CAMBRIDGE LIBRARY COLLECTION

Books of enduring scholarly value

Earth Sciences

In the nineteenth century, geology emerged as a distinct academic discipline. It pointed the way towards the theory of evolution, as scientists including Gideon Mantell, Adam Sedgwick, Charles Lyell and Roderick Murchison began to use the evidence of minerals, rock formations and fossils to demonstrate that the earth was older by millions of years than the conventional, Bible-based wisdom had supposed. They argued convincingly that the climate, flora and fauna of the distant past could be deduced from geological evidence. Volcanic activity, the formation of mountains, and the action of glaciers and rivers, tides and ocean currents also became better understood. This series includes landmark publications by pioneers of the modern earth sciences, who advanced the scientific understanding of our planet and the processes by which it is constantly re-shaped.

The Natural History of Igneous Rocks

Alfred Harker (1859–1939) was a prominent petrologist who spent his career at St John's College, Cambridge, lecturing on and researching rock formations and related geological activity. He was elected a fellow of the Royal Society in 1902, and was president of the Geological Society from 1916 to 1918. He used his Cambridge lectures as the foundation for this book (first published in 1909), offering an introduction to the development of rocks and related volcanic activity. With more than one hundred diagrams of various aspects of geological formations, this work also provides a visual guide to the location and formation of igneous rocks. Over the course of the work, he covers the themes of vulcanicity, rock structure, crystallization, the role of magma and the principles of rock classification, giving a broad picture of the field of petrology around the beginning of the twentieth century.

Cambridge University Press has long been a pioneer in the reissuing of out-of-print titles from its own backlist, producing digital reprints of books that are still sought after by scholars and students but could not be reprinted economically using traditional technology. The Cambridge Library Collection extends this activity to a wider range of books which are still of importance to researchers and professionals, either for the source material they contain, or as landmarks in the history of their academic discipline.

Drawing from the world-renowned collections in the Cambridge University Library, and guided by the advice of experts in each subject area, Cambridge University Press is using state-of-the-art scanning machines in its own Printing House to capture the content of each book selected for inclusion. The files are processed to give a consistently clear, crisp image, and the books finished to the high quality standard for which the Press is recognised around the world. The latest print-on-demand technology ensures that the books will remain available indefinitely, and that orders for single or multiple copies can quickly be supplied.

The Cambridge Library Collection will bring back to life books of enduring scholarly value (including out-of-copyright works originally issued by other publishers) across a wide range of disciplines in the humanities and social sciences and in science and technology.

The Natural History of Igneous Rocks

Alfred Harker

Cambridge
University Press

CAMBRIDGE UNIVERSITY PRESS

Cambridge, New York, Melbourne, Madrid, Cape Town,
Singapore, São Paolo, Delhi, Tokyo, Mexico City

Published in the United States of America by Cambridge University Press, New York

www.cambridge.org
Information on this title: www.cambridge.org/9781108028134

This edition first published
This digitally printed version 2011

ISBN 978-1-108-02813-4 Paperback

THE NATURAL HISTORY OF IGNEOUS ROCKS

VESUVIUS IN 1888: SHOWING AN EXPLOSION OF THE 'STROMBOLIAN' TYPE

FROM A PHOTOGRAPH BY DR. TEMPEST ANDERSON

THE NATURAL HISTORY OF IGNEOUS ROCKS

BY

ALFRED HARKER

M.A., F.R.S.

FELLOW OF ST. JOHN'S COLLEGE
LECTURER IN PETROLOGY IN THE UNIVERSITY OF CAMBRIDGE

WITH 112 DIAGRAMS AND 2 PLATES

METHUEN & CO.
36 ESSEX STREET W.C.
LONDON

First Published in 1909

PREFACE

THE following pages are, in substance, a course of lectures delivered at Cambridge; and, in offering them to a larger audience, the general plan of treatment has been retained.

Igneous action and igneous rocks are first considered from a purely geological standpoint. In emphasizing this aspect, I have been influenced by a conviction that the subject has not yet been accorded its due place as an integral part of historical geology.

The middle portion of the book deals with the crystallization of igneous rock-magmas, regarded as complex solutions. Although the subject is one which might be more skilfully handled by a professed chemist, a presentation of it as it appears to a petrologist may be not less useful to the student. I have in the main followed Professor Vogt, whose writings in the last few years have thrown so much light upon the matter.

Some of the questions briefly and tentatively touched in the concluding chapters are of a more speculative character, but so fundamental in their scope that they could not be passed over in silence.

For the two plates I am indebted to the kindness of Dr. Tempest Anderson. One of them, which has already appeared in the *Geographical Journal*, is reproduced by permission of the Council of the Royal Geographical Society.

A. H.

ST. JOHN'S COLLEGE, CAMBRIDGE
March, 1909

CONTENTS

CHAPTER V

MUTUAL RELATIONS OF ASSOCIATED IGNEOUS ROCKS

CHAPTER VI

IGNEOUS ROCKS AND THEIR CONSTITUENTS

CHAPTER VII

ROCK-MAGMAS

CHAPTER VIII

CRYSTALLIZATION OF ROCK-MAGMAS

CHAPTER IX

SUPERSATURATION AND DEFERRED CRYSTALLIZATION

CHAPTER X

ISOMORPHISM AND MIXED CRYSTALS

LIST OF DIAGRAMS

LIST OF PLATES

From photographs by Dr. Tempest Anderson.

THE NATURAL HISTORY OF IGNEOUS ROCKS

CHAPTER I

IGNEOUS ACTION IN RELATION TO GEOLOGY

Petrology as a branch of geology.—Cosmogony affords no sound basis for a theory of igneous action.—Petrology and the geological record. —Chronological distribution of igneous rocks.—Igneous action and crust-movements.—Geographical distribution of the younger igneous rocks.—Cycles of igneous activity.

Petrology as a Branch of Geology.—The petrology of igneous rocks, like other special departments of natural science, can be approached from more than one side. That which has received most attention is the descriptive side—*petrography* in the proper sense—which possesses a large and rapidly growing literature, its popularity dating from the introduction of microscopical methods some fifty years ago. With this we shall be only incidentally concerned in the present work. Most of the later chapters will be occupied with *petrogenesis*—*i.e.*, the study of igneous rocks as the products of consolidation of molten magmas. Discussion concerning the genesis of igneous rocks has hitherto been, and to some extent still is, of a speculative nature; but the recent developments of physical chemistry indicate at least the general lines upon which a more systematic treatment may be attempted, and laboratory research has already begun to lay an experimental foundation for such treatment.

It is, however, the *geological aspect* of petrology that will first engage our attention. Here we have to consider igneous action in its relation to other geological events, and with reference to the part which it has played in the history of the globe. Among the subjects which will properly find their place here are the distribution of igneous rocks in time and space, and their connection with tectonic geology; their relations to one another and to other rocks, as displayed in their mutual associations and relative ages, and also in the morphology and anatomy (as distinguished from the histology) of igneous rock-bodies. In the physical interpretation of the facts, what we have called the geological side comes in contact with the petrogenetic, and (again borrowing terms from a sister science) we are confronted by the difficult problems of the physiology of igneous processes and the evolution and phylogeny of different rock-types.

It is, as we have said, the descriptive side of petrology that has received the largest share of attention. Indeed, the accumulation of facts in this department has so far outrun the collation and interpretation of them, that the work done loses much of its immediate value. This is, doubtless, the chief reason why, in the minds of many geologists, petrology has come to be identified with petrography, and regarded as a science apart, or at least a study for specialists, having no direct bearing on geology in general. The writings of Scrope and Darwin, of von Richthofen and Lossen, of Suess and Marcel Bertrand, of numerous workers in the Alpine regions of Europe and the High Plateaux of North America, go far to prove that igneous action is, on the contrary, closely bound up with the geological history of the globe; but they have not yet availed to modify very materially the attitude of mind which we have designated.

That this branch of geology has not yet been accorded its due place as an integral part of the science is sufficiently apparent on a glance at the geological literature of the day. In most manuals of instruction the divorce between petrology and the physical and historical parts of geology is almost complete. In memoirs dealing with particular districts it is still a common practice to touch only cursorily on igneous action in the chronological treatment which occupies the body of the work. If further details are given, they are of a purely descriptive kind, are relegated to a more or less perfunctory appendix, and are usually supplied by a specialist who knows the rocks only from specimens. More striking is the evidence of geological maps. Here the sedimentary formations are ranged in due order in the marginal index; while below, as if of no particular age, are inserted the igneous rocks, distinguished usually according to lithological characters only, and often set down in random order. There are, of course, many exceptions to this reproach; but it is still generally true that, in respect of the igneous rocks, our maps are coloured on the 'geognostic' principle, as they were a hundred years ago for all rocks. A student can scarcely fail to receive the impression that, while the accumulation of successive sediments is an ordered process, telling a story full of significance, igneous action is a meaningless interpolation, without relation to other episodes in the geological record. It is therefore our first business to show that igneous activity, standing in close connection with the tectonic development of a region, is very intimately and fundamentally related to other parts of its geological history.

Cosmogony affords no Sound Basis for a Theory of Igneous Action.—A formal treatment of our subject on systematic lines might perhaps be expected to begin with a discussion of the origin and cause of igneous action. If any apology be needed for adopting a different course in this volume, it will be found in the author's conviction that the present state of knowledge does not warrant us in undertaking such a discussion with any prospect of profit. Further, this seems to be no matter for great regret, if we remember that the older geological theories, which proceeded on those lines, have, one and all, proved barren of any results germane to our present subject; while such positive contributions as we have to record are the outcome, in every case, of direct appeal to geological history, with a due apposition and correlation of the facts elicited.

The question of the ultimate origin of igneous action, including the source of the magmas and the nature of the motive power, is indeed inseparably bound up with other questions—the origin of our planet and the present physical and thermal condition of its interior parts—which at present we must be content to leave open. Cosmogony assuredly can afford no firm foundation for *a priori* reasoning. The Nebular Hypothesis in Laplace's form, if not discredited, has at least been shown to involve great difficulties, to which no answer is yet forthcoming; the Meteoritic Hypothesis, resting from the first upon a more precarious basis, is involved practically in the same damaging criticism; and the Planetesimal Theory has as yet scarcely emerged from the tentative stage. As regards the present physical condition of the interior of the globe, the most diverse theories still find champions. The old conception of a comparatively thin solid crust floating on a liquid interior may perhaps be considered obsolete. Lord Kelvin, however, never withdrew his hypothesis of a solid globe having a rigidity comparable with that of steel, though the supports of that doctrine have, one by one, fallen away. The fundamental assumption upon which the *a priori* argument for the solidity of the deep interior rests—viz., the assumption that the melting-points of minerals continue to be raised indefinitely by increased pressure—can no longer be taken for granted since the publication of Tamman's researches. Fisher's doctrine of a fluid substratum over a solid nucleus is, from this point of view, merely a variant of the solid-earth theory. Arrhenius believes the greater part of the globe to be gaseous, passing through a liquid layer to a solid crust some forty miles thick. As regards the physical assumptions made, this theory has at least as much plausibility as Kelvin's, and it accords in a somewhat striking manner with the indications derived from seismic phenomena. All these conflicting theories assume, more or less explicitly, the Nebular Hypothesis of the birth of our planet. On the Planetesimal Hypothesis, as developed by Chamberlin,[1] the globe may be supposed effectively solid throughout, but with liquid tongues and threads, which emanate from places of local fusion in the middle zone, and work their way outward.

The various hypotheses to which allusion has been made agree in presupposing some more or less defined present distribution of temperatures within the Earth, which has resulted from the primitive physical and thermal conditions of our planet. The theories which have been entertained by different authorities concerning the origin and cause of igneous action proceed, avowedly or by implication, on similar lines. The energy invoked is energy in the form of *heat*—the residuum of the original heat, derived from the parent incandescent nebula, from the collision of meteorites, or from compression in the interior of a mass composed of aggregated planetesimals, according to the particular reading of cosmogony adopted. Recent developments in chemistry, however, invite us to contemplate the existence within the Earth of a vast store of energy in another form—viz., the *chemical*, an energy very gradually set free by those ultra-atomic changes which seem to be indicated by the remarkable phenomena of radio-activity. In particular, Strutt,[2] after estimating the amount of radium contained in various rock-specimens, has arrived at the conclusion that the observed temperature-gradient in the accessible part of the Earth's crust can be wholly accounted for in this way. Such a conclusion, if fully admitted, manifestly carries far-reaching consequences in the class of questions at which we are glancing. The temperature-gradient merely proves that the Earth is losing heat by conduction outward; but if there be a continual generation of heat in the interior, we have no longer any good reason for believing that the Earth is cooling.

[1] 'Fundamental Problems in Geology,' *Carnegie Institute Yearbook*, No. 3 (1905), pp. 195-258. Chamberlin and Salisbury, *Geology*, vol. ii. (1906), pp. 38-81, 99-106.

[2] 'On the Distribution of Radium in the Earth's Crust, and on the Earth's Internal Heat,' *Proc. Roy. Soc.* (*A*), vol. lxxvii. (1906), pp. 472-485, and lxxviii. (1906), pp. 150-153.

Important as the genesis of our planet is in relation to a complete view of the world's history—and, in the opinion of the present writer, cosmogony is at least as much in the province of the geologist as in that of the astronomer—the present situation is so unsettled that one who is not directly led to discuss these large questions may legitimately adopt towards them an attitude frankly agnostic. He will at least, if prudent, refuse to build on so insecure a foundation, so long as other ground is open. The data furnished by geological observation, it is true, do not as yet suffice to lay down a complete working theory of the nature and operation of igneous action; but we shall find in them a suitable starting-point for that side of the inquiry which can be most profitably discussed by the geological student.

Petrology and the Geological Record.—The relation between rational (as distinguished from descriptive) petrology and historical geology is one of mutual dependence. On the one hand, any geological account of the globe, or of any selected region, is incomplete if it does not recognise adequately the essential part which has been played by igneous action at many different epochs. On the other hand, the materials for a study of igneous action must be derived mainly from an examination of the geological record. In this study, indeed, our method of inquiry is necessarily different from that which has produced such brilliant results in other branches of physical or 'dynamical' geology. The operations of erosion and sedimentation, for example, may be seen in progress at the present time, and the part of geology dealing with these processes has thus been built up on strictly Lyellian principles. For reasons which are sufficiently obvious, this method is not applicable to igneous action.

In the first place, the operations to be studied are, with very partial exceptions, subterranean, and therefore concealed from view. The visible manifestations of modern igneous activity, as displayed in volcanoes, are of too limited and specialised a kind to afford more than vague clues to the essential character of igneous action in general. The much more important hypogene processes, which may or may not give rise to volcanic outbursts as one episode in their development, are hidden from observation, and are conducted under conditions of which we have no practical knowledge. Not only are the circumstances different from any that can be witnessed in nature or imitated in the laboratory, but a further difference is imported by the time element. It cannot be doubted that the differentiation of rock-magmas, and their intrusion and extrusion in ordered sequence—in short, the evolution of any great connected suite of igneous rocks—must be a long-protracted process: a cycle embracing a varied succession of events, and not always comprised within the limits of one geological period. We are living in a time which coincides with a particular phase of one of these great cycles of igneous activity; and it is probable that, even if direct observation were possible, observation extending over a few centuries only would not suffice to throw much light on the larger problems of the origin and distribution of igneous rocks. The data for the solution of these problems must be sought in the records of the past, as displayed in eroded regions of former igneous activity.

The superficial volcanic phenomena of the present time illustrate in themselves the necessity of this appeal to the geological record. Although active volcanoes have been studied from an early time, and in some respects with great thoroughness, some of the most fundamental facts concerning them are very imperfectly elucidated. For instance, the arrangement of volcanic vents is in general a linear one, giving place in some cases to a disposition in groups; and the lines are often related to coast-lines, while the groups are often island groups. On these and other data some have based theories of the situation of vents on hypothetical lines of fracture, and others have supposed an accession of sea-water to be the proximate cause of volcanic outbursts; but such speculations rest on a very slender basis of ascertained fact. Other questions are suggested by an examination of the ejected products. Thus, throughout the vast belt of the Andes all the volcanic rocks are closely related in characters, and usually depart but little from a central type; while in the small group of the Æolian Isles we find varieties of lava of widely diverse and highly-specialised types. To such problems a study of modern volcanoes alone can bring no satisfactory solution. We are led to inquire whether some empirical laws, which seem to hold good, are not rather parts of larger principles, and whether apparent anomalies may not be explained as survivals of a pre-existing state of things.

It appears, then, that, for the purposes of the inquiry before us, the maxim of the school of Hutton and Lyell must be reversed. Instead of applying a knowledge of processes now going on around us to elucidate the record of past ages, we must seek rather to use the history of the past to explain the phenomena of the present. Geology may thus repay part of the debt which she owes to Physical Geography.

Chronological Distribution of Igneous Rocks.—It is a familiar fact that igneous action has been especially characteristic of certain geological periods. Thus, in Europe, and, indeed, over a large part of the world, we find igneous rocks of Tertiary and Recent age in great variety and in many countries, and others, again, of Palæozoic and pre-Palæozoic age; while in the intervening time the record is almost a blank. With a few exceptions in the Triassic (representing the dying out of Palæozoic activity), and again in the Cretaceous (the prelude to Tertiary activity), there was a general cessation of igneous manifestations throughout the Mesozoic periods.

It is this fact that has lent colour to the doctrine, so long and tenaciously held by many Continental geologists, of a fundamental difference between 'older' (pre-Tertiary) and 'younger' (Tertiary and Recent) igneous rocks. On the one hand, the older have naturally suffered more in most cases from the effects of secondary changes, and have become in some measure disguised; different rock-types being thus affected in different degrees, and certain volcanic rocks being usually transformed beyond superficial recognition. On the other hand, in most of the best-known districts of Europe, denudation has not progressed so far as to expose the Tertiary plutonic rocks in any great force, while well-preserved volcanic products of this age cover considerable areas. The alleged differences between older and younger rocks are thus partly the differences between the altered and the fresh, partly those between plutonic and volcanic. This relic of the Wernerian dogmas is no longer seriously entertained, though it is unfortunately preserved in the nomenclature of igneous rocks current in Germany and France. We know that a given rock-type — say, a hypersthene-andesite—may be Ordovician, or Devonian, or Permian, or Eocene, or Recent, and that exceptionally the Palæozoic examples are as well-preserved as the Tertiary. We are no longer invited to believe that the laws of physics and chemistry in Palæozoic times were different from those which govern the consolidation of rock-magmas to-day.[1]

While geological age thus ceases to obtrude itself into systematic petrography—or will cease when the inertia of conservative habit is exhausted—it does not follow that the age of an igneous intrusion or extrusion is a matter of indifference. Indeed, the discarding of the age-factor from classification and nomenclature enables us to approach the important subject of the chronology of igneous action unhampered by arbitrary preconceptions. We then perceive that the cessation of igneous activity during the greater part of Mesozoic time is not a unique fact: pauses of equal significance, and perhaps of no less duration, divided distinct periods of activity in Palæozoic and earlier times. It is true that the earlier chapters of the geological record are less complete, and sometimes less clearly read, than those dealing with the later periods; but this necessary imperfection does not impair the validity of the general statement. In no other country is the history of past igneous action so full as in the British Isles; and here the alternation of periods of activity with periods of quiescence is very clearly

[1] On this question see *Science Progress*, vol. ii. (1894), pp. 48-63.

discerned.[1] A very brief summary will suffice to recall facts which are familiar to all students of British geology.

The Lewisian rocks are, with relatively trifling exceptions, wholly igneous—viz., a complex of plutonic gneisses, granites, pegmatites, etc., intersected by numerous dykes and sheets. They point to an important period of igneous activity at the beginning of the known geological history of the British area. The Torridonian, which succeeded, was a period of quiescence.[2] In the younger pre-Cambrian districts of England and Wales we again meet with evidence of igneous action, which was probably on a somewhat extensive scale. In this case the products which remain, and can be examined, are chiefly volcanic rocks.

A pause ensued, coinciding with the accumulation of the Cambrian and the older Ordovician strata. Then followed the Ordovician outbreaks (mainly of Llandeilo and Bala age) in numerous parts of England, Wales, Scotland, and Ireland. Here both volcanic and plutonic rocks are well represented, besides minor intrusions. The Silurian was a period of quiescence, though some belated volcanic outbursts are indicated at one or two scattered centres, in the south-west of England and the west of Ireland.

In Lower Old Red Sandstone times the northern half of the British area was the theatre of igneous action on a large scale. To this age belong not only very considerable accumulations of volcanic rocks in the midland belt of Scotland, the Cheviots, and Co. Tyrone, but also a large part at least of the 'newer granites,' etc., of Scotland with other granite masses in the English Lake District, besides an important suite of minor intrusions. The Upper Old Red Sandstone, following the Lower after a long hiatus in the stratigraphical succession, presents a strong contrast with it, in that there

[1] Data for a chronology of igneous action in Britain may be found in Sir Archibald Geikie's Presidential Addresses to the Geological Society in 1891-2 (*Quart. Journ. Geol. Soc.*, vols. xlvii., xlviii.), and in his *Ancient Volcanoes of Great Britain* (1897).

[2] The series which occupies so large a part of the Scottish Highlands, the Dalradian of Sir A. Geikie, is omitted, its chronological place being still undecided. It includes a system of intrusive sills.

was an almost complete cessation of igneous manifestations throughout the British area. Sir A. Geikie finds only two isolated districts, in which occurred comparatively feeble revivals of activity which are referred to this age—the one near Limerick and the other in the Orkney Isles. After this pause came another widespread outbreak of igneous action in Lower Carboniferous times, taking the form of volcanic eruptions of more than one type and the intrusion of an important series of sills and dykes. The evidence of this is found in many parts of midland and southern Scotland, in the Isle of Man, Derbyshire, the Bristol district, and the Limerick basin. In England and Ireland activity was confined to a limited range in the Carboniferous Limestone age, but in Scotland it was prolonged to the beginning of the Coal-Measures time. Sir A. Geikie has described volcanic rocks occurring in the New Red Sandstone of Ayrshire and Dumfries, which may be regarded as representing a revival of the Carboniferous igneous activity, upon a much smaller scale and with a greatly reduced areal distribution.

After a prolonged pause, covering practically the whole of Mesozoic time, came the great Tertiary outbreak, the last in the history of our country. It included volcanic outpourings of enormous extent, plutonic intrusions at numerous centres, and a long series of minor intrusions, collectively of even greater bulk. These last at least affected most of Scotland and the northern parts of England, Wales, and Ireland. The volcanic rocks are of Eocene age, but the intrusions may perhaps be referable in part to some later division of Tertiary time.

Summarily, the great periods of igneous activity in Britain, separated by longer or shorter intervals of quiescence, are six in number: two of pre-Palæozoic age, one Lower Palæozoic, two Upper Palæozoic, and one Tertiary. In order to be able to treat the British area as in some degree a geological unit, we have omitted Cornwall and Devonshire, an area the petrological affinities of which are with the European continent rather than with the rest of Britain. Whether the igneous rocks in the Ordovician of Cornwall include true contemporaneous lavas seems to be still uncertain. Volcanic rocks are well represented in part of the Middle and Upper Devonian of Devonshire, and some of the intrusions in Cornwall and Devon are to be referred to a Devonian age. In the Carboniferous we again have evidence of igneous activity, both in volcanic rocks interstratified in the Culm and in a series of intrusive sills. The granites of Cornwall and of Dartmoor, and the various dykes ('elvans' and lamprophyres) related to them, are of a later age—either late Carboniferous or post-Carboniferous. Finally, in the Exeter district and at Cawsand, there are igneous rocks intercalated in the Permian. In this south-western part of England, therefore, there have been at least three or four periods of igneous activity, divided by quiescent intervals, though the record is not quite parallel to that of the rest of Britain.

It may be remarked that the history of igneous action in this country in no wise bears out any theory of a general decline of energy with the lapse of ages. The latest of the active periods which we have recognised was not inferior, so far as can be ascertained, to any that preceded it, either in the character of the manifestations, the duration of the activity, or the extent of area involved. Setting aside the earliest chapters in the Earth's history, concerning which we have little or no direct information, we may say that, while there have been alternations of activity and repose, indicating a certain rough periodicity, there is nothing to suggest a secular waning.

Igneous Action and Crust-Movements.—Not less significant than the limitation of igneous action to certain periods is the restriction of its manifestations, in any one of these periods, to certain defined regions of the Earth's crust. It is not necessary to enlarge upon a fact which is apparent on a glance at the geological map of any extensive area. Taking now the chronological and the geographical distribution together, we are led to observe a relationship of a very fundamental kind—viz., the general correspondence, as regards both time and space, of igneous action with important movements of the Earth's crust. Since this relationship is the starting-point of many considerations which are to follow, it demands a closer scrutiny.

The idea of a connection between igneous action and disturbances of the solid crust is not a new one, but has figured in geological literature in various forms. Von Buch's 'elevation theory' of volcanoes has now only a historic interest; but we know from the phenomena of laccolites that igneous intrusion may be the immediate cause of folding and faulting. In other typical cases which might be cited, it would be more natural to regard igneous intrusion as the consequence, rather than the cause, of displacements of the solid crust. In a broader view, however, the distinction between apparent cause and apparent effect ceases to be applicable, and the two sets of phenomena—igneous action and crust-movement—are seen as co-ordinate effects of the same ultimate cause. The strains set up in the crust of the globe in various ways, cosmic and regional, develope local stresses, which may increase until they reach the limit of resistance. That limit passed, the stresses must find relief in a readjustment of the crust; and such readjustment may be effected by relative displacements of the solid rock or of fluid rock-magma, or, more generally, of both together. Igneous operations (intrusion and extrusion) and crust-movements thus represent two different, but closely related, ways of relieving crustal stress and restoring equilibrium. They are in some degree complementary to one another; and, in view of the greater mobility of the fluid magma, it is easy to understand that the correspondence between the two sets of phenomena is one of general agreement, not exact coincidence. Accordingly, we find that igneous action has prevailed at those periods of geological history and in those regions of the Earth's crust which have also been characterized by relative elevation and depression, folding, faulting, and overthrusting of the solid rocks. Regarded more closely, the extrusion and intrusion of molten magmas have not always been strictly simultaneous with the main displacements of the solid crust, nor has igneous activity always been localised exactly on the principal lines of mechanical disturbance.

The general geological structure of the great Continental masses is known as regards its main features, and a masterly presentation of this knowledge is contained in Suess' great work, *Das Antlitz der Erde* (1885-1908).[1] From this side, therefore, the data are at hand for a comparison of the chronological and geographical distribution of crust-movements on the one hand, and igneous action on the other. The data concerning the ages and areal distribution of igneous rocks must be gathered from many sources. For the European area a comparative study on the lines indicated was attempted in 1888 by Marcel Bertrand;[2] and, though such a first essay is necessarily incomplete, it is of value, both in itself and as an example. Bertrand discusses the distribution in Europe of the chief groups of igneous rocks of various ages in the light of their relation to four main systems of folding—the Huronian, the Caledonian, the Hercynian, and the Alpine. A more detailed examination of the subject would doubtless lead to the recognition of more than four systems of crust-movements, each having its attendant train of igneous rocks. In this place we shall be content to glance very briefly and generally at the distribution over the globe of the 'younger' igneous rocks in its relation to the latest great crust-movements.[3] The data are naturally more complete here than in the case of any of the older suites of igneous rocks.

Geographical Distribution of the Younger Igneous Rocks.—Beginning with South America, which, as Suess remarks, "presents in a higher degree than any other part of the world all the features of a homogeneous structure," we have in the Andes an example on the grandest scale of a belt of igneous activity following a system of folds which has, in general plan, a simple linear extension. The structure of

[1] Of this there are French and English translations, the former annotated; *La Face de la Terre* (1897-1908), and *The Face of the Earth* (1904-1908).

[2] 'Sur la distribution géographique des roches éruptives en Europe,' *Bull. Soc. Géol. Fra.* (3), vol. xvi. (1888), pp. 573-617.

[3] See the writings of Suess, Bertrand, de Lapparent, Michel-Lévy, and others.

the range, due to movements directed towards the Pacific, is not in reality simple, there being in most parts three or four parallel chains. Folding, beginning at a date between Cretaceous and Eocene, has been renewed at later epochs, and some movement is probably still in progress. Darwin's conclusions concerning very considerable recent upheaval have been combated, but frequent earthquakes attest the instability of the western border of the mountains. Igneous activity, in like manner, has been renewed from time to time, and throughout a great part of the long line is not yet extinct. It is a feature of the Andes that the great volcanoes are ranged along the main orographic line itself. Abyssal and hypabyssal rocks, belonging to earlier epochs of this system of igneous activity, have been exposed by erosion in many places.

In Columbia the various branches of the Andean system diverge, and the belt of igneous activity follows the westerly branch by Panama into Central America, where numerous active volcanoes still survive. The most easterly branch, which curves away to north-east and east to connect with the Antillean chain, is not accompanied by Tertiary igneous rocks in Venezuela and Trinidad, but the evidences of igneous action are met with again in the Antilles. The late outbursts in St. Vincent, Martinique, and Dominica are a survival or revival of Pleistocene activity, and in Guadeloupe and Antigua eruptions began as far back as the Miocene. In the Virgin Islands occur late Cretaceous and Eocene plutonic rocks of the same types which are found in the Andes.[1] Along this great curve the volcanoes are situated not on the crest, but on the inner border. In Hayti the Antillean chain bifurcates, one branch passing through Jamaica to Guatemala, and the other through Cuba to Yucatan. Central America makes in some sense a break in the great Cordilleran system, the main axes following a transverse direction, and, as Suess points out, this is the only part of the Pacific border where the coast is not determined by folded mountain-chains.

[1] Högbom, *Bull. Geol. Inst. Upsala*, vol. vi. (1905), pp. 214-233.

The igneous activity which has affected the Mexican plateau from Miocene times to the present day is related to crust-movements of a different class from those which express themselves in folded mountain-chains. In the distribution of active and recent volcanic vents, for instance, no orderly arrangement is obvious, though some approach to allignment is discovered by those geologists who would connect the outbreaks with important transverse fissures.[1] The Rocky Mountains begin only in New Mexico.

The mountain-system of the western part of North America is much more complex than the Andes. Not only has movement recurred at different epochs, but the effects of folding are complicated by great vertical displacements superposed thereon. The Rocky Mountains are faced by the range of the Sierra Nevada to the west, the formation of which seems to have taken place about the close of Jurassic time, and was accompanied by great intrusions of plutonic rocks. Between the two opposed ranges lies the plateau region of the Great Basin, bounded and intersected by enormous faults. This, as King[2] remarks, "has suffered two different types of dynamic action—one, in which the chief factor evidently was tangential compression, which resulted in contraction and plication, presumably in post-Jurassic time; the other of strictly vertical action, presumably within the Tertiary, in which there are few evidences or traces of tangential compression." The topographic features of the country—notably the great cañons—prove that, in its later part at least, the vertical movement has been one of upheaval relatively to sea-level.[3] That relative displacement continues in this area to the present time is attested, not only by the deformation of the old shore-lines of 'Lake

[1] See, *e.g.*, Felix and Lenk, *Zeits. deuts. geol. Ges.*, vol. xliv. (1892), pp. 303-323, pl. xix.

[2] *U.S. Geol. Explor. 40th Parallel*, vol. i., p. 744.

[3] Suess here pushes to the extreme his thesis that apparent uplift is only differential subsidence. "The elevations present the characters of 'horsts,' and even the immense High Plateaux of Utah must be regarded as results of unequal subsidence. In this region is sunk the Cañon of the Colorado."—*Das Antlitz der Erde*, vol. i. (1885), p. 757.

Bonneville,' but by recent earthquakes and demonstrable movement along the old fault-lines.[1] As regards igneous action, intrusions and extrusions which must be connected with the building of the Rocky Mountains occur at numerous places along the eastern border of the range, from western Texas to Montana. Where their age can be verified, they are found to be late Cretaceous, Eocene, Oligocene, or Miocene at different centres. The lavas of the Sierra Nevada are younger—Miocene and Pliocene, possibly extending into early Pleistocene. In the High Plateaux country volcanic activity has been manifested in different places at several epochs from Miocene, or even Eocene, to a very late date, but the great outpourings of lava which flooded vast areas were of Pliocene age. If we were to regard only volcanoes now active, we should see in this part of North America between Central Mexico and Alaska a break in the great 'Circle of Fire' which girdles the Pacific Ocean. But at numerous places throughout the region volcanic cones and craters not yet destroyed by erosion, geysers, hot springs, and other significant indications prove that, if igneous activity has ceased, its extinction is at least a thing of the near past. Russell[2] remarks that this apparent exhaustion of activity is found in the region where the belt of country affected broadens out from a narrow strip to a width of sometimes as much as 1,000 miles. Where the belt narrows again, in Alaska, volcanoes still active reappear, and the line continues by the Aleutian Isles into Asia. Very noticeable is the total absence of igneous rocks among the relatively undisturbed strata of the interior part of the United States. In the east the Appalachian belt of folding, of much greater geological antiquity, has its own suite of ancient volcanic and other igneous rocks.

The eastern coast of Asia, with its island outposts, is defined, like the western coasts of the American continents, by strongly-marked mountain-chains, folded towards the

[1] Gilbert, *2nd Ann. Rep. U.S. Geol. Sur.* (1882), pp. 169-200; *Amer. Journ. Sci.* (3), vol. xxvii. (1884), pp. 49-53.

[2] *Volcanoes of North America* (1897), p. 131.

ocean; but instead of long parallel lines, we see here a series of pronounced curves, with convexity facing the ocean, which succeed one another with a 'festoon' arrangement apparent even on a glance at the map. This structure characterizes the whole western border of the North Pacific, until the great 'Malay arc,' running through Sumatra, Java, and Flores, marks the limit in this direction of the Pacific régime. Almost everywhere the main orographic lines are accompanied by geologically young igneous rocks, and in most parts they are marked by active volcanoes. The precise age of the Tertiary igneous rocks is not known in every case, but it is evident that there have been pauses and renewals of activity. In Japan, as in some of the North American chains, there are intrusions of Cretaceous age. The Tertiary volcanic sequence here belongs to the Miocene and Pliocene, and there are post-Miocene plutonic rocks like those of the Andes; finally came a revival of volcanic activity in Quaternary and Recent times. Similarly in Java there was a great outbreak of vulcanicity about the close of the Oligocene, continued in the earlier part of the Miocene, and a revival at the end of Tertiary time, prolonged to the present day. Only in Formosa, Further India, and Borneo has there been no important outbreak of igneous action.

As is well known, most of the island groups in the Pacific are composed largely of volcanic rocks of late age, and they include (in Hawaii) the greatest active volcanoes in the world. The relation of igneous action to crust-movements is necessarily here a matter of inference and speculation. In the far south 'younger' igneous rocks and great volcanoes are known in South Victoria Land. The circle of the Pacific is completed by New Zealand and Australia. In New Zealand volcanic activity broke out in the North Island probably in the Eocene, and is not yet extinct. Tertiary volcanic rocks, though of very different types, occur also in the Dunedin district. In Tasmania, Victoria, New South Wales, and part of Queensland extensive outpourings, especially of Tertiary basalts, mark the course of what Suess terms the 'Australian Cordillera.' According to Gregory,[1] the older basalts of Victoria may have begun as far back as the Cretaceous, while the newer, which have been assigned to the Pliocene, "probably lasted till the human occupation of Victoria." In Queensland, too, the earliest eruptions have been ascribed to a Cretaceo-Eocene age, while the latest are represented by recent ash-cones, still recognisable. Only in New Caledonia are no evidences of late igneous activity known. Summarily, we may say that during Tertiary times the borders of the Pacific have been peculiarly affected by crust-movements, largely of the accentuated type which results in folded mountain-chains, and in many places there are indications that these movements are not wholly exhausted. We see that during the same time this part of the world has been, and still continues to be, the theatre of igneous action.

Turning to the other side of the globe, we find another important area of Tertiary igneous activity in the North Atlantic and Arctic regions. It includes the northern half of the British Isles, the Färöer, Iceland, the greater part of explored Greenland, Spitzbergen, Franz Josef Land, and possibly a still wider extent in the far north. Activity began in the Eocene with the pouring out of great floods of basalt, followed by a system of intrusions probably still more voluminous. In Britain the record is comprised probably within the first half of Tertiary time; but in some parts, notably in Iceland, igneous action is still in force. This region is not one of folded mountain-chains, but of broken plateaux, and the crust-movements indicated are of the vertical kind. The linear element being lacking or inconspicuous, the connection of the igneous processes with displacements of the crust is less obvious, but it is not less real. The Inner Hebrides, for instance, which have been the chief theatre of activity in the British part of the region, are parts of a sunken area, dropped by great faults between the older masses of the Scottish mainland and the Outer Isles, and broken by many other faults.

[1] *Proc. Roy. Soc. Vict.* (2), vol. xiv. (1902), p. 213.

Differential subsidence, as contrasted with lateral displacement, is a general feature of the latest crust-movements over a great part of the Atlantic region, and the relation of igneous action to tectonic development is therefore less sharply defined than in the case of folded mountain-chains. The numerous volcanic centres which rise from the ocean between the Azores and Tristan d'Acunha are distributed, not in lines, but in groups. There are, however, parts of the world where the linear element enters very conspicuously in connection with movements of differential subsidence. The most remarkable example of this is the 'Great Rift Valley,'[1] which extends from the Jordan Valley to Lake Nyassa. The long trough-like depression has been produced by differential vertical movements at different epochs from the Eocene to the later Pleistocene. The attendant 'plateau' eruptions took place in the Cretaceous and Miocene, and activity was prolonged through the Pliocene and Pleistocene at various places along the line.

In Central and Southern Europe we see a great development of igneous rocks of late geological age connected with the Alpine system of crust-movements. The relation is not strictly analogous with that which we have noted on the borders of the Pacific Ocean, for igneous action has not broken out along the main orographic lines, but between and outside them. The volcanic districts fall into two main belts, an inner and an outer, as shown on the accompanying map (Fig. 1).[2] The inner belt includes numerous districts lying in the loops made by the curving mountain-chains. It is divided into two groups by the line of the Apennines. The western (Iberian-Tyrrhenian) group has its limits partly defined by the Apennines, Atlas, and Pyrenees. In Spain Calderon recognises three lines running N.E. to S.W.—one following the Mediterranean coast, and including several

[1] See especially Gregory, *The Great Rift Valley* (1896).

[2] The Spanish, South Italian, and Ægean districts lie outside the limits of the map. The main orographic lines are marked by small v v v, the points of which indicate the direction of thrust.

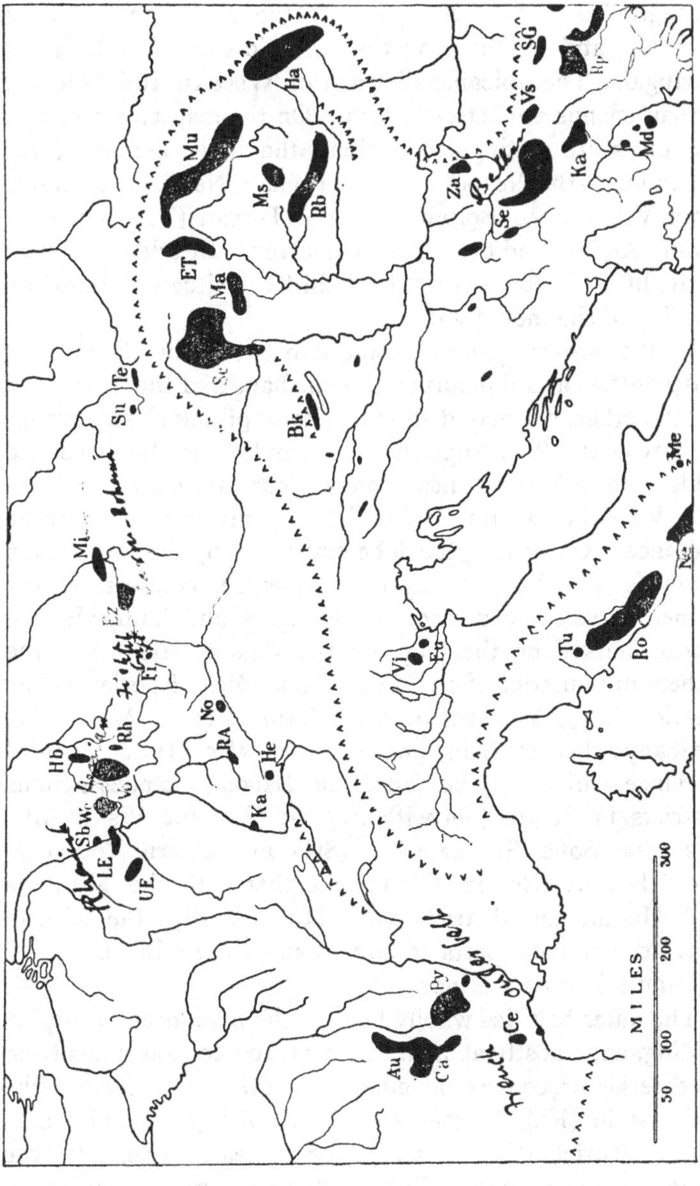

FIG 1.—DISTRIBUTION OF THE YOUNGER VOLCANIC ROCKS IN THE EUROPEAN AREA.

districts, from Gerona to Cabo de Gata; another passing through the middle of Spain by the Ebro fault and La Mancha; and the third on the Atlantic side, in Galicia and Portugal. The volcanic districts of Western Italy and the Italian islands are better known. On the map two principal provinces are distinguished along the inner border of the Apennines—the Neapolitan and Roman (Ne, Ro), extending from Vesuvius to Bolsena, and the Tuscan (Tu), including Monte Amiata and other minor centres. In addition, there is the little isolated district of Melfi (Me), situated almost on the line of the mountains.

In the eastern group, embraced between the Swiss Alps, Carpathians, and Balkans on the one hand and the Apennines on the other, we may distinguish four principal sub-groups or provinces. The eruptions have avoided the closely packed folds of the Alps; but near Padua occur the small Euganean and Vicenzan districts (Eu, Vi), constituting an Adriatic province. Occupying the lobe embraced by the Carpathian chain is the large Hungarian province, comprising the numerous volcanic districts of Hungary and Transylvania. Those marked on the map are the Bakony Mts. (Bk), the Schemnitz district (Sc), Matra Mts. (Ma), Eperjes-Tokay district (ET), Munkacs district (Mu), Meszes Mts. (Ms), Rezbanya district (Rb), and Hargitta Mts. (Ha). A third province, which may be called the Balkan, includes various districts in Servia (Se) with Zajecar (Za), the Visker Mts. (Vs) near Sofia, Sredna Gora (SG) in Eastern Roumelia, and Rhodope (Rp) in Thrace, together with the Karatova and Macedonian districts (Ka, Md). Lastly, the Ægean province embraces parts of Greece and the Archipelago, and continues into Asia Minor.

The outer belt lies wholly beyond the main area of Alpine folding, and mostly at a distance of 100 to 300 miles from its exterior or convex boundary. A few minor centres—the Kaiserstuhl (Ka), Hegau (He), Rauhe Alb (RA), and Nordlingen district (No)—range along a nearer line. In the main outer belt we may distinguish four principal provinces. The French includes the Cevennes district (Ce), Le Puy

(Py), Cantal (Ca), and Auvergne (Au). The Rhenish province includes the Upper and Lower Eifel (UE, LE) and the Siebengebirge (Sb); the Hessian province the Westerwald (We), Vogelsberg (Vo), Rhön (Rh), and Habichtswald (Hb). The Fichtelgebirge district (Fi) stands rather apart. The Saxon-Bohemian province embraces the volcanic districts of the Erzgebirge (Ez) and Mittelgebirge (Mi); and finally, we may note less important centres in the Sudetic Mts. (Su) and the neighbourhood of Teschen (Te).

Abyssal and hypabyssal rocks of Tertiary age are exposed in Elba and the adjacent islands, in the mountains of Servia and Southern Hungary, at Ditro in Transylvania, and in the Bohemian Mittelgebirge. To these we may probably add important groups of intrusions in the Adamello Mts., the Monzoni district, and elsewhere.

Very noteworthy is the total absence of igneous rocks of Tertiary age in those parts of the European continent situated outside the influence of the Alpine system of crust-movements. No younger igneous rocks are known in Western and Northern France, Belgium and the Netherlands, North Germany, Denmark, and Scandinavia, nor anywhere in the vast expanse of Russia outside the southern mountain border formed by the Crimea and Caucasus. The Ural chain belongs to an earlier (Hercynian) system, and, like the old mountain-districts of North Germany and France, has associated with it a different suite of igneous rocks.

Cycles of Igneous Activity.—Setting aside for the present operations conducted in hypothetical intercrustal magma-basins, or generally in the unknown depths of the Earth's crust, we recognise the actual manifestations of igneous action chiefly in the forcing outward of molten magmas from a lower to a higher level within the crust or through the crust to the surface. Such flow takes place, of course, in accordance with the fundamental hydrodynamical law, from a place of higher to a place of lower pressure. If we could assume the total volume of liquid and solid to remain constant, it is clear that any uprise of magma must be compensated by a corresponding settling down of some part of the solid crust. Expansion and contraction attendant on changes of temperature, or on fusion and crystallization, modify this rough statement, but it must hold good approximately for any particular act of intrusion or extrusion. The correlative movement of the solid crust may, however, be distributed over a wide space, and, in describing the effects, it is on the displacement of the mobile rock-magma that we naturally fix attention. The movement of bodies of magma is, then, in the main, one of uprise; but with this is connected in most cases lateral movement, taking the form of quasi-horizontal injection within the crust, or the spreading of extravasated lava at the surface.

Confining our attention still to the dynamical (as distinguished from the thermal) aspect of igneous action, we may distinguish *intrusion* and *extrusion*, according as the displaced magma is raised from a lower to a higher level within the Earth's crust or is raised to the surface and poured out as subaërial or subaqueous lavas. The conditions which determine one or the other result are not of an accidental but of a significant kind. This appears from ample evidence, which proves that in general intrusion and extrusion characterize distinct stages in the history of a region, and must be interpreted as representing different phases of the development of igneous activity. Nor does this twofold division exhaust the distinctions to be recognised. We have further to distinguish *two intrusive phases:* the one characterized by large bodies of plutonic rocks, the other by the injection of numerous smaller bodies, which assume the forms of dykes, sills and other sheets, laccolites, 'plugs,' etc. The rocks forming these smaller bodies have in general the characters which are implied in Brögger's term 'hypabyssal.' They include the dyke-rocks (Ganggesteine) of Rosenbusch, but also other types, such as quartz-porphyry and dolerite, which are otherwise placed in his classification. We shall recognise then a phase of minor intrusions as contrasted with that of large plutonic intrusions. It has sometimes been styled the 'dyke phase,' but in many actual cases stratiform injections (sills, etc.) play a more important part than dykes. Summarily, then, we have *three distinct phases of igneous action*, their respective characteristics being volcanic extrusions, large plutonic intrusions, and minor intrusions.

It has been intimated that these different manifestations belong to different chapters in the history of a given region, and so by implication would seem to represent different stages in the development of igneous activity. It is proper, therefore, to inquire whether such development proceeds on definite lines, constituting a *regular cycle*, in which the several phases follow in a fixed order. The answer to this question is a qualified affirmative. In most cases in which the record is complete the order is found to be the same—viz., that in which the three phases are enumerated above—and, despite some exceptions, we are warranted in ascribing to it a fundamental significance. Provisionally, then, at least, we may accept as the normal cycle that in which igneous action manifests itself successively under three different phases—(i.) volcanic, (ii.) plutonic, (iii.) minor intrusions. As a single illustration, we may take the igneous rocks of Lower Old Red Sandstone age in Scotland.[1] The plutonic intrusions of this period disrupt and metamorphose the volcanic rocks, wherever these are preserved in the neighbourhood, while the dykes and sheets intersect the plutonic as well as the volcanic rocks. The evidence is very clearly presented at the Cheviot centre (Fig. 2), where the succession is as follows:

(i.) *Volcanic Phase.*—Pyroxene-andesite lavas (silica percentage, 59-64), with some subordinate tuffs.

(ii.) *Plutonic Phase.* — A rather basic pyroxene-bearing granite (66-67).

(iii.) *Minor Intrusions.*—Dykes (and some sheets), falling into two main groups: (*a*) acid rocks—quartz-felsites and granophyres (70-73); (*b*) intermediate rocks—mica-porphyrites (61). (N.B.—The Acklington dyke, which crosses the southern part of the area, belongs to a later (Tertiary) age.)

The British Tertiary cycle shows the same sequence, with

[1] *Cf.* Kynaston on the Lorne Area, *Summary of Progress Geol. Sur.* for 1897, p. 87.

a longer record of events. Without multiplying examples of the rule, we will consider some exceptions, and apparent exceptions, to it. The absence of one or other phase in some cases is susceptible of more than one explanation. In the Lewisian system we find preserved only extensive

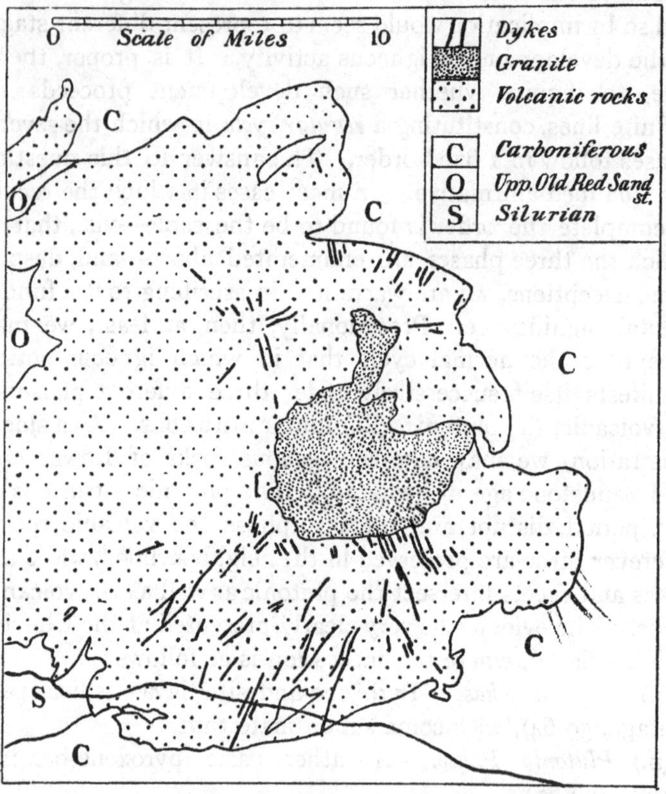

FIG. 2.—SKETCH-MAP OF THE CHEVIOT DISTRICT, ILLUSTRATING THE THREE SUCCESSIVE PHASES OF IGNEOUS ACTION.

Compiled from the maps of the Geological Survey. Scale, 1 inch to 5 miles.

plutonic masses intersected by numerous dykes and sheets. Here it is possible that volcanic rocks once overlying have been removed by pre-Torridonian erosion, and pebbles of rhyolite in the Torridonian sediments lend some support to this supposition. The Lower Old Red Sandstone cycle shows in Scotland the normal sequence; but in the English Lake District the volcanic phase is not represented. In this outlying area the only igneous rocks of the period are the granites of Shap, Skiddaw, and Eskdale and the dykes belonging to them (acid rocks and mica-lamprophyres). In the British Carboniferous we find only volcanic rocks and minor intrusions, the plutonic phase being wanting. The suppression or non-appearance of one of the three phases, when it is not due merely to the imperfection of the geological record, is a departure from what we have laid down as the normal sequence in a complete cycle, but it does not import any reversal or confusion of order. An exception of a different kind is the recurrence of extrusion as the latest phase of waning activity, or, rather, the revival of activity in this form after a more or less prolonged interval of quiescence. Examples of this are perhaps frequent enough to suggest that our provisional rule needs to be supplemented or modified to cover this case. Thus, in the British succession, the Silurian lavas and tuffs of Gloucestershire and Co. Kerry may be regarded as an appendix to the Ordovician cycle; the volcanic rocks of Upper Old Red Sandstone age in the Orkneys and the Limerick district seem to stand in like relation to the cycle of the Lower Old Red Sandstone; and the Permian volcanic rocks described by Sir A. Geikie in Ayrshire and Dumfries may be attached in the same way to the Carboniferous cycle. Other instances might be cited from the continent of Europe. Thus, Bertrand regards the Triassic eruptions of the Monzoni district[1] and elsewhere as the latest waning phase of Hercynian (Permo-Carboniferous) activity. There has been, then, in several past periods a *final reversion to the extrusive phase* of action, following what we have regarded as the normal cycle, and divided from it by a considerable interval of time. This recrudescence of vulcanicity

[1] The evidence seems to indicate that only the interbedded volcanic rocks are of Triassic Age, the associated intrusions being Tertiary. See Salomon, *Tsch. Min. Petr. Mitth.* (2), vol. xvii. (1897), pp. 109-283; Mrs. (Ogilvie) Gordon, *Quart. Journ. Geol. Soc.*, vol. lv. (1899), pp. 560-633, and other papers.

has always been relatively feeble, and has operated within a much-restricted area, or broken out at a few isolated centres.

To the widespread volcanic activity of the present age no such subordinate part can be assigned, and we have to inquire what is its true significance. Is it to be regarded as a survival of Tertiary activity, or as the opening phase of a new cycle? Probably both interpretations may be maintained with reference to different parts of the globe. There is abundant evidence, both in the past and in the present, that the development of igneous action on what we have considered the normal lines may be more rapid in one area than in another. Thus, in Britain the latest great cycle ran its course, including the three phases, and came to an end long before the close of the Tertiary era; but in Iceland vulcanicity, initiated at about the same time (Eocene), has continued, apparently with an unbroken record, to the present day. On the other hand, in many parts of the world vulcanicity at the present time seems to be, not a survival, but a revival, of the activity of an earlier epoch, being divided from it by a more or less prolonged interval. In some parts of the great Pacific circle the distinction is emphasized by a difference in the mode of action, the early Tertiary and mid-Tertiary outbreaks taking the form especially of fissure-eruptions, while those of Pleistocene and recent age are of the central type. The geological record of the near past being necessarily imperfect in a special sense, and that of the future wanting, we shall forbear to offer any decisive judgment on the point raised; and, since Tertiary and Recent volcanic operations have affected for the most part the same regions of the Earth's crust, we may for many purposes group all the 'younger' igneous rocks together.

For the Devonian igneous rocks of the Christiania district Brögger has deduced a succession which does not seem to harmonise with what we have regarded as the normal cycle of events. The rocks fall into natural families, designated by the principal abyssal type in each—essexite, larvikite, etc. Although Brögger does not state the conclusion in definite shape, his account seems to imply that all the rocks of the essexite family — volcanic, abyssal, and hypabyssal — were extruded or intruded before the next succeeding family made its appearance, and so for the rest in order. Instead of one cycle, there would thus be a number of minor cycles. The field-evidence on which this interpretation rests has not yet been clearly set forth. Other districts more or less closely comparable with that of Christiania, such as Essex County in Massachusetts and the 'Monteregian Hills' in Canada, seem to conform to the normal sequence—at least, so far that the minor intrusions fall into a distinct phase posterior to all plutonic intrusions.

CHAPTER II

VULCANICITY

The place of vulcanicity as one phase of igneous action.—Intercrustal magma-reservoirs.—Two contrasted types of vulcanicity.—The function of water in the mechanism of volcanoes.—Fissure-eruptions. —Central eruptions.

The Place of Vulcanicity as one Phase of Igneous Action. —The actual phenomena of volcanic eruptions are of so striking a kind, and their consequences to human life and property have in some cases been so momentous, that from an early time they have excited an interest by no means confined to professed students of science. Being, moreover, in the popular experience, by far the most remarkable manifestations of terrestrial igneous action, they probably represent in the minds of most who are not geologists the sum-total of such manifestations; though even among superficial phenomena there are some, such as hot springs, which suggest a more continual and widespread operation of subterranean igneous energies. It is unfortunate that this partial and misleading view is countenanced by the language employed by many geological writers, who habitually use the terms 'vulcanicity,' 'volcanic rocks,' etc., in a comprehensive sense to include the operations and products of igneous action in general.

The point here raised is not merely one of terminology. The idea, whether present or not in the mind of the writer, is inevitably suggested to that of the reader, that the important results of igneous action are those which take effect in outbursts at the Earth's surface, and that such operations as have their *locus* in the crust itself are merely subsidiary and ancillary to these superficial eruptions. A few geologists there are who would maintain the thesis roughly indicated, and might defend, on that ground, the comprehensive usage of the term 'volcanic.' Taking a wider view of the province and functions of igneous action, we shall avoid confusion by employing the word 'volcano' and its derivatives only in application to actual subaërial or submarine outbursts. Further, for molten rock-material in general we shall adopt the term 'magma,' reserving 'lava' for flows of such molten matter extravasated at the surface, and for the rocks formed by the consolidation of these flows under superficial conditions.

Almost any theory of cosmogony requires us to believe that, as regards the greater part of their solid mass, the Earth and other members of the Solar System are composed of igneous rocks consolidated under plutonic conditions. It is probable, as suggested by Tschermak,[1] that the requisite conditions for superficial igneous outbursts are realised only during a certain stage in the history of a planet—a stage through which the Earth is now passing, while the Moon has outlived it. In any case, the volcanic phase of igneous action must be, from the cosmic point of view, subsidiary and of quite minor importance as regards permanent consequences. Confining our attention, however, to that part of the Earth which is accessible to direct observation, we may still assert confidently that intruded rock-bodies bulk far more largely in the composition of this outer crust than extruded material. Great outpourings of lavas, it is true, are known in various parts of the world. The 'Deccan traps' of India are estimated to cover an area of 200,000 square miles, with a thickness varying from 200 to 6,000 feet; though it is probable[2] that this thickness has been much exaggerated by reckoning as lava-flows the intercalated intrusive sills. These and other such great lava-fields belong to the later

[1] 'Ueber den Vulcanismus als kosmische Erscheinung,' *Sitz. k. Akad. Wiss. Wien.*, *Math.-Nat. Classe*, vol. lxxv. (1877), pp. 151-176.

[2] Fermor, *Rec. Geol. Sur. Ind.*, vol. xxxiv. (1901), p. 161.

chapters of geological history; and analogy from other areas would lead us to infer that these imposing outpourings of lava have been accompanied by even more voluminous subterranean intrusions, not yet revealed by erosion. In peninsular India itself ancient crystalline rocks, in the main plutonic intrusions, occupy a larger area than the Deccan lavas, with an unknown depth. The vast tracts of granitoid and gneissoid rocks in Scandinavia, Canada, Brazil, and other countries must greatly outweigh the known volcanic rocks of all ages, and each such tract is only a part of a larger concealed mass.

Setting aside the question of the relative bulk of intruded and extruded rock-magmas, it is not difficult to see that intrusion rather than extrusion must be the more immediate phase of igneous action; and the geological record shows that the former is more directly related than the latter to the larger geological conditions which determine this class of effects. The one represents the general and the other a special case. Igneous action, in its dynamic aspect, consists in the moving of a body of magma from one situation to another, in response to differences of fluid pressure; but whether or not some part of the magma is forced out at the surface depends upon conditions of the second order of importance. Intrusion accordingly may, and does, occur without extrusion, but the converse case never. Any extravasation of magma at the surface implies, of course, an intrusion of the same magma through the subjacent crust, and a residual portion necessarily occupies the channels of uprise in the form of dykes, 'plugs,' or the like. But, apart from this, it seems to be invariably the case, wherever the natural processes of dissection have progressed far enough to reveal the complete history, that volcanic outbursts are accompanied by intrusions of abyssal or hypabyssal character, which are in no sense subsidiary to the surface eruptions, but independent, and in general of later date.

Summarily, as already laid down in the preceding chapter, volcanic eruptions occupy a more or less definite place in the succession of events which goes to make up a complete igneous cycle, being characteristic of, and usually confined to, a certain stage of development; and it even appears that the volcanic phase may be in some cases wholly suppressed.

Intercrustal Magma-Reservoirs.—A systematic discussion of igneous action, which we shall not attempt in any formal shape, would involve the consideration of two questions, distinct but not wholly separable—viz., the source of igneous rock-magmas, and the nature of the forces by which they are displaced and brought into action in the manner of intrusion and extrusion. The close relation between these two questions, involving at once thermodynamical and hydrodynamical problems, makes the subject in its entirety one of peculiar complexity.

As regards *the source of the magmas*, it is clear that a fundamental inquiry would take us back into a region of speculation from which we have already turned away as unprofitable. It does not appear, however, that ignorance concerning the condition of the interior of the Earth as a whole need materially embarrass the pursuit of the object immediately before us. There are at least considerations which may serve to reassure us that the problems of petrology are in great measure independent of the speculations of cosmogony.

One of these considerations, tending to remove the immediate subject of study from the wider and more hypothetical questions which we have excluded, may be briefly stated. Even if we assume the terrestrial mass to have been at some past time wholly fluid, and make whatever supposition may be deemed most probable as to the persistence to the present time of a liquid condition in some portion of the interior, it cannot be supposed that the igneous intrusions and extrusions of which we have actual knowledge have been fed directly from any such reservoir of residual liquid. This remark applies equally to the general fluid interior conceived, for instance, by Arrhenius, to the fluid substratum of Fisher's theory, or to the liquid portions, honeycombing an otherwise solid globe, which form a part of Lord Kelvin's hypothesis. The evidence of volcanoes now active is in itself sufficient to negative the supposition that they draw *directly* upon a common source. For instance, between the summits of the lava-columns in Mauna Loa and Kilauea, some twenty miles apart, there is a difference of level of nearly 10,000 feet. The simple hydrostatic law plainly forbids the supposition of any continuous fluid connection between the two vents, and it is manifest that their immediate sources must be, in a sufficiently definite sense, independent. The same conclusion is strongly enforced when we have regard to the wide differences of composition sometimes found between the products of even closely neighbouring active volcanoes, as in the group of the Lipari Isles. It would be easy to multiply examples, and to show that what is true of volcanic eruptions holds good of igneous action in general. The facts warrant us in rejecting decisively the conception of an immediate common stock-reservoir.

What has been urged concerning a certain sufficiently defined independence even between adjacent volcanic vents must, however, be considered in connection with another class of facts, already alluded to, which indicate a community of relationship, more or less close, extending even throughout a large region which is at some period the seat of igneous action. Considering still the volcanic phase of activity, as being more immediately in question, it is well known that modern volcanoes in a given district sometimes show a marked sympathy of behaviour, as illustrated, *e.g.*, by the outbursts in Martinique and St. Vincent in 1902. Further, the products erupted in such cases are often closely similar in composition. More generally, we see that, in the past and the present, volcanic conditions may be realised contemporaneously (in a broad sense) throughout a large but still defined region of the Earth's surface, and that within this region the magmas erupted may show a similarity of composition which can only be explained by community of origin.

The facts here alluded to will receive closer attention in the following chapters. They are briefly cited here to show that the reservoirs of molten magma, which must be postulated within the Earth's crust in a region of igneous activity, are of different orders of magnitude and different degrees of permanence. While we picture each distinct volcanic centre as possessing its own proper reservoir of lava, it is necessary to suppose that such reservoir is of relatively small dimensions and of temporary status. In a large class of volcanoes, at least, there are strongly marked epochs of eruption separated by intervals of quiescence, which may last for years, or even for centuries. We may suppose that by each great eruption, or connected succession of eruptions, the local reservoir becomes exhausted, in the sense that the magma which has not been extravasated becomes wholly or mainly solidified. A new eruption after a more or less prolonged pause implies, therefore, a renewal of the local source of magma; or in the case of a volcano of steadier activity, a more or less continual renewal is implied. This rejuvenescence, whether recurrent or continual, might conceivably be brought about in either of two ways—by the melting (or remelting) of solid rock at the volcanic focus, or by an accession of fresh rock-magma from some other source. Considerations sufficiently obvious enable us to discard at once the former alternative. Apart from the impossibility of accounting for the generation of an indefinite supply of heat at a particular place within the Earth's crust, what is known of the composition of lavas clearly forbids us to suppose that they arise from local melting of the crust within a relatively small compass. We must conclude, therefore, that the local reservoir of an individual volcano is supplied by drafts from some much larger body of rock-magma with which it is from time to time in communication. In accordance with the general law of uprise, we must suppose the large parent reservoir to be situated at a greater depth than the local springs of the volcanoes connected with it. Where throughout a considerable tract we find what may be accepted as evidence of a community of origin among the volcanic eruptions, we must conceive *a very extensive intercrustal magma-basin, or a train of such basins*, underlying the tract in question. It should be observed that community of origin by no means necessitates identity of composition in the products at the several centres. We shall see later that a large body of magma in such a reservoir as is here considered may undergo such 'differentiation' that drafts from different parts of it will differ widely in composition.

Various lines of argument go to show that the small local reservoirs of individual volcanoes are not necessarily situated at any very great depths below the surface. For instance, De Lorenzo[1] calculates that for the volcanoes of the Campi Phlegræi the depth is from 1,000 to 2,500 metres, and for Mte. Somma not more than 3,000 metres. At a depth of a mile or two miles the normal temperature of the Earth's crust is still a comparatively low one, and a great amount of heat would be required to raise the solid rock to the temperature of fusion and melt it. At the depth at which we suppose a large magma-reservoir to be situated the conditions are very different. It is easy to see that in any (relatively) *permanent* intercrustal reservoir the magma must have nearly the same temperature as the solid rocks which enclose it. Unless we assume a very important difference of composition between the magma and the rocks with which it is in contact, any noteworthy superheating of the former above the temperature of the latter is impossible, just as it is impossible to heat water in a vessel made of ice. The magma in the reservoir and the contiguous solid crust must be approximately at the same temperature—viz., the temperature of fusion[2] under the pressure which obtains at that depth. Any accession of heat or relief of pressure will cause melting of part of the enclosing solid rock, and any loss of heat or increase of pressure will cause crystallization of part of the magma. Gain or loss of heat will not alter the temperature, and change of pressure will merely adjust it to the new temperature of fusion.

[1] 'Considerazioni sull' origine superficiale dei volcani,' *Atti. R. Acc. Sci. fis-mat. Napoli* (2*a*), vol. xi., No. 7.

[2] The term 'temperature of fusion', as applied to a rock composed of different minerals, is a loose one, but does not require explanation in this place.

Since the first requisite in any general theory of the *causes of igneous action* must be the provision of some source for the molten magmas which are intruded and extruded, we here touch as closely as it is our intention to do upon this fundamental question. As remarked, the parent magma may be provided in some deep-seated part of the Earth's crust either by an accession of heat or by a relief of pressure. It is not easy to frame any adequate explanation of the subterranean development of heat in quantity and its localisation in the manner postulated. Mallet's theory of the mechanical generation of heat by the crushing of rock-masses has been destructively criticized by Fisher and others. We conceive the seat of igneous action to be situated at depths where crushing of the kind contemplated cannot be effective. Mallet's theory, again, was propounded with special reference to volcanic action; but, as we shall see, in the more important type of vulcanicity the crust-movements concerned are not of a kind involving any important crushing of rock-masses. More recently Dutton[1] and others have put forward hypotheses depending on a chemical generation of heat by radioactivity. Here the connection between igneous action and crust-movements, which we have seen reason to regard as of fundamental significance, finds no place. Data are not yet obtainable for testing such a theory on its merits; but it may be remarked that, according to Strutt's conclusions, little radium is to be expected in the deeper parts of the Earth's crust.[2] Dutton believes volcanic action to have its seat wholly within a very shallow depth of the crust (1 to 2½ miles)—a supposition which we must reject for reasons already given.[3] It appears, then, that we must seek the immediate cause of igneous action, not in the generation of heat, but chiefly in *relief of pressure* in certain deep-seated parts of the crust where solid and molten rock are

[1] 'Volcanoes and Radio-activity,' *Journ. Geol.*, vol. xiv. (1906), pp. 259-268.

[2] See also v. d. Borne, *Jahrb. f. Radioakt. u. Electron.*, vol. ii., pp. 77-108.

[3] *Cf.* Louderback, *Journ. Geol.*, vol. xiv. (1906), pp. 747-757.

approximately in thermal equilibrium. We are thus led by an independent line of reasoning to the principle already enunciated, which connects igneous action primarily with crustal stresses, and so secondarily with crust-movements.

The idea of large intercrustal magma-basins occupies an important place in that conception of igneous action which seems to accord best with the facts of geology, as will be more fully apparent when we come to discuss 'petrographical provinces' and the course of events comprised within a complete cycle of activity in a given tract of the Earth's crust. We conceive the parent magma-basin as underlying such tract at a very considerable depth, and having, with reference to the cycle of events, a considerable degree of permanence. It is not necessarily coextensive with the tract in question, for the magmas drawn from it may not only ascend, but also travel laterally. The degree of permanence postulated implies that the parent reservoir is available as a direct or indirect source of molten magma at those epochs of the complete cycle when intrusion or extrusion takes place. It does not exclude the possibility of the basin becoming partially (or even totally) frozen at particular stages and again remelted. In particular, we may not improbably imagine it divided, in the waning stages of activity, into several smaller isolated reservoirs. Further, we are free to contemplate the possibility of subsidiary reservoirs being formed at higher levels within the crust and having a less degree of permanence, these being mainly fed in the first place from the larger basin below, but becoming in their turn sources of intrusion and extrusion.

We have no data for determining with any degree of confidence the depths at which large intercrustal magma-reservoirs of the kind described may exist. Almost any plausible theory of the present state of the interior of the globe leads to the conclusion that at some depth, measured probably by decades of miles, the rocks must be closely at the point of fusion, but any calculation of the depth depends necessarily upon precarious assumptions. Nor can we tell to what further depth the condition thus reached may be supposed to continue. On Chamberlin's hypothesis, the greater part of the Earth's interior is supposed to be at or near the temperature of fusion.[1] If large molten reservoirs can be formed only at depths of at least 30 or 40 miles, it can scarcely be possible for the great bodies of plutonic rock, which must mark their situation, to be exposed by erosion. In that case the largest igneous rock-masses known to geologists may represent subsidiary reservoirs, which were only off-shoots from the main basin and at a higher level. In confirmation of this, it is found that such masses very generally give evidence of distinctly intrusive relations, and are sharply separable from the contiguous rocks, which, though highly metamorphosed, show no sign of fusion.

Two Contrasted Types of Vulcanicity.—Distinct from the question of the source of the molten rock-magmas, though, as has been remarked, closely bound up with it, is that of the mechanics of igneous action. We have already seen that the cause of the displacement of bodies of rock-magma is to be sought in readjustment of the Earth's crust, which has become subjected to powerful stresses as a consequence of deformation. This, like the former question, would, if discussed on systematic lines, carry us back to speculations concerning the genesis of the Earth and the present condition of its interior as a whole. As before, we may avoid this by approaching the subject from the geological side. The readjustment takes effect in displacement of parts of the solid crust, as well as of fluid magma. The effects of crust-movement, as seen in faulting, folding, etc., thus afford a picture of the forces which have also been responsible for igneous intrusion and extrusion within the same region of the Earth's crust and at the same period.

It follows that an analysis of crust-movements from the dynamical point of view would lead to a systematic classification, from the same aspect, of the movements of rock-magmas—*i.e.*, of the material operations of igneous action. In this place we shall notice only one distinction, which seems to be of a very fundamental kind. On the broadest

[1] Chamberlin and Salisbury, *Geology*, vol. i. (1905), p. 541.

view *crust-movements are of two types*, according as the displacement is essentially in the vertical or the horizontal sense. We shall term these for convenience *plateau-building* and *mountain-building* movements respectively. In the former there is a relative upheaval or depression of great blocks of the Earth's crust; deformation and tilting of the blocks themselves being subsidiary to the bodily displacement, and any folding or other effects due to lateral thrust being only a secondary result. Whether the absolute displacement (relative to the centre of the Earth) is upward or downward is immaterial for our purpose: in Suess' view the movement is always one of subsidence, but in different degrees. In the mountain-building type the movement is primarily in the horizontal sense, and results from lateral thrust. Any vertical displacement, such as the relative upheaval of the core of the Alpine chain, is a secondary and local effect. Since lateral thrust and horizontal movement necessarily have some definite direction in azimuth, there is in this case a linear (curvilinear) disposition of the region principally affected, which is wanting in the plateau type. The tectonic features characteristic of the two contrasted types are sufficiently familiar. In a plateau region we find extensive tracts of relatively undisturbed strata, often with uniform and gentle inclination, or quasi-horizontal, and, as a characteristic incident, normal faulting or its analogue, monoclinal folding. In a mountain region we have, on the other hand, high dips, anticlinal or isoclinal folding, reversed faulting or over-thrusting, and all the well-known features of Alpine architecture.

Corresponding with the two principal categories of crust-movements, and the two characteristic tectonic types which result from them, we have to recognise two contrasted modes in which volcanic action manifests itself. We shall distinguish these as *fissure-eruptions* and *central eruptions*; and they are the expression, as regards igneous action at the surface, of plateau-building and mountain-building forces respectively. A like distinction will be made later in the case of igneous intrusions.

The close relation of the eruptions to the crust-movements is usually less obvious in a plateau than in a mountain region—a consequence of the wide extent and simple nature of the displacements in the former case—but it is not less real and essential. A classical example is that part of Southern Norway lying west and north of the Christiania Fjord, which has been the subject of exhaustive study in the hands especially of Professor Brögger.[1] Here a tract, having an area of nearly 4,000 square miles, is composed of Lower Palæozoic strata and igneous rocks of Old Red Sandstone age. It is surrounded by Archæan gneisses, and the boundary between these and the Lower Palæozoic rocks is everywhere a faulted one. The actual relations, as mapped out by Brögger, are somewhat complex. In places there is one great fault, with a throw of over 3,000 feet; in other places a group of faults produces a like total effect; but the general effect is a relative depression of this tract, some 150 miles long, as compared with the adjacent country. It was within this sunken area that igneous action broke out. There were volcanic outpourings on a very extensive scale and a great suite of intrusions, the larger bodies taking the general habit of laccolites. The relation of the igneous rocks to the crust-movements is apparent in respect of time as well as of areal distribution. The relative displacement along the main lines of faulting has been effected by repeated movements, and the more important of these were contemporaneous, in a broad sense, with the igneous activity. In some districts—notably that of the Langesundsfjord—actual simultaneity of crust-movement and igneous intrusion has given rise to effects of a remarkable kind.

The actual distribution of stresses in the crust of the Earth, and the visible displacements which result, are necessarily of a more complex kind than can be expressed in a brief formal statement, and what has been said of plateau and mountain regions must be understood with this qualification. Indeed,

[1] 'Ueber die Bildungsgeschichte des Kristianiafjords,' *Nyt. Mag. for Naturv.*, vol. xxx. (1886), pp. 99-231; also, for the igneous rocks, *Zeits. f. Kryst.*, vol. xvi. (1890), and later memoirs.

it results from the spherical form of the Earth that vertical and horizontal movements can never be theoretically independent. Displacements of the plateau-building type may be accompanied only by very subsidiary movements connected with lateral thrust, but it appears that effects of the latter kind may also acquire a certain importance. Of such nature is the overriding of a relatively depressed block by an adjacent tract, upon which Suess[1] has laid stress. A good example is that of the Lower Old Red Sandstone volcanic district of Glencoe,[2] which constitutes a sunken and faulted area in the midst of the Moine series (quartzites, etc.). Along part of its course the bounding fault is one of overthrust, and along its outer edge is a belt of intrusions approximately contemporaneous with the faulting.

The relation of central volcanic outbursts to crust-movements of the mountain-building type is in general sufficiently evident; but we may remark that it is not always of the same kind. In what may be regarded as the simplest case, the volcanic vents break out along the main orographic axis itself, and there results a *serial* arrangement which is highly characteristic. It is illustrated on a grand scale in the Andes, and again in the Caucasus. Another case is exemplified by the Tertiary and Recent volcanic districts of Central and Southern Europe, as related to the powerfully overthrust mountain-chains of the Alpine system. Here the eruptions have found vent, not along the orographic lines themselves, but within the loops embraced by the various divergent curves, and the linear disposition is lost in a *grouped* arrangement.

The Function of Water in the Mechanism of Volcanoes.—Volcanoes differ greatly in magnitude, in the degree of violence of their eruptions, and in the nature of their liquid and solid ejectamenta; but a constant characteristic common to all volcanic outbursts is the emission of large volumes of steam, with other gases. It is therefore natural to see in the force of these imprisoned gases the proximate cause, at

[1] *Das Antlitz der Erde*, vol. i. (1885), pp. 181-187.

[2] Bailey, *Summary of Progress Geol. Sur.* for 1905, pp. 96-99.

least, of the eruptions themselves, and this factor has occupied a prominent place in most speculations dealing with the subject. The exponents of these various views—not in all cases professed geologists—have usually regarded vulcanicity as an isolated phenomenon, and those theories which have taken explicit form are, for the most part, not theories of igneous action, but merely of volcanic outbursts.

We have been led to regard vulcanicity as merely one phase—and not the most characteristic phase—of igneous action, and igneous action itself, not as an incident, but as a very essential part of the economy of the terrestrial globe, closely bound up with deformation of the solid crust. No other view is possible if we believe that, at a sufficient depth, such conditions of temperature and pressure prevail that solid and liquid rock are in approximate thermal equilibrium. Any local relief of pressure within that region, connected with a redistribution of stress in the crust, must then give rise to melting and the formation of an intercrustal reservoir of rock-magma. We can easily conceive that the same purely mechanical causes may bring about the transference of part of the magma to higher levels within the Earth's crust, and even its extravasation at the surface. That this mechanical factor does at least control and direct extrusion as well as intrusion is manifest from the intimate association of volcanic eruptions with obviously mechanical processes such as the folding of strata in mountain-chains. Eruptions of the plateau type, in the nature of the case, have no such evidently significant localisation; but we may often see in connection with them equally convincing indications of vertical movement in the sense of subsidence of the solid crust. Granted a sufficiently extensive local reservoir of magma at a moderate depth below the surface, it would seem possible to explain *fissure-eruptions* by gravitational readjustment alone—*i.e.*, by the mere pressing outward of the fluid magma through fissures concurrently opened, with a corresponding settling down of the heavier solid crust which overlay the magma. A similar view of the mechanics of massive extrusions, and of the analogous sill-formed intrusions, was long ago put forward by Hopkins,[1] but apparently with reference to a general intratelluric body of magma rather than the local and temporary reservoir, constituted *ad hoc*, which is here contemplated.

The purely hydrostatic hypothesis, which seems to cover the main facts with reference to fissure-eruptions, leaves to the dissolved water (or steam) in the magma only a subsidiary part. Free fluidity is essential to the process described. Most of the great massive outpourings of which we have knowledge are of basic lavas, for which a relatively low degree of viscosity is a specific property; but the presence of water, and perhaps of more potent volatile fluxes, in any magma doubtless causes a notable reduction of viscosity.

In *central eruptions*, and especially in those of a paroxysmal kind, a much more important rôle must be assigned to the elastic force of steam and other gases imprisoned in the magma. These volatile constituents exist, in solution under pressure, in all rock-magmas; and a portion of magma forced up to a higher level, there to constitute a smaller and temporary reservoir, carries with it its share of them. It doubtless carries something more than its proportionate share, for these volatile substances, even in solution, must tend to concentrate to some extent at the higher levels in any large continuous body of magma. With transference of the magma to a higher level within the Earth's crust, the confined gases may thus be brought into the condition for an explosive outburst, and an orifice perforated through the overlying rocks. The drilling of a cylindrical aperture in this manner is illustrated by the interesting experiments of Daubrée[2] on the mechanical effects of high explosives, which likewise throw light on the comminution of rocks to make volcanic dust and on other geological phenomena.

[1] 'Researches in Physical Geology,' *Trans. Camb. Phil. Soc.*, vol. vi. (1835), pp. 1-84; also *An Abstract of a Memoir on Physical Geology; with a Further Exposition* . . . (1836).

[2] *Comptes Rendus*, vol. cxi. (1890), pp. 767-774, 857-863, and vol. cxii. (1891), pp. 125-136, 1434-1490, 1890-1891; also *Bull. soc. géol. Fra.* (3), vol. xix. (1891), pp. 313-354.

Communication with the surface once established, the conditions are changed. The sudden relief of pressure, causing the disengagement of much of the gaseous constituents of the magma, and consequently a great expansion of volume, may occasion not only an uprise of magma in the conduit which has been opened, but a violent expulsion of a portion of it through the vent. To what extent the uprush of steam and other gases carries the liquid lava with it will depend on the viscosity of the magma, the width of the volcanic conduit, and other circumstances. In exceptional cases the gases alone find exit, and the only permanent record of such an eruption may be a tuff or breccia wholly of non-igneous material, torn from the walls of the conduit. More often lava is blown out, giving rise to volcanic bombs, ashes, or dust. We may presume that an individual eruption is brought to an end, not by exhaustion of the local magma-reservoir, but by the choking of the conduit with magma which has attained a prohibitive degree of viscosity, a consequence mainly of its becoming depleted of its gaseous constituents. The plug may be blown out by a new explosion, when the imprisoned gases below have gathered sufficient strength. In short, it is easy to see that the course of events at a given volcanic centre and the character of the eruptions must depend on a number of variable factors; but among these the content of water and other volatile substances in the magma is necessarily of prime importance.

It is proper to make a few remarks relative to *the source of the water* which takes part in volcanic eruptions. The hypothesis of oceanic waters obtaining access to the heated interior of the globe, and there being absorbed into molten rock-magmas, is, at the first glance, a plausible one. It was entertained at an early time by von Buch and von Humboldt, was adopted with more or less modification by Scrope, Dana, and others, and has more recently been advocated by Arrhenius.[1] Those who have embraced this theory have urged the influence of water in lowering the melting-points

[1] 'Zur Physik des Vulkanismus,' *Geol. Foren. Förh. Stockholm*, vol. xxii. (1900), pp. 395-419.

of minerals, and therefore in bringing about fusion; and they have pointed to the constant presence among volcanic products of steam, and often of other substances which may conceivably be derived from sea-water. It has often been insisted, also, that most modern volcanoes are situated at no great distance from the sea; but this generalisation at once loses its force when we include past as well as present volcanoes in our survey. Indeed, it is not difficult to see that the relation so frequently observable between coast-lines and belts of volcanic activity is not an essential but a secondary one, arising from the fact that both are related directly to the same orographic lines.

It is certain that at a very few miles beneath the surface of the Earth—viz., at the limit of what the American geologists term the '*zone of fracture*'—any open fissure in the rocks, however narrow, is impossible. No downward flow of water, in the ordinary sense, can therefore be postulated. Further, a consideration of the vapour-pressure of water at different temperatures, and of the actual temperatures at which lavas are erupted, shows that it is not with liquid but with gaseous water that we are concerned. The theory would therefore require us to believe that steam can force its way, against enormous pressure, through capillary channels in the deeper parts of the Earth's crust. The latter, as Arrhenius remarks, must act as a semi-permeable membrane, with pores wide enough to admit liquid or gaseous water, but not wide enough for the passage of the other constituents of a rock-magma; and he confesses that, to make this hypothesis intelligible, water must have quite different properties from those generally attributed to it.

On the other hand, we know that water and various gases are present in all igneous rocks, even the most deep-seated, which have been examined. The water amounts on the average to about 1½ per cent., a proportion quite sufficient to endow the molten rock with the properties displayed in volcanic eruptions. Some rocks contain a very much larger amount, various hypogene processes conspiring to concentrate the water in particular magmas. A fresh hypabyssal pitchstone has usually 5 to 10 per cent. or more; while a surface-lava of the same general nature (obsidian) seldom contains so much as 1 per cent., owing to the escape of steam during eruption. The other volatile constituents found in igneous rocks are in much smaller amount by weight: Tilden's[1] experiments would indicate an average of about 0·17 per cent. He found that hydrogen and carbon dioxide are the most abundant gases, while carbon monoxide, methane (CH_4), and nitrogen also occur. It is possible that some part of the lighter gases found may be formed by reactions in the laboratory. To these free gases we must add those which occur combined in common minerals of igneous rocks—viz., boric, hydrofluoric, hydrochloric, and hydrosulphuric acids, with others. Our knowledge of the gases actually given out through volcanic vents is derived almost wholly from fumeroles, which are found to emit different mixtures of gases at different stages of their development;[2] but it is clear that these agree generally with the list just given, and not with the composition of the atmosphere. Gautier[3] has contended that most of the gases contained in rock-magmas are generated there by various chemical reactions; but the conditions of his experiments are not such as we can suppose to be realised in nature. In any case, there seems to be no reason to postulate a meteoric source for the water and gases which figure in volcanic eruptions; and this gratuitous supposition reverses the causal relations which are suggested by a simple survey of the facts. If we may speculate so far on the past history of the globe, it would seem, not that the sea is the source of the volcanic water, but that vulcanicity (in the broad sense of direct communication between the heated interior and the exterior of the globe) is the original

[1] *Proc. Roy. Soc.*, vol. lx. (1897), pp. 453-457. See also Gautier, *Comptes Rendus*, vol. cxxxii. (1901), pp. 58-64.

[2] For a summary of the facts see Clarke, *The Data of Geochemistry, Bull. No.* 330 *U.S. Geol. Sur.* (1908), pp. 212-220.

[3] Various memoirs in *Comptes Rendus*, vol. cxxxi.-cxliii. (1900-1906), and general discussion in *Ann. des Mines* (10), vol. ix. (1906), pp. 316-370.

source of the oceanic waters, and is slowly adding to them.[1] A like volcanic origin, in a broad sense, may be predicated for the atmosphere, although the composition of the latter is greatly modified by continued reaction with the Earth's crust and with organic matter, living and dead.

Fissure-Eruptions.—Active volcanoes of the central type are numerous at the present time in many parts of the world; and, in particular, well-characterized examples of this type have been intermittently in action for many centuries in the Mediterranean basin, the home of the western civilisation. These have naturally attracted much observation, and we possess very full information concerning the phenomena which they exhibit. To conclude that the mode of eruption represented by Etna and Vesuvius is the more natural, or the more important geologically, or that it is in any sense the normal type and the other exceptional, would, however, be a hasty and unwarranted generalisation. We have here an illustration of a remark made in our first chapter: that, as regards igneous action, even in its superficial and overt manifestations, observation limited to the actual phenomena of the present epoch is necessarily partial and misleading. The geological record tells a different story. Since von Richthofen,[2] forty years ago, drew attention to the vast outpourings of lava in the form of 'massive' or fissure-eruptions in the Western States of America and elsewhere, it has come to be recognised that this type of vulcanicity has played the leading part at numerous epochs, and sometimes over very large regions of the Earth's surface.

The extensive basaltic tract of the Deccan, assigned to a late Cretaceous age, has already been mentioned. In South Africa voluminous outpourings of a like nature have occurred at more than one epoch; and others are indicated

[1] Gregory has maintained that the deep-seated springs in the interior of Australia derive their water in great part directly from an intratelluric source (*The Dead Heart of Australia* (1906), p. 339).

[2] 'The Natural History of Volcanic Rocks,' *Mem. Calif. Acad. Sci.*, vol. i., part ii. (1868); also *Zeits. Deuts. Geol. Ges.*, vol. xx. (1868), pp. 663-726, and vol. xxi. (1869), pp. 1-80.

along the course of the Great Rift Valley of Eastern Africa, where Gregory distinguishes two epochs of plateau-eruptions —one Cretaceous, and the other probably Miocene. In the Pliocene period there were immense floods of lava, mainly from fissure-eruptions, over large stretches of country to the west of the Rocky Mountains. Russell estimates the extent of the 'Columbia lavas' of Washington, Oregon, and Idaho at 200,000 to 250,000 square miles, and the maximum thickness at more than 4,000 feet. Of the Eocene basalt plateaux of the British Isles, Iceland, etc., only relics are preserved; but there is good reason for believing that at that time a continuous lava-field extended from Antrim to far within the Arctic Circle, a distance of at least 2,000 miles. Sir A. Geikie estimates that in Iceland the thickness exceeds 5,000 feet. In that part of the region activity has continued until the present time, and our knowledge of the phenomena of modern fissure-eruptions is derived largely from accounts by Thoroddsen, Helland, and others of the Icelandic area. A good summary has been given by Geikie,[1] who has enforced the geological importance of this type of vulcanicity, especially with reference to the Tertiary igneous rocks of Britain. Dutton[2] has recorded many instructive particulars concerning the fissure-eruptions of Utah and Arizona, where some of the flows are sufficiently recent to retain much of their original characters.

The essential features of this type of volcanic action, so far as regards visible manifestations, may be gathered from these examples. The eruptions take effect, not through a single orifice, but more or less continuously, *along an extensive fissure or group of parallel fissures*, or, at least, at very numerous points distributed along such fissures. These in a given region have certain definite directions, which we may suppose

[1] *Ancient Volcanoes of Great Britain* (1897), chap. xl., with references. For photographs see Tempest Anderson, *Volcanic Studies* (1903). For a somewhat different interpretation see von Knebel, *Zeits. Deuts. Geol. Ges.*, vol. lviii. (1906), pp. 59-76.

[2] *Report on the Geology of the High Plateaux of Utah* (1880), and *Tertiary History of the Grand Cañon District* (1882).

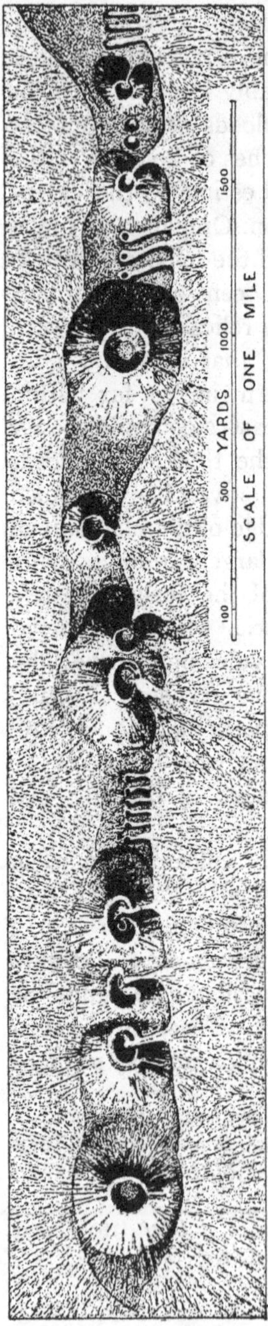

Fig. 3.—Part of the Laki Fissure, Iceland, showing Allignment of Small Cones and Craters formed in the Eruption of 1783. (After Helland.)

The figure reproduces on a reduced scale about one-sixth of Helland's long folding ground-plan.

to be related to crustal strains of a large order. In Iceland one set runs north-east to south-west, and another north to south. The fissures appear at the surface as long open rents (Icelandic 'gja') a few feet wide, and of great depth. The Eldgja, near the Mýrdals Glacier, is more than 18 miles long, with a depth ranging to more than 600 feet. It appears, however, from Dutton's account that the fissures themselves sometimes fail to reach the surface, and are then only indicated by a linear arrangement of numerous small cinder-cones. Such small cones, of slag, scoriæ, or heaps of blocks, are usually formed when the eruption is localised at particular points along a line of fissure; but they are only an incidental feature, and Dutton notes that—*e.g.*, on the Markágunt Plateau—great floods of basalt have been extruded without any formation of cones.

The *volume of lava* extravasated in some fissure-eruptions is enormous, as compared with the ejectamenta of the more familiar central volcanoes. The most considerable eruption re-

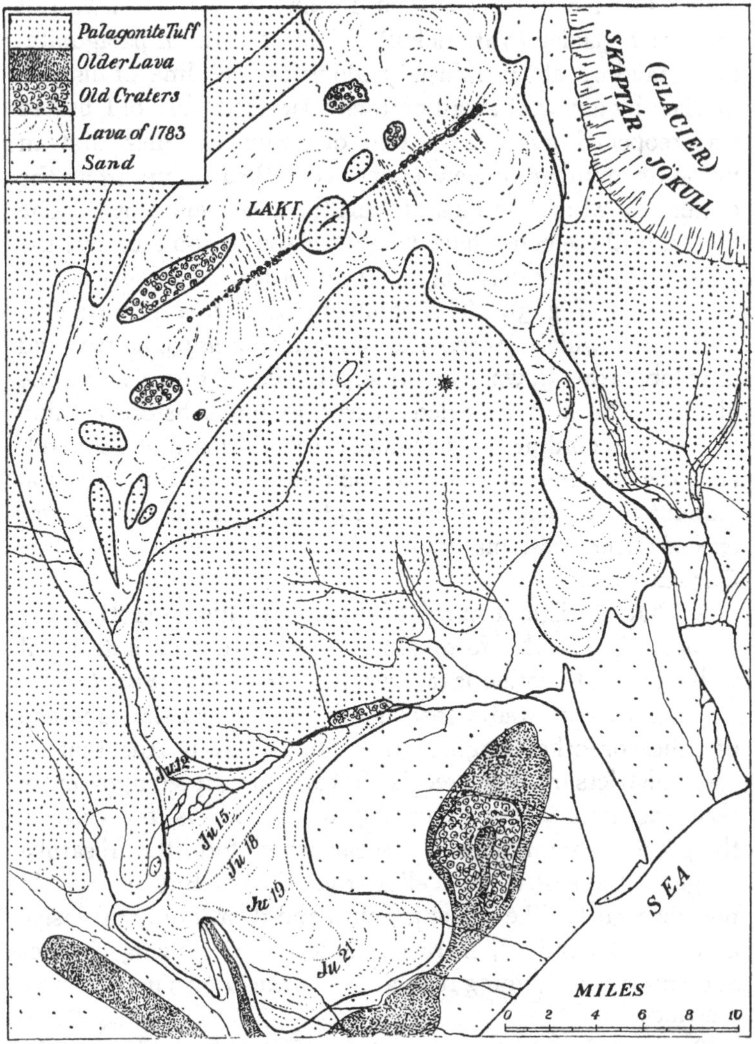

FIG. 4.—MAP TO SHOW THE EXTENT OF THE LAVA-FLOOD FROM THE ERUPTION OF LAKI IN 1783. (AFTER HELLAND.)

In the southern part of the map dotted lines mark the front of the stream at different dates.

corded in historic times is that which took place in 1783 at Laki,[1] situated in the south of Iceland, to the south-west of the great glacier (Vatnajökul). Laki is a hill of palagonite-tuff, standing about mid-way along an old line of fissure, marked by a row of small craters. In 1783 this old fissure was reopened along an extent of some 20 miles, and an immense number of new cones were thrown up, from the craters of which streams of basaltic lava welled out. The largest of these cones have heights of 100 to 300 feet, but the great majority, closely set in linear order, are of very trivial dimensions (Fig. 3). The confluent lava-streams formed a vast flood, which overflowed the surrounding country, and sent long arms down two principal valleys (Fig. 4). In one of these the lava travelled nearly 40 miles before coming to rest. According to the conservative estimate of Thoroddsen, the volume of material poured out in this eruption amounted to nearly 3 cubic miles, which doubtless far exceeds the total material ejected by Vesuvius since the Christian era.

A characteristic feature of fissure-eruptions is that the outbreak takes the form almost exclusively of a tranquil welling out of molten lava, with *little or no intervention of the explosive element*, and, consequently, with a general absence of ashes or other fragmental products. It is true that in some districts massive breccias have been erupted apparently along fissures. This, according to Hague,[2] is the case in the great accumulations of breccias which form the Absaroka range, to the east of the Yellowstone Park. But breccias do not necessarily, like volcanic ashes and dust, imply explosive action; and, indeed, intrusive breccias, occupying fissures and sometimes forming regular dykes, are known in numerous districts.

Connected with the fluid nature of the ejectamenta is the most obvious negative characteristic of fissure-eruptions, the

[1] Helland, *Lakis kratere og lavaströmme* (1886, Universitetsprogram, Christiania).

[2] See especially *Presid. Address Geol. Soc. Washington*, 1899, and *Compte-Rendu VIII. Congr. Géol. Internat.* (Paris, 1900), pp. 364, 365.

absence of any important cone of the Vesuvian type. Although the surface of any lava-stream has a certain inclination, depending mainly on the viscosity and the rapidity of cooling, the actual gradient is very slight in the case of a stream of large volume. This is seen in the flat profiles of the great lava volcanoes of Hawaii, and still more in the voluminous outpourings of lava now considered. It comes about thus that a region of fissure-eruptions is typically *a plateau country in surface topography* as well as in geological structure.

As bearing on the subject of the mechanics of extrusion of the plateau type, it is especially noticeable that the fissures have been determined without regard to pre-existing faults or other details of geological structure. This is very strikingly shown on the High Plateaux of Utah and Arizona. Dutton remarks further of the same region that the fissures have broken through without reference to the bold surface-relief of the ground, so that they are situated with indifference on the top of a butte, in the bottom of a valley, or even on the brink of a precipice.

At the close of an eruption the channel of uprise, filled by a portion of the magma, becomes a dyke; and systems of parallel dykes are a conspicuous feature of any district of former fissure-eruptions which has been sufficiently dissected by erosion to lay bare its inner structure.

Central Eruptions.—The general characteristics of volcanic eruptions of the central kind are too familiar to need description in this place. It will suffice to call attention to the wide range of variety exhibited by the phenomena at different volcanic centres, with the corresponding variety shown in the nature of their products and in the forms assumed by the cones, which are so prominent a feature of this class of volcanoes. These things are related to one another, and connect themselves especially with the degree to which the explosive element enters into the eruptions. On this depend most of the differences which are to be observed between the eruptions of different volcanoes, or to a less extent between different eruptions of one volcano, or even different stages of one eruption. As already remarked, the degree of violence of an outburst is controlled by several factors, but especially by the content of dissolved water and other gases in the magma and by its viscosity. The latter being in part a specific property, we can sometimes trace a connection between the composition of the magma and the charactor of the eruptions.

In what we may conveniently call the *Hawaiian type*, represented by the giant volcanoes of the Sandwich Isles, the lavas, of basaltic composition, are exceptionally fluid. The violently explosive element is lacking in the eruptions, and there is consequently an absence of fragmental products. The volcanic mountain itself, built up entirely by successive

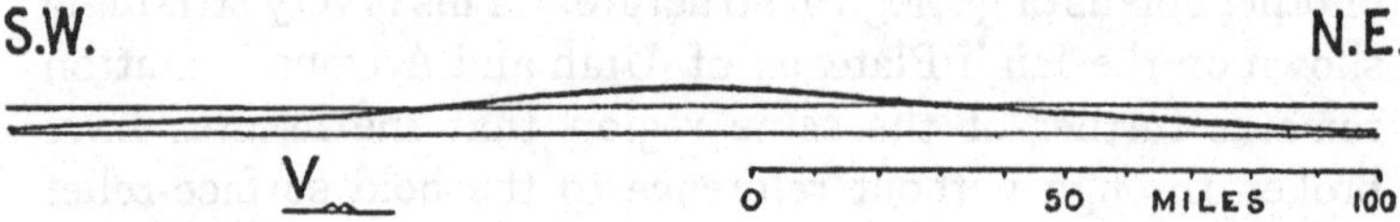

FIG. 5.—IDEAL SECTION OF MAUNA LOA, TO TRUE SCALE. (AFTER DUTTON.) The upper of the two horizontal lines represents the sea-level. Vesuvius is shown on the same scale at V.

outflows of very fluid lava, presents very gently inclined slopes, and, as illustrated by Kilauea and Mauna Loa, may attain enormous dimensions. The latter mountain has a height, reckoned not from sea-level but from the sea-floor, of nearly 30,000 feet, with a base covering an area about 160 by 130 miles (Fig. 5). It is true that Dutton [1] attributes some part of the elevation to actual uplift, but no positive evidence is adduced in support of the suggestion. The Hawaiian type of eruption is an extreme one among central volcanoes. In the great volume of the lava poured forth and the tranquil mode of its emission we may recognise an approach to the characteristics of fissure-eruptions.

In the ordinary eruptions of most of the better-known volcanoes the operations are on a smaller scale, but relatively of a more vigorous order. Mercalli has distinguished two

[1] *Hawaiian Volcanoes*, *4th Ann. Rep. U.S. Geol. Sur.* (1884), pp. 75-219.

STROMBOLI ON 20 APRIL, 1904: SHOWING AN EXPLOSION OF THE 'VULCANIAN' TYPE

FROM A PHOTOGRAPH BY DR. TEMPEST ANDERSON

chief types—the Strombolian and the Vulcanian—and to these Lacroix has added the Peléan. The differences seem to be related primarily to different degrees of viscosity in the magmas. In the *Strombolian type* the magma possesses a considerable measure of liquidity, though more viscous than the Hawaiian lavas, and the extravasation of lava-flows is a characteristic feature as well as the ejection of fragmental material. The latter consists typically of partially solidified bombs and lapilli, large or small, thrown out in an incandescent state by the gaseous outbursts. Of ashes in the usual sense there is commonly very little, the clouds of steam which are shot up being free from any noteworthy charge of finely divided solid matter. Our Frontispiece is

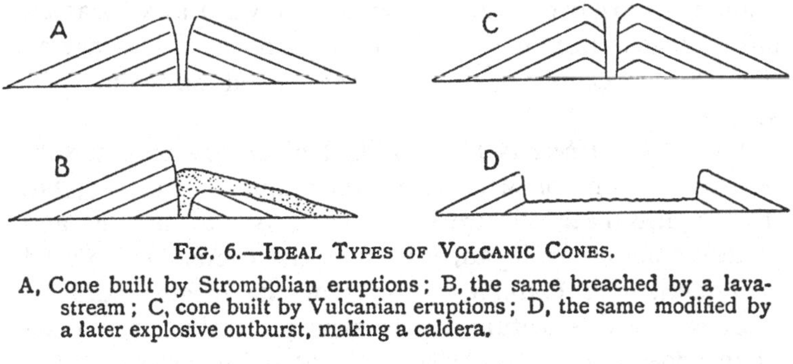

Fig. 6.—Ideal Types of Volcanic Cones.

A, Cone built by Strombolian eruptions; B, the same breached by a lava-stream; C, cone built by Vulcanian eruptions; D, the same modified by a later explosive outburst, making a caldera.

from an instantaneous photograph of a Strombolian explosion from the crater of Vesuvius, and shows ejected masses of pasty lava in the air. The ideal form of cone, built up by repeated fragmental discharges of the kind described, is that shown in Fig. 6, A. Such eruptions, however, are usually varied by flows of lava, which breach the crater-wall, and modify the symmetrical structure of the cone (Fig. 6, B); and these may become a prominent feature of a series of eruptions.

In the *Vulcanian type*, as exemplified in the eruptions of Vulcano in 1888-89, the magma is notably more viscous. In the case cited it was of dacitic composition, doubtless specifically more viscous than the basalts of Stromboli, but the temperature of the magma and its content of water must also be taken into account. Owing to its viscosity, the lava tends constantly to obstruct the crater and conduit of the volcano, thus giving rise to violent explosions, accompanied by pulverisation of the rock and the production of abundant ashes. The bursts of steam shot out from the vent are densely charged with solid matter in a state of fine division, and part of this may fall at a distance from the vent. Plate I. shows an explosion of the Vulcanian type at Stromboli. The 'cauliflower' form of the dark cloud seems to be characteristic of a rush of heated gas carrying fine powder in suspension. On the right, however, the ashy material is beginning to drop out of the cloud. The characteristic structure of an ash-cone built up by Vulcanian eruptions is that of Fig. 6, C, with an inward as well as an outward slope from the crater-rim. Outflows of lava are on the whole exceptional, and do not extend far from their source.

The *Peléan type* has been studied especially by Lacroix[1] in the eruptions of Mt. Pelée, Martinique, in 1902-03; but it has also been illustrated in volcanoes of dacitic and andesitic nature in Japan, Java, and elsewhere. Its special characteristic is extreme viscosity of the magma, and its most remarkable feature is the slow protrusion of a mass of incandescent lava too viscous to flow. At certain stages ashes and lapilli are shot upward in explosive bursts of steam; but a more striking phenomenon is the emission of incandescent clouds, which, in consequence of the choking of the orifice, are not projected vertically, but roll down the flanks of the cone. A cloud of this kind consists of steam densely charged with dust, lapilli, and blocks, and possessing a very high temperature. It gathers speed as it descends, at the same time expanding and gradually dropping its load of solid material. A hot blast of this kind destroyed the town of S. Pierre on May 8, 1902, and numerous other outbursts occurred in the succeeding months.

The crater of Mt. Pelée was occupied by a great mass of

[1] *La Montagne Pelée et ses Éruptions* (1904), and *La Montagne Pelée après ses Éruptions* (1908).

viscous lava in continual agitation. From this gradually rose what was variously described as an obelisk, or spine, or needle (Fig. 7)—a mass of lava (pyroxene-andesite), solidified on the outside but still viscous within. In the May of the following year it had attained a height of nearly 1,500 feet, but it was of ephemeral nature. It is generally interpreted as a plug thrust up through the orifice of the volcano, after the manner of soft metal in the process of wire-drawing.

The eruptions of some volcanoes now extinct appear to

FIG. 7.—THE SPINE OF MT. PELEE, APRIL 12, 1903. (AFTER LACROIX.)

have been comparable with the Peléan type. Here belong the trachytic domes of Auvergne, more particularly of the latest (Quaternary) volcanoes which constitute the chain of the Puys, and, again, the phonolitic cupolas of the Velay and the Hegau. In all of these the lava has evidently been extremely viscous, and has piled up about the volcanic orifice instead of flowing. Michel-Lévy has supposed that the protrusions in such cases took place under a covering of blocks and other fragmental accumulations, and therefore partook somewhat of the nature of intrusions. The protruded mass of lava in this type of eruption has an endogenous structure, contrasting with the exogenous growth of the ordinary ash-cone. Reyer[1] has imitated it by forcing pasty plaster of Paris, variously coloured, through a hole in a board (Fig. 8).

The eruptions at a given centre are not always of one type. Vesuvius and Etna are volcanoes which have exemplified at different times the Strombolian and Vulcanian types, and the

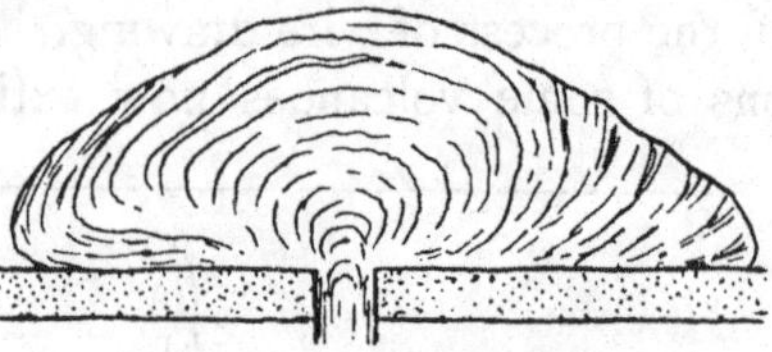

FIG. 8.—PROTRUSION OF VISCOUS LAVA IMITATED WITH PLASTER OF PARIS. (AFTER REYER.)

alternation of lava-flows and fragmental ejections has built up a cone of composite structure. In relation to the cone, the more violent eruptions of the Vulcanian type have not a constructive, but a destructive effect, and the bulk of the volcanic mountain is by no means increased as a result of every eruption (Fig. 9). In particular, a succession of

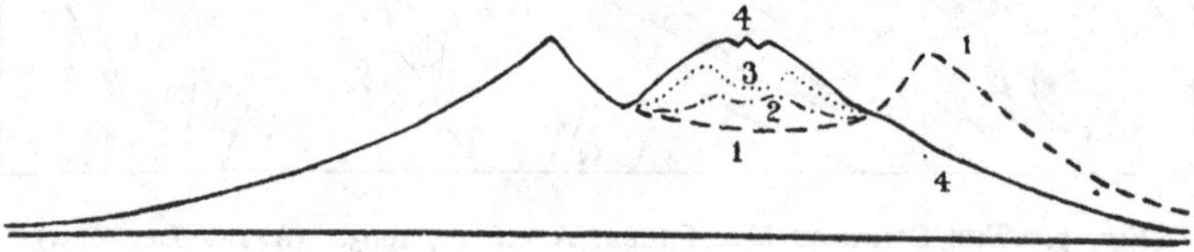

FIG. 9.—OUTLINES OF VESUVIUS AT DIFFERENT PERIODS OF ITS HISTORY. (AFTER JUDD.)

1, Somma, before A.D. 79; 2, Vesuvius iu the sixteenth and seventeenth centuries; 3, after the great eruption of 1822; 4, after the eruption of 1872.

explosions, or a single great catastrophic outburst, may not only expel the materials accumulated in the crater, but also blow away all the central part of the cone, producing a *caldera* (Fig. 6, D), the steep walls of which show in section the structure of the old cone. The floor of the caldera is covered with loose material. It may become a 'crater-lake,' as in the case of Albano and other Latian volcanoes. Later

[1] *Theoretische Geologie* (1888), p. 152.

eruptions may build up a new cone standing within the caldera, as Vesuvius stands within the broken ring of Mte. Somma.

The *Krakatoan type*, representing the extreme of violence, is illustrated especially by the eruption which, in 1883, altered the map of the Sunda Straits, and projected a volcanic dust so fine as to float for months in the higher strata of the atmosphere.

CHAPTER III

IGNEOUS INTRUSION

Geological and morphological classification of intrusive rock-bodies.—Concordant intrusions in plateau regions.—Dykes in plateau regions.—Intrusions in regions of mountain structure.—Plutonic intrusions of irregular habit.

Geological and Morphological Classification of Intrusive Rock-Bodies.—The forms assumed by intruded bodies of rock-magma, solidified as igneous rocks, exhibit a wide range of variety; and the evident tendency of bodies intruded in like circumstances to exhibit like forms makes it clear that these forms are of geological significance, as being dependent on conditions which can in part be predicated.

It is possible to classify igneous intrusions with reference to form alone, as has been done explicitly by Daly;[1] but it will be more in accordance with the point of view here adopted to take account likewise, and primarily, of geological relations. Bearing in mind the intimate connection between igneous action and crust-movements, which we have taken as our guide, we must expect to find *different forms of intrusive bodies related to different types of crust-movement.* Although the data at hand do not warrant any attempt to work out such relation in detail, it is at least clear that, of the two main types of crust-movement which we have distinguished, the *plateau-building* and the *mountain-building*, each has its own forms of intrusive rock-bodies, which are in a large degree distinctive. This we shall take as the primary basis of classification. Comparing actual areas which exemplify the two

[1] *Journ. Geol.*, vol. xiii. (1905), pp. 485-508.

contrasted tectonic types, such as the Isle of Skye with North Wales, or the High Plateaux of Utah with the Appalachian region, we cannot fail to notice the difference in the characteristic forms of the intrusive rock-masses. Where the crustal stresses, as shown by the displacements, have been of more mixed character, the contrast may be obscured, and intermediate forms may occur; but, though this may make the application of the distinction more difficult in exceptional cases, it does not detract from the significance of the principle. It cannot be doubted that the different behaviour of the intruded magmas in the two general cases distinguished is a consequence of different mechanical conditions. Roughly stated, in the one case the stresses relieved by the intrusions were of a large order, and initially of approximately uniform character; the localisation of individual intrusions depended on factors beyond our knowledge, probably comparatively small inequalities in the distribution of stress; and the minor disturbances of the strata observable are consequences of the intrusions themselves. In the other case intrusion has followed (in the causative sense) disturbances of the strata due to definite external forces, commonly of the nature of lateral thrust; and the situation and form of the intruded bodies have been determined by these same causes.

Under each of the two categories thus recognised are comprised intrusions of widely different habits. The most obvious differences are seen in the diverse postures assumed by the intruded rock-bodies and their attitude towards the 'country'-rocks in which they are intruded. It is evident that we are still dealing with the effects of different distributions of crustal stress, partly modified, however, by pre-existing structures in the country-rocks. As regards the relation of intrusions to these latter, we may distinguish more or less sharply the *concordant* and the *transgressive*. In the one case the intruded magma has been guided by surfaces of structural weakness, such as particular bedding-planes, in the rocks invaded. Typical examples are presented by the sill, the true laccolite, and other lenticular forms sometimes termed laccolites; but it must be observed that the general guidance premised does not preclude some degree of irregularity of behaviour. In the second case the intrusions have systematically broken across the bedding or other directional surfaces of the country-rock. Typical examples of this are the dyke and the plug. As a rule, the plane of greatest extension of the intruded mass makes in the former case a low angle and the latter a high angle with the horizontal—at least, where the country-rocks are not highly disturbed; and the distinction pointed out therefore corresponds in some measure with one based on the *different postures* of intruded bodies, whether quasi-horizontal or quasi-vertical. The correspondence is, however, an imperfect one, and where the country-rock is one (such as granite) without directional structures, the terms 'concordant' and 'transgressive' cease to be applicable. The distinction based on what we have styled posture is in one respect the more fundamental, in that it stands in close relation with the character of the provocative crustal stresses.

Another kind of distinction might be drawn which, however, will not figure directly in our classification. Intrusions may differ *functionally*—that is, as regards the parts which they play in the general economy of igneous action. For instance, there are independent or self-contained intrusive bodies, such as laccolites and stocks, which we may conceive as forced in a fluid state into a certain situation and there consolidated. Contrasted with these are in-filled channels, which have served as conduits for the more or less prolonged flow of molten magma, and are filled merely by the latest-risen material; the dyke-feeders of fissure-eruptions or of sill intrusions are of this kind.

The subdivisions of the classification must be made with reference to the minor factors which have contributed to determining the form and habit of igneous intrusions. Among intrusive bodies which fall into the same primary category, having something of the same general habit, and sustaining like relations to the country-rock, there may be differences of form (as between the sill and the laccolite) sufficiently pronounced to demand recognition in terminology. They are, however, differences in degree rather than in kind, and the ideal types are connected by intermediate forms. As regards the controlling conditions, such differences seem to depend on certain factors not yet mentioned, especially the *depth of 'cover'* at the time of the intrusion (implying the pressure under which the magma was intruded) and the *nature of the igneous rock* itself. Since the rocks are intruded as fluid magmas, the only specific properties which can enter in this connection are those proper to the liquid state. They are, firstly, the degree of fluidity or viscosity of the magma, and, secondly, its density.

The various forms of intrusive bodies to be discussed in the following sections of this chapter may be grouped as below:

Plateau District.	*Mountain District.*
Concordant { sill, laccolite, bysmalite.	Concordant, phacolite.
Transgressive, dyke.	Transgressive { sheet, dyke.

In addition, it will be necessary to say something of those intrusive bodies, usually of deep-seated origin and large dimensions, which cannot without undue straining be brought into the scheme just set forth (batholites).

Concordant Intrusions in Plateau Regions.—The *sill*, which is the most typical form of intrusion in plateau regions, is characterized by a wide lateral extension as compared with its vertical thickness. Its regularity of behaviour in following a defined bedding-plane is often very remarkable, and the thickness is sometimes very uniform for long distances. In some plateau regions sills attain an enormous development, exhibiting at the same time a very general uniformity of petrographical characters. A great group of this kind forms part of the British Tertiary suite of igneous rocks. The thick piles, in places exceeding 3,000 feet in thickness, which build the 'basaltic plateaux' of Ulster and the Inner Hebrides, consist to the extent of about one-half of basalt lavas, while the remainder is made up by a great succession of dolerite sills belonging to a later epoch. These vary in thickness from a few feet to 100 feet or more, and individual sills can be followed for many miles. The upper ones were in general intruded earlier, and a newer injection of magma has in many cases spread along the lower surface of an older sill, with no intervening country-rock. The sills extend down into the well-bedded Jurassic strata of Skye and Eigg, but not into the massive Torridonian sandstones or the unstratified Cambrian dolomites.

The best-developed and most regular sills are usually of basic rocks; which has been attributed to the greater fluidity of basic as compared with acid or felspathic magmas. Another question relates to the depth below the surface at which sills are intruded. Russell[1] has maintained that they are more numerous in the upper parts of the Earth's crust, and many observations go to support this generalisation. There must, of course, be a minimum depth for sills intruded beneath a land-surface; for, if the pressure above be insufficient, the magma will burst out. Under submarine conditions, where the pressure of a column of water is added to the weight of the rocks, sills may be intruded close to the sea-floor. Probably some are forced in along sediments actually in process of deposition in deep water, the distinction between intrusion and extrusion being in such circumstances of little significance. Here belong many of the rocks distinguished by what Sir A. Geikie[2] has styled 'pillow structure.'

The *laccolite* was first recognised as an important type of intrusive body, and provided with a distinctive name, by Gilbert.[3] It is illustrated in its ideal development in the High Plateaux region to the west of the Rocky Mountains, and is reproduced with modifications in the Rocky Mountain belt itself.[4] The typical laccolite has a flat base and a domed

[1] *Journ. Geol.*, vol. iv. (1896), pp. 177-194.

[2] *Ancient Volcanoes of Great Britain* (1897), vol. i., pp. 193, 201, 240, 252.

[3] *Report on the Geology of the Henry Mts.*, 1879; 2nd ed., 1880.

[4] Cross, *The Laccolitic Mountain-Groups of Colorado, Utah, and Arizona*, *14th Ann. Rep. U.S. Geol. Sur.* (1894), II., pp. 165-241.

upper surface, though the regularity of this plano-convex form may be modified by circumstances, and in particular by the proximity of another laccolite of earlier intrusion (Fig. 10, A). In the Henry Mts. of Utah the laccolites vary in diameter from ½ to 4 miles; the ground-plan is an oval not very different from a circle; while the height averages about one-seventh of the horizontal diameter, and never exceeds one-third. We may reasonably include with the ideal type varieties modified in some particulars; but the name has sometimes been farther extended to embrace the lenticular intrusions proper to folded regions, which are essentially different in significance, and these we shall exclude.

The difference between the laccolitic and sill forms is one of degree rather than of kind. A sill, although it may maintain a nearly uniform thickness over a considerable lateral extent, thins out towards its edge. In the laccolite the thinning out from centre to periphery is more rapid, and this type therefore differs from the other merely in having a more restricted areal extension and a greater relative thickness in the centre. The elevation of the overlying strata, instead of being extended and of small amount, is localised and correspondingly accentuated. This seems to be connected with a higher degree of viscosity in the intruded magma, and also with a greater depth of intrusion. The most typical laccolites are not of such rocks as dolerites, but of more acid rocks or of felspathic types rich in alkalies, and it is probable that the magmas were somewhat viscous when intruded. As regards depth of 'cover,' the fact that the strata forming the roof of the laccolitic chamber have often been sharply arched without fracture clearly points to great pressure and therefore great depth. In the Henry Mts. the several occurrences range through a vertical height of more than 4,500 feet, and Gilbert estimates that the highest were intruded under a cover of from 7,000 to 11,200 feet of strata. From petrographical considerations, also, Cross was led to postulate a great depth for the laccolitic instrusions of the High Plateaux in general, but the argument is not quite convincing. There is reason to believe that in some regions

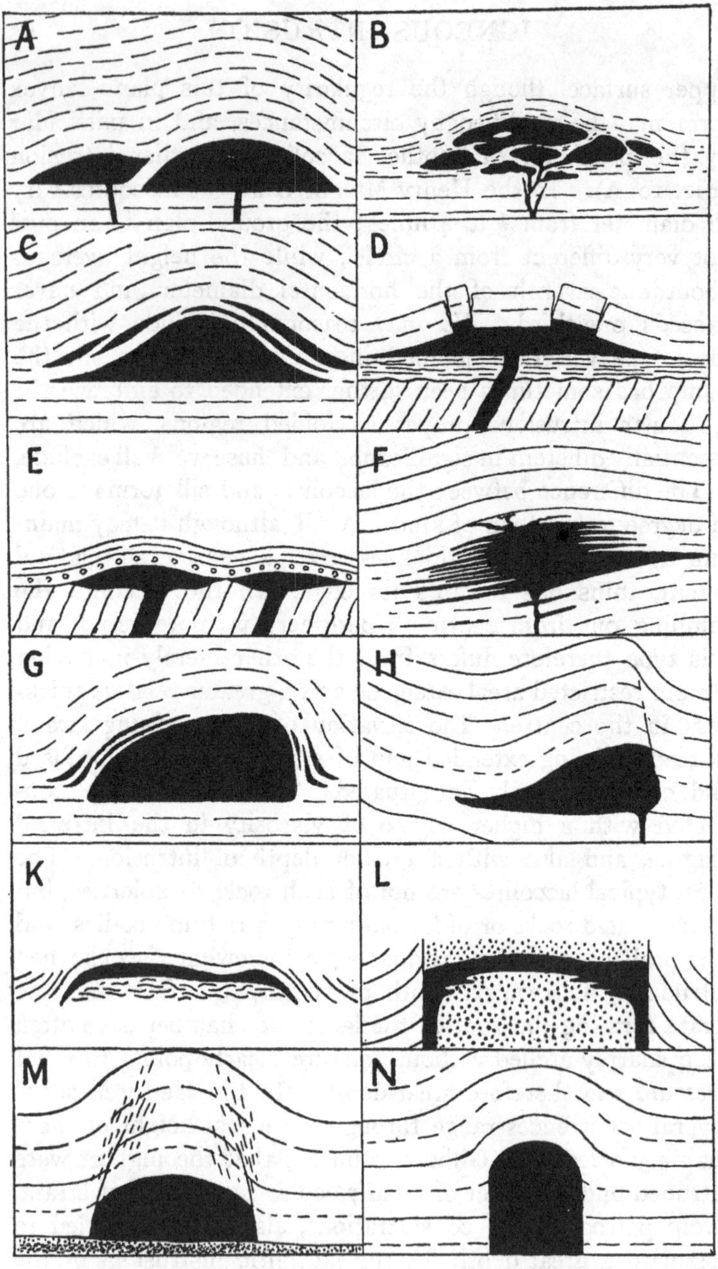

Fig. 10.—Laccolitic and Allied Types of Intrusions.

A, Typical laccolites: Mt. Holmes, Henry Mts., Utah (after Gilbert); B, group of intrusions separated by stratified rocks, but collectively forming large bodies of plutonic rock, of somewhat irregular habit but with generally laccolitic relations, have been intruded at depths by no means extreme.

It is not necessary to suppose that a laccolitic chamber is always filled by a single influx of magma, and in the larger intrusions of this class the reverse is doubtless the case. Sometimes successive injections of magma have taken effect at slightly different horizons in a well-bedded series of strata, the result being a cluster of separate intrusive bodies, divided by partitions of stratified rocks, but making up collectively what may be regarded as a compound laccolite (Fig. 10, B). More usually the later accessions of magma, rising through the same channel as the earlier, have spread along the same bedding-plane; and the heterogeneity of the resulting laccolite, taking the form of a certain quasi-stratified arrangement, may be made apparent by some petrographical differences. Again, a modification of the ideal laccolitic shape is found in some cases, in which the intrusive mass fringes out at its margin into a number of tapering sheets, intruded along different bedding-planes. This 'cedar-tree' form, as distinguished from the ideal 'mushroom' shape, was described by Holmes[1] in the La Plata Mts. of Colorado (Fig. 10, F). It is met with in other large intrusive masses having the general habit of laccolites, such as the gabbro of the Cuillin Hills in Skye. The tapering sheets, which run out into the neighbouring rocks, represent so many distinct injections, and these may often be distinguished by differences of texture or composition

[1] *9th Ann. Rep. U.S. Sur. Territ.* (1877), Pl. XLV., Fig. 2.

a compound laccolite: El Late Mts., Colorado (after Cross); C, laccolite with subsidiary sheets: Judith Mts. type, Montana (after Pirsson); D, laccolite with broken cover: Ragged Top Mt., Black Hills, South Dakota (after J. D. Irving); E, laccolitic intrusions at horizon of unconformity: near Deadwood Gulch, Black Hills (after Irving); F, compound laccolite (cedar-tree type): La Plata Mts., Colorado (after Holmes); G, abruptly protuberant laccolite: Mt. Hillers, Henry Mts. (after Gilbert); H, asymmetric laccolite: Mt. Marcellina, West Elk Mts., Colorado (after Cross); K, intrusion in Little Rocky Mts., Montana (after Weed and Pirsson); L, intrusion in volcanic vent: Kilchrist, Isle of Skye; M, bysmalite: Mt. Holmes, Yellowstone Park (after Iddings); N, ideal section to illustrate Russell's conception of a 'plutonic plug.'

even where they are in juxtaposition in the interior of the mass.

The sharp arching of the strata in the roof of a typical laccolite involves a lateral elongation and tension. Under great pressure the strata stretch without fracture; but if the tension has to be relieved by actual disruption, the fissures formed are concurrently occupied by molten magma, so that

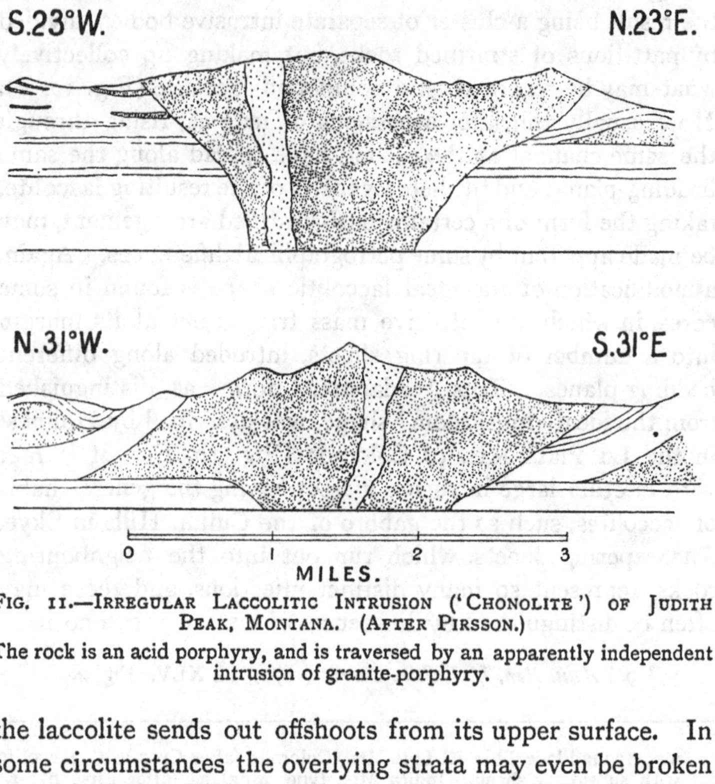

FIG. 11.—IRREGULAR LACCOLITIC INTRUSION ('CHONOLITE') OF JUDITH PEAK, MONTANA. (AFTER PIRSSON.)

The rock is an acid porphyry, and is traversed by an apparently independent intrusion of granite-porphyry.

the laccolite sends out offshoots from its upper surface. In some circumstances the overlying strata may even be broken into blocks and displaced, as shown in J. D. Irving's[1] diagrammatic section through the laccolitic intrusion of phonolitic rock at Ragged Top Mountain in the Black Hills of Dakota (Fig. 10, D). Over the marginal parts of a simple laccolite, where the curvature is concave, there is a

[1] *Ann. N.Y. Acad. Sci.*, vol. xii. (1899), p. 218.

certain relief of pressure and a consequent tendency of the bedding-planes to open. There may be an injection of magma at such places, forming lenticular sheets subsidiary to the main body (Fig. 10, C), and this is especially the case when the main laccolite is of unusually protuberant shape (Fig. 10, G).

The *channel of supply* of a laccolitic intrusion is doubtless in general represented by a dyke or group of dykes; but, in the nature of the case, this can rarely be demonstrated as regards the larger laccolites. Sometimes the feeders may be more accurately described as sills in highly-inclined strata, as in Fig. 10, E, where the laccolites themselves are intruded along a surface of unconformity. Exceptionally, intrusive bodies having many of the characters of laccolites have been fed from the circumference. An example of this is the large granite-porphyry mass of the Little Rocky Mountains[1] (Fig. 10, K). It is intruded in an unconformity, and in connection with a sharp uplift; though it is not clear how far the latter is the cause, how far the effect, of the intrusion. Near Loch Kilchrist, in Skye, is a remarkable granophyre, enclosing half-digested gabbro material, which occupies part of an old volcanic vent. The magma has risen along the cylindrical walls and, at a certain level, spread out as a thick horizontal sheet in the volcanic agglomerate (Fig. 10, L).

The essential characteristic of the laccolite, as distinguished from certain other intrusive bodies which may resemble it in form, is that *the intrusion is the cause, not the consequence, of the concomitant folding*. Subject to this criterion, we may include as irregular laccolites some generally lenticular bodies which sustain the same geological relations, although their boundaries are in part transgressive. The irregularity is carried far in some of the larger examples, such as that of Judith Peak in Montana[2] (Fig. 11), where, however, the geological function of the intrusion and its association with simple laccolites

[1] Weed and Pirsson, *Journ. Geol.*, vol. iv. (1896), pp. 409-412.
[2] Pirsson, 18*th Ann. Rep. U.S. Geol. Sur.*, II. (1898), pp. 533-538.

suffice to place it under the present head. Daly[1] includes this example under his type 'chonolite,'[2] which, however, is defined without reference to those geological relations which we have regarded as of fundamental significance.

In the case of the ideal laccolite, intruded in connection with crustal stresses purely of the plateau-building kind, the precise location of the intrusions will depend upon intratelluric conditions beyond our cognisance. If, however, there be any intervention, as a minor factor, of forces tending directly to produce folding of the strata, this subsidiary element will determine the situation of intrusions which in all essential characters may be ranked as true laccolites. Such is the case of the gabbro of the Cuillin Hills, already mentioned, and other Tertiary plutonic intrusions in the British area. In other cases, not only the situation, but the form is affected, and comes to be determined by the resultant of the regional plateau-building forces and local forces of the other category. In this light we may regard the *asymmetric laccolites*, which occur in the Western States of America and elsewhere. Mt. Marcellina, in Colorado, is a typical example of this transitional form (Fig. 10, H). The diagram shows on one side the tapering form of the regular laccolite, but on the other a steep boundary, involving abrupt displacement of the strata. The curvature on this side is replaced by a fault of peculiar nature, passing gradually into a fold in the upward direction, but terminating suddenly downward at the base of the intrusive body. Again, it appears that, where an intrusion of generally concordant habit is forced in beneath a cover of no great thickness, the disturbance of the strata produced by the intrusion itself may sometimes be so great as to obscure entirely the simple relations proper to the ideal laccolite. Remarkable instances have been described in the neighbourhood of Piatigorsk,[3] on the northern border of the Caucasus.

[1] *Journ. Geol.*, vol. xiii. (1905), pp. 498-501.

[2] 'Chonolith' of Daly; but we adopt uniformly the English spelling in preference to the German.

[3] Vera de Derwies, *Récherches géologiques et pétrographiques sur le laccolithes des environs de Piatigorsk* (1905), Geneva.

Here the disturbance amounts in places to total inversion (Fig. 12). Although the district lies outside the main belt of folding, it can scarcely be doubted that so wide a departure from the regular habit is connected with the intervention of mountain-building forces. The 'sphenolite' described by Philippi[1] at Parroquias in Mexico seems to be of somewhat similar nature.

To the sill and the laccolite we may add the *bysmalite* of Iddings,[2] making a third term in this series of types of plateau intrusions. It is an extreme type, and one not yet widely recognised. The narrower localisation of the intrusion and of the resulting uplift, which distinguishes the laccolite from the sill, is here carried much farther. The peripheral boundary is so steep as to approach the vertical,

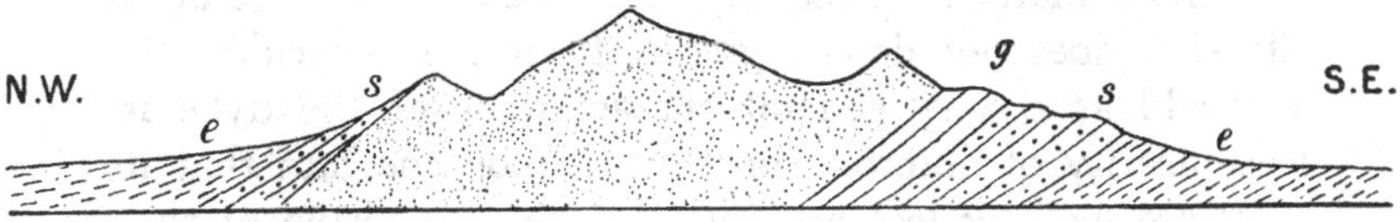

FIG. 12.—SECTION THROUGH BESCHTAOU, NEAR PIATIGORSK. (AFTER MLLE. DE DERWIES.) SCALE, NEARLY 1 INCH TO A MILE.

A quasi-laccolitic intrusion of quartz-porphyry has given rise on the south-east side to complete inversion of the strata. *g*=Gault; *s*=Senonian; *e*=Eocene.

and the discontinuous disruption of the contiguous strata, which was noted as an incident in the asymmetric laccolite, becomes here the rule (Fig. 10, M). We may, indeed, conceive the intrusive body as bounded by a circular fault, which merges into a dome-like uplift above, but terminates abruptly downward at the level of the base, below which the strata are undisturbed. Iddings' typical example is taken from the Yellowstone Park, but he would include under the same head a group of intrusions in the Black Hills of Wyoming and South Dakota, described by Russell[3] under the name 'plutonic plug.' Russell's conception is that of a cylindrical plug of igneous rock thrust through the strata,

[1] *Centralbl. für Min.*, 1907, p. 456.
[2] *Journ. Geol.*, vol. vi. (1898), pp. 704-710.
[3] *Ibid.*, vol. iv. (1896), pp. 23-43.

terminating in a dome-like uplift above, but extending indefinitely downward without any base (Fig. 10, N). This raises obvious mechanical difficulties, and has been criticized by more than one writer. According to Jaggar,[1] the remarkable cylindrical hill named Mato Tepee, taken as a typical example of such a plug, is really an outlier of a laccolite resting on unbroken strata, and owing its isolated appearance to erosion in conjunction with vertical jointing.

Dykes in Plateau Regions.—The main characteristics of the dyke, the most familiar of all types of intrusion, are two. Firstly, it is bounded by nearly parallel surfaces, and has an approximately constant width—small in comparison with the longitudinal and vertical extension; in other words, it fills a rent or fissure in the rocks, which has been only slightly and uniformly enlarged. Secondly, its posture at the time of its intrusion does not depart greatly from the vertical.[2] We may add as a very general characteristic of the dyke its rectilinearity of outcrop, consequent upon the fact that its walls are more or less accurately plane. In virtue of their quasi-vertical posture, dykes communicate between different levels in the crust of the earth. Some dykes represent the conduits of sills or laccolites, or the channels of fissure-eruptions at the surface; others are to be regarded as independent and self-contained intrusive bodies, not subserving any ulterior end. These functional differences, though not always easily discriminated in practice, imply relationships which have their significance in the development of igneous activity in any region.

The dyke type of intrusion is found in folded mountain districts as well as in regions of plateau-structure, and many of the characteristic features are common to dykes of all kinds. A general discussion of the *mechanics of dyke-intrusion* would be a subject of some complexity. Hopkins seems to have considered that, a dyke-fissure being once initiated and

[1] *21st Ann. Rep. U.S. Geol. Sur.*, III. (1901), pp. 264, 265, Pl. XXXVIII.

[2] Certain geologists have rejected this criterion; see Daly, *Journ. Geol.*, vol. xiii. (1905), pp. 488, 489.

occupied by fluid magma, hydrostatic pressure, tending to separate the walls of the fissure, will be sufficient to cause the propagation of the dyke directly onward. The tendency to indefinite extension and the characteristic rectilinearity of course would thus be properties inherent in the dyke itself. This is certainly a very inadequate view of the matter. Many dykes have undoubtedly been injected in fissures already prepared, such as joints and faults. In unbroken rock-masses the formation of fissures must often be the most direct way of relieving the crustal stresses, which we postulate as the prime cause of igneous intrusion, and the arrangement of such fissures must be in relation with the disposition of stresses to which they give relief. Pre-existent structures may also have an influence. Thus, in rocks already affected by well-marked bedding-planes, no great tension could be developed in the direction perpendicular to those surfaces Sills may be injected along the bedding-planes, but any new fractures formed will tend to be at right angles to them, and this is clearly an important factor in determining the direction and hade of dykes. Among gently inclined strata in a plateau region dykes may often be observed to have a hade which suggests that they have been tilted in common with the strata; but it can sometimes be verified that such dykes were intruded after the tilting, and the only explanation of this is the tendency to perpendicularity alluded to. Similarly, a younger group of dykes may have the same direction as an older group, and then not infrequently follow the same fissures, producing double or multiple dykes; but if a new direction is taken, this tends to be at right angles to the former.

The *direction of flow* of the fluid magma within a dyke-fissure is by no means always vertical.[1] It is often indicated, especially at the margin of a dyke, by various fluxional and taxitic structures, and may have any direction between the vertical and the horizontal. It appears that dykes may originate in the depths within comparatively limited spaces,

[1] *Tertiary Igneous Rocks of Skye*, *Mem. Geol. Sur.* (1904), pp. 294, 337; and *Geology of the Small Isles* (1908), p. 150.

the linear extension observed at the outcrop depending on the propagation of the dykes in a lateral direction. In

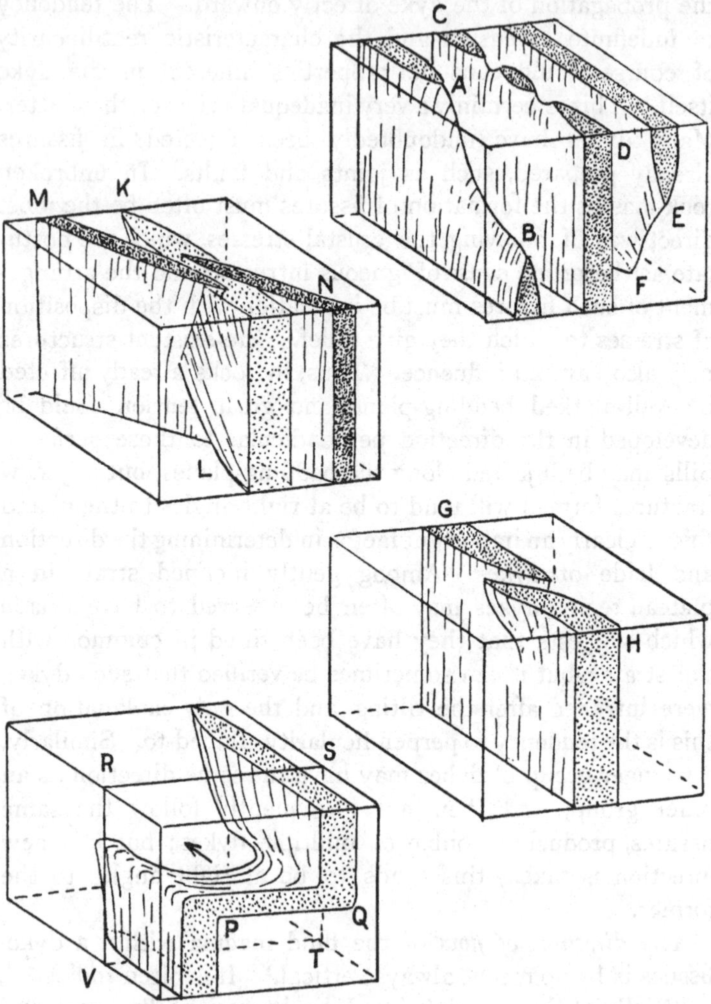

FIG. 13.—DIAGRAMS TO ILLUSTRATE IRREGULARITIES IN THE BEHAVIOUR OF DYKES.

general we may conceive a dyke as terminating along an inclined and curved line, such as A B in Fig. 13. Then the outcrop on a horizontal plane dies out laterally (A), but in a vertical section it dies out upward (B). If the terminal edge has a general horizontal direction, but with undulations, the outcrop may consist of detached portions (C D). If the edge has an overhanging position (E F), a vertical section may show the dyke dying out downward. Again, the outcrop of a dyke is sometimes seen to be interrupted, and shifted to a parallel course at some little distance (G H). Doubtless a section at a lower level would make the real continuity apparent. In like manner, where an earlier dyke terminates in an overhanging edge (K L), a later one (M N) may pass from one side to the other without actually cutting through the first. Among well-bedded strata a dyke is sometimes seen to stop abruptly at a particular bedding-plane (P Q), follow it for some distance as a sill, and then resume the dyke habit as abruptly. If the direction of flow makes a considerable angle with the vertical, the connecting sill lags behind the two portions of the dyke in their propagation laterally (in the direction of the arrow), and a vertical section along such a plane as R S T V will then show the two parts of the dyke, one terminating upward and the other downward at the same bedding-plane, with no connecting sill.[1]

Considering now more particularly the *dykes in plateau regions*, we have to remark that the crustal stresses relieved by the intrusions are of relatively simple type, which may be almost uniform over large tracts. Accordingly, the dykes may form extensive systems, and tend to parallelism in a direction which is nearly constant, or changes only gradually. Thus the British Tertiary dykes, which affect an area embracing about half of the British Isles, rarely depart much from a bearing between north-west and north-north-west. They belong to numerous distinct groups, intruded at different epochs, and serving different geological functions, but the same *law of parallelism* governs all. It is only locally modified, in the direction of a radial disposition, about certain special centres of disturbance.

[1] Gilbert, *Report on the Geology of the Henry Mountains* (1879), pp. 28, 29; Harker, *Geology of the Small Isles* (1908), pp. 151-153.

Although the dykes of this great system are in some districts extremely numerous, it is often observable that they are not equally abundant in different kinds of country-rock,[1] and dykes which freely traverse one rock may terminate suddenly on encountering another. They do not, however, suffer any deviation in passing from one rock to another, and they are usually quite indifferent to faults or other geological accidents. These characters are not found in dyke-intrusions of every kind, and they seem to indicate that the opening of the dyke-fissures was, in such cases, *a sudden act*.[2] There are, indeed, in connection with many systems of dykes, circumstances which suggest that the injection of the fissures with molten magma must have been very rapidly accomplished.[3] Since no noteworthy superheating can be assumed in the magma as intruded, a very moderate fall of temperature would render it highly viscous, and additional cooling would cause it to solidify. Nevertheless, dykes of slender width have often penetrated thousands of feet of cool rocks. The inference is that intrusion was effected too rapidly to permit any considerable cooling during the process.

Intrusions in Regions of Mountain Structure.—We have next to review briefly those characteristic types of intrusion which are associated especially with folded strata. As before, the intrusions are to be regarded as affording relief to crustal stresses, and their form and habit must therefore stand in relation to the distribution of stresses by which they are provoked. Here, however, the stresses are in general of a very complex order, and the problem has not yet been brought within the province of strict mechanical analysis. Nor is it in any wise elucidated by the so-called 'experiments in mountain-building,' which present only a rough geometrical illustration, not a mechanical analogy, of

[1] *Tertiary Igneous Rocks of Skye* (1904), pp. 292-294; *Geology of the Small Isles* (1908), pp. 143, 144.

[2] Compare Cross, *14th Ann. Rep. U.S. Geol. Sur.* (1895), p. 240.

[3] Compare Barrell, *Geology of the Marysville Mining District, Montana, Prof. Paper No.* 57 *U.S. Geol. Sur.* (1907), pp. 157-159.

the processes of nature; for it is impossible to reproduce in a small model the actual conditions which have been realised in the Alps or the Scottish Highlands.[1]

Taking first *concordant intrusions*, in which pre-existing structures in the country-rocks have exerted a directing influence in conjunction with the mountain-building forces, it is to be remarked that the typical sill is not found among folded strata.[2] The reason is sufficiently obvious. A bedding-plane, which was originally a surface of weakness throughout its extent, ceases to be so after folding of the strata. In the ideal case of a system of undulatory folds (Fig. 14) there is

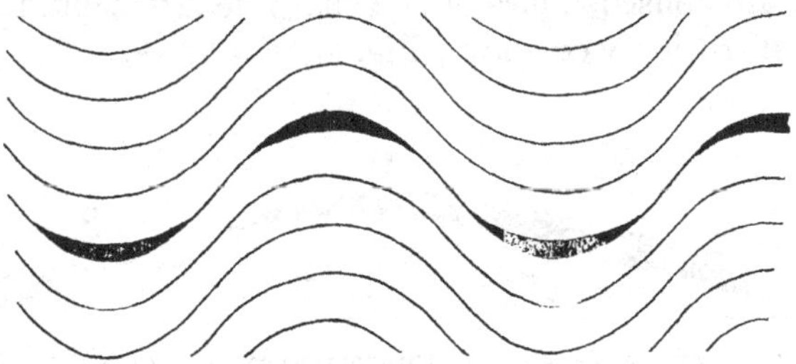

FIG. 14.—DIAGRAM TO ILLUSTRATE PHACOLITES INTRUDED IN CONNECTION WITH FOLDING.

increased pressure and compression in the middle limbs of the folds, but in the crests and troughs a relief of pressure and a certain tendency to opening of the bedding-surfaces. A concurrent influx of molten magma will therefore find its way along the crests and troughs of the wave-like folds. Intrusive bodies corresponding more or less closely with this ideal case are common in folded districts. Since some distinctive name seems to be needed, we may call them *phacolites*. The name laccolite has often been extended to include such bodies, but this is to confuse together two things radically different. The intrusions now considered are not, like true laccolites, the cause of the attendant

[1] This follows from Newton's Principle of Similitude, which is habitually disregarded in this connection.

[2] We except, of course, the case of subsequent folding, affecting the sills in common with the stratified rocks.

folding, but rather a consequence of it. The situation, habit, magnitude, and form of the phacolite are all determined by the circumstances of the folding itself. In cross-section it has not the plano-convex shape of the laccolite, but presents typically a meniscus (Fig. 15), or sometimes a doubly convex form. Except where the folding has the character of a dome, a phacolite does not show the nearly circular ground-plan of a laccolite, but has a long diameter in the direction of the axes of folding. As regards the mechanical conditions of its injection, the phacolite resembles rather the small subsidiary intrusions which sometimes accompany a laccolite, and are consequences of the sharp flexure caused by the primary intrusion (see above, Fig. 10, C and G).

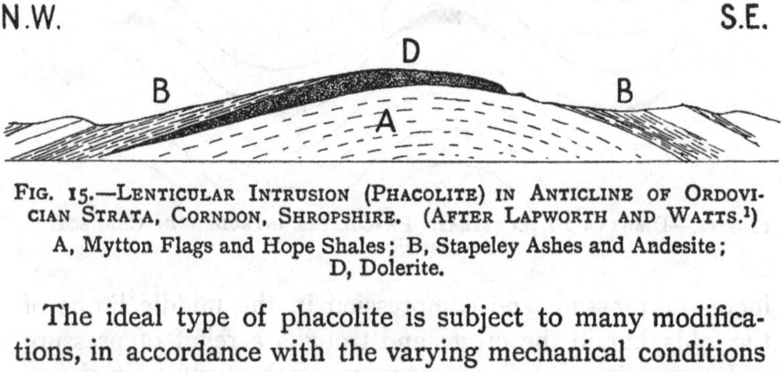

FIG. 15.—LENTICULAR INTRUSION (PHACOLITE) IN ANTICLINE OF ORDOVICIAN STRATA, CORNDON, SHROPSHIRE. (AFTER LAPWORTH AND WATTS.[1])
A, Mytton Flags and Hope Shales; B, Stapeley Ashes and Andesite; D, Dolerite.

The ideal type of phacolite is subject to many modifications, in accordance with the varying mechanical conditions of intrusion. Some bodies of this nature, in the Alps and elsewhere, attain large dimensions. According to Baltzer,[2] the Aletsch mass is 18 or 19 miles long and 2 miles broad, with a visible thickness of 2,600 to 3,200 feet, while the St. Gotthardt mass has a length of 45 miles and a breadth of 2 or 3 miles. In another place Baltzer[3] shows how, in smaller intrusive bodies of the same general nature ('Rücken'), the originally concordant relation may be obscured, owing to the igneous rocks becoming *involved in later folding*, to which they have yielded in some measure, but less than the surrounding rocks (Fig. 16).

[1] *Proc. Geol. Assoc.*, vol. xiii. (1894), p. 342.

[2] *Compte Rendu IX. Congr. Géol. Internat.* (1904), II., pp. 787-798.

[3] *Neu. Jahrb.*, Beil. Bd. xvi. (1903), pp. 292-324.

Turning now to *transgressive intrusions*, the stratiform injections, which we shall distinguish as *sheets*, related to mountain-building differ fundamentally from the sills of plateau regions. In the first place, their areal distribution

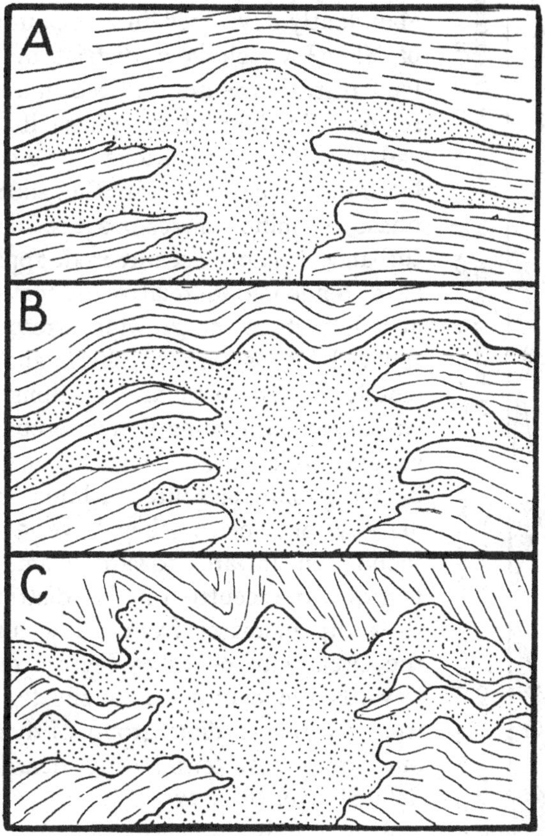

FIG. 16.—DIAGRAM OF SUCCESSIVE STAGES IN THE FORMATION OF 'RÜCKEN.' (AFTER BALTZER.)

is in more or less evident relation with special foci of activity, often marked by earlier plutonic intrusions. Secondly, the direction of the sheets is, in the general case, not determined by structural planes in the rocks invaded, but related to the system of crustal stresses which was the

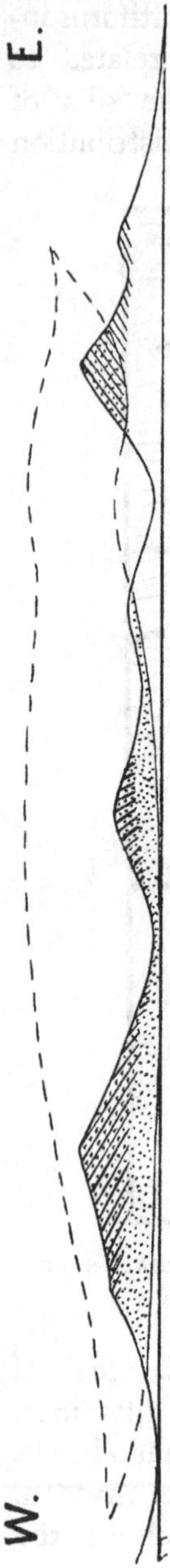

FIG. 17.—SECTION ACROSS THE GABBRO LACCOLITE OF THE CUILLIN HILLS, SKYE, SHOWING THE DISTRIBUTION AND DIP OF THE INCLINED SHEETS OF DOLERITE WHICH INTERSECT IT. SCALE, $\frac{3}{5}$ INCH TO A MILE.

immediate precedent cause of the intrusions. The actual disposition may, therefore, vary much in different cases.

A comparatively simple case is illustrated by the group of dolerite sheets so abundantly developed in the Cuillin Hills of Skye,[1] which is a district of special local disturbance interpolated in a plateau region. In the plutonic phase it was invaded, as already mentioned, by a complex laccolite of gabbro. In the succeeding phase of minor intrusions the local disturbance became accentuated, apparently as a tendency to relative upheaval, and at a late stage the growing stresses were relieved by a system of sheets and dykes closely related to these conditions. The sheets are extraordinarily numerous within an area which coincides nearly with that of the gabbro laccolite. Almost parallel at any one place, they have a steady inclination inward, towards some point beneath the centre of the district, and their dip increases from a low angle at the circumference to 45 degrees or more towards the centre. The sheets, which were fed by a radiate group of dykes, die out downward, terminating nearly at a defined level (Fig. 17).

As a highly complex example we may select that of the Monzoni district, as described by Mrs. Ogilvie Gordon[2] (Fig. 18). In the following quotation 'sill' is replaced

[1] *Tertiary Igneous Rocks of Skye*, *Mem. Geol. Sur.* (1904), pp. 366-369.

[2] The Geological Structure of Monzoni and Fassa, *Trans. Edin. Geol. Soc.*, vol. viii. (1903), Special Part (pp. 141, 175).

by 'sheet' for reasons given above: "In the first instance, 'Monzoni' was a simple wedge-shaped sheet-mass with sheet-fingers intercalated in the strata, but during the progress of the Judicarian deformational movements that sheet was added to, became a *sheet-complex*, which bulked more largely, pushed vertically and laterally, and *evoluted* itself as an intrusive mass thrusting itself within the crust. Space was provided for it by tears in the roof of strata, by partial absorption, lateral incrush, and involution of the sedimentary

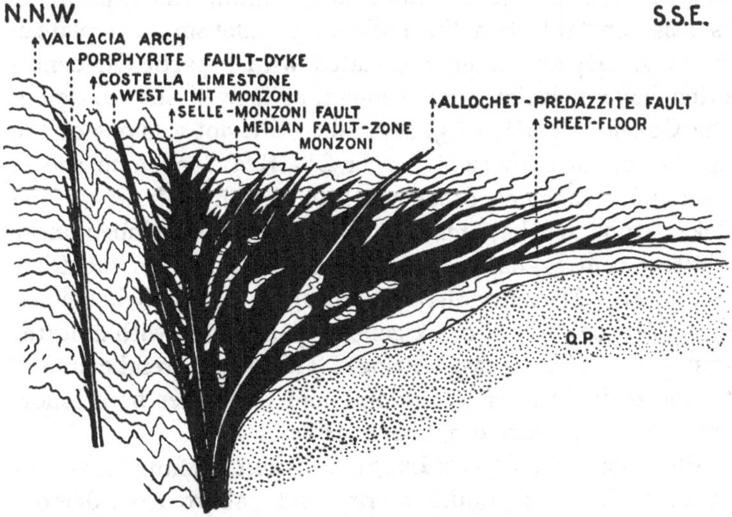

FIG. 18.—GENERALISED SECTION OF THE MONZONI COMPLEX, TIROL. (AFTER MRS. OGILVIE GORDON.)

strata. Thus the 'Central Massive' type of the present Monzoni group is a natural evolute, or overthrust, form assumed when molten rock-material intermittently ascends from a lower horizon to a higher at a locality of fault-intersection, and consequently of crust-weakness."

As regards *dykes*, a mountain district differs from a region of plateau structure, not so much in any special features of the individual intrusions as in the distribution of groups of dykes and the arrangement of the members of a group. The dykes affect certain parts, or are most numerous in certain parts, of the area, and are gathered especially about certain centres, which have often been marked out at an earlier stage by plutonic intrusions. About a given centre there may be several groups of dykes, distinguished by their petrographical characters. The disposition of the dykes, as regards direction, being related to the crustal stresses which found relief in the intrusions, varies with the conditions of the case. Where there are definite axes of folding, there is often a tendency for the dykes to run at right angles to the axes, and they show, therefore, a parallel disposition, which, however, is less constant than the regional parallelism in a plateau country. Dykes closely associated with a particular centre often have a radiate arrangement, as is well shown, *e.g.*, in the Cheviot district (Fig. 2, p. 26). Cheviot was one of the special foci of activity of Lower Old Red Sandstone age in Scotland. In some other parts of the area, however, the dykes, though they may be numerously developed near plutonic centres, have no radiate disposition. In the Blair Atholl and Pitlochry district, and again about Ballachulish and Glencoe, they show a general parallelism with a north-east or north-north-east bearing. This is the trend common to the main axes of movement, and the dykes here follow the direction, not of dip, but of strike.

Plutonic Intrusions of Irregular Habit.—Many intrusions of such rocks as granite, diorite, and gabbro have demonstrably the habit of laccolites, sheets, or dykes, though their behaviour in detail is sometimes less regular than that of hypabyssal rocks in smaller intrusive bodies. But we have also to recognise many extensive masses of plutonic rocks which have been only partially exposed by erosion, without revealing unequivocal evidence of their true geological relations. These we may conveniently call *batholites*, using this term of Suess in a purely descriptive sense, without implication of the various theoretical views which he has at different times attached to it.[1] Such intrusive bodies are often clearly transgressive as regards their visible boundaries, but the

[1] *Das Antlitz der Erde*, vol. i. (1885), pp. 218-220; Engl. Trans., pp. 168, 169.

form and relations of their deeper parts are beyond direct observations.

On the *assimilation hypothesis*, still supported by some French geologists, an igneous rock-magma is supposed to be capable of melting and incorporating freely the solid rocks which it encounters, so that an intrusion of this kind may to a great extent replace, rather than displace, sedimentary and other solid rocks. Michel-Lévy[1] ascribes these effects to a rise of the isogeotherms in the neighbourhood of the intrusion, and an intense circulation of mineralising fluids. A large body of granite is pictured as having a pyramidal or conical shape, expanding indefinitely downward and possessing no other base than itself.

The difficulties—mechanical, thermal, and chemical—involved in such a theory have been sufficiently set forth by Brögger and others. The geological evidence seems to prove very clearly that these large igneous rock-bodies, like the smaller ones, are essentially *intrusive* as regards their relation to the environing rocks. Concerning the actual mechanism of intrusion there is still considerable divergence of opinion. Suess has supposed that the injection of a granitic intrusion at a high temperature must necessarily be preceded by the formation of a corresponding cavity. To provide this he invokes the action of tangential thrust in the upper part of the Earth's crust, causing one part to be lifted off or thrust away from another. The formation of empty cavities is quite irreconcilable with the strength of the materials of the crust and the large dimensions of many batholites, and the opening of any large chamber must certainly imply the concurrent filling of it with fluid rock-magma. Suess' treatment of the question is nevertheless instructive, as reminding us that, when very large bodies of magma are intruded under a considerable cover, their form and disposition may be determined mainly by the distribution of stress which thus finds relief, with little regard to the structure of the encasing rocks. A very general form will be one with quasi-horizontal extension,

[1] See especially *Contribution à l'étude du granite de Flamanville*, *Bull. carte géol. Fra.* (1893).

its vertical thickness being only a fraction of its length and breadth. This will have something of the laccolitic shape, though determined by quite other mechanical conditions.

Brögger[1] has discussed the nature of batholites, with special reference to the Drammen granite and other large plutonic intrusions of Southern Norway. His interpretation is based on the known facts with reference to laccolites; and his general conclusion is that the laccolitic form, however disguised by local irregularities, is the natural habit of large plutonic intrusions in regions not affected by dynamic metamorphism—or, in our phrase, in regions of plateau structure—like the faulted and sunken Christiania district. The term 'laccolite,' here employed in a generalised sense, may be understood to imply a rock-body which is distinctly intrusive, occupying a space opened concurrently with its injection, and having a thickness presumably much less than its lateral extent. The essential point is that such a mass must possess *an under as well as an upper surface*; and if strata are truncated by the visible upper surface, we must suppose that their prolongation still exists beneath the intruded mass. The actual upper surface of contact may be of more or less irregular shape, owing especially to portions of the roof having become detached and sunk in the magma (compare Fig. 19). This action has by some geologists been accorded a foremost place in the mechanism of batholitic intrusion. It is the process which Daly[2] styles 'overhead stoping.' He supposes blocks to be wedged off by veins of the magma penetrating fissures in the roof, aided by the local stresses set up by unequal expansion; and he considers that a magma, at the time of its intrusion, is not too viscous to permit the sinking of such detached blocks. This view assigns to the magma itself a merely passive rôle in the act of intrusion, which, as a mechanical process, consists in the removal of solid rock from the roof of the chamber and its transference

[1] *Eruptivgesteine des Kristianiagebietes*, II. (1895), pp. 116-153.

[2] 'The Mechanics of Igneous Intrusion,' *Amer. Journ. Sci.* (4), vol. xv., pp. 269-298, and vol. xvi., pp. 107-126 (1903); also *The Geology of Ascutney Mountain, Vt., Bull.* 209 *U.S. Geol. Sur.* (1903).

by gravity to the base. In other words, there is a gradual working upward, but no enlargement of the magma-chamber. For the latter, however, Daly supposes, in addition, a melting of the detached blocks as they sink, and their incorporation into the magma. His theory of 'abyssal assimilation,' since developed farther in a highly hypothetical paper,[1] seems open to many of the objections which apply to the marginal assimilation of Michel-Lévy.

Barrell[2] has applied Daly's hypothesis to the large batholites of Boulder and Marysville in Montana (Fig. 19). The form of the upper surface accords well with the supposition that large blocks have been detached from the roof of the chamber and

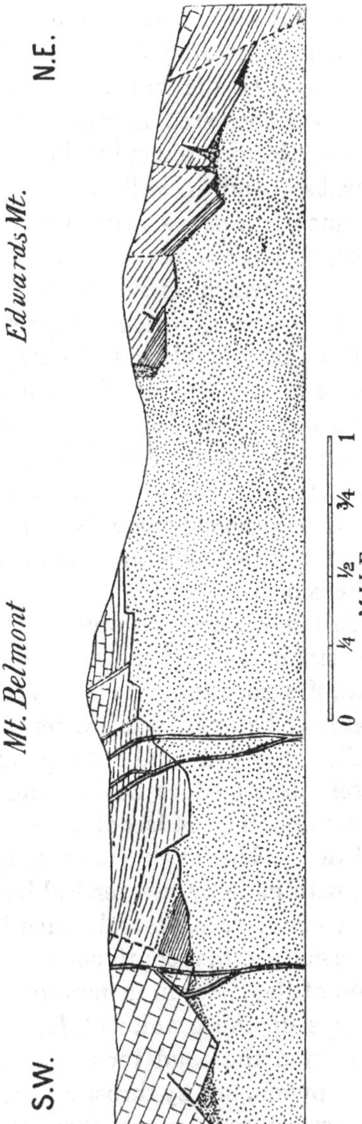

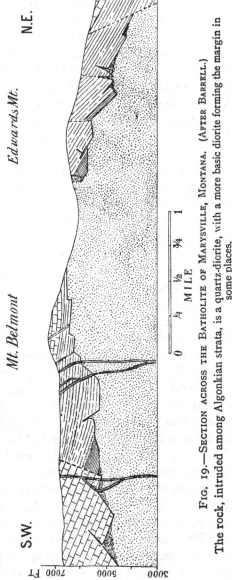

Fig. 19.—Section across the Batholite of Marysville, Montana. (After Barrell.)
The rock, intruded among Algonkian strata, is a quartz-diorite, with a more basic diorite forming the margin in some places.

[1] *Amer. Journ. Sci.* (4), vol. xxii. (1906), pp. 195-216.

[2] *Geology of the Marysville Mining District, Montana, Prof. Paper No.* 57, *U.S. Geol. Sur.* (1907).

sunk into it, but there is nothing to suggest that the igneous mass extends indefinitely down into a region of 'pressure-solid magma.' If the breaking away of portions of the roof is to be considered as more than a subsidiary factor, there is an obvious difficulty in setting a limit to the action invoked, and it will not be easy to explain why, although such intrusions may approach somewhat near to the surface, they never break through. The *minimum depth of cover* under which large plutonic rock-bodies may be intruded is, perhaps, often exaggerated. According to Barrell, the great Boulder batholite, covering more than 2,000 square miles, had at Elkhorn a cover of 2,000 feet or perhaps less. Some of the British Tertiary intrusions of gabbro and granite must have been intruded at less depths than this. For the Drammen granite Brögger estimates an original cover of about 2,000 feet, and for some intrusions of augite-syenite, nepheline-syenite, and nordmarkite only a few hundreds of feet.

An intrusion working its way up through solid rocks by 'overhead stoping' must, if this action be sufficiently continued, acquire something of the vertical cylindrical form which seems to be implied in the term *boss*. Although most of the bodies of granite and other plutonic rocks which have been loosely described as bosses, and so rendered in ideal sections, are doubtless of laccolitic or other stratiform shape; some, not of the largest dimensions, appear to have a plug-like form, with more or less vertical boundaries. The space which they occupy can scarcely have been provided by the thrusting aside of the contiguous rocks, assisted by some contraction of the latter in metamorphism. They appear, as Barrois[1] says of the granite of Rostrenan in Brittany, to have penetrated the solid rocks by cutting a way through them like a punch, not by thrusting them aside like a wedge. In such cases we may suppose that 'stoping' has played an important part, but there is no evidence of concomitant assimilation. An interesting example is afforded by the

[1] *Ann. soc. géol. Nord.*, vol. xii. (1885), p. 105.

gabbros and granites of Skye.[1] In part of the area these intrusions assume the habit of bosses, and where they traverse the Cambrian dolomitic limestones the relations are very clearly exhibited (Fig. 20). The junctions are every-

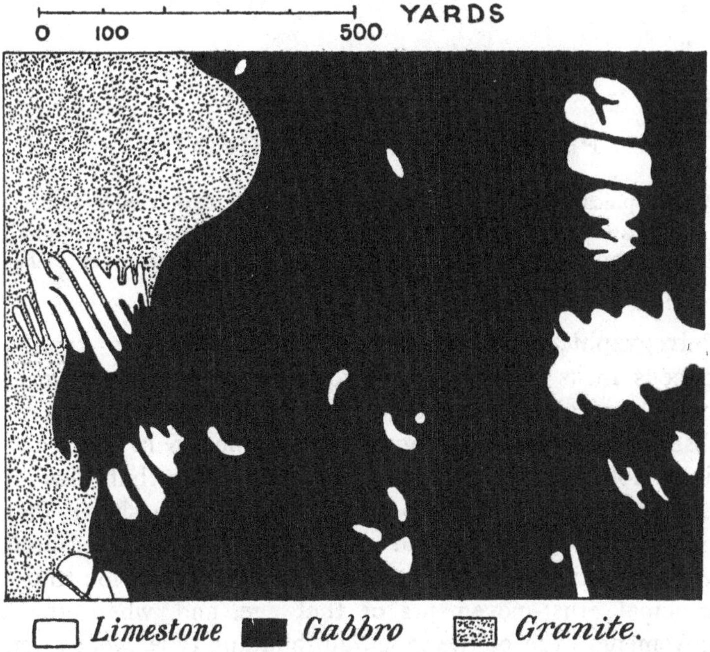

FIG. 20.—SKETCH-MAP OF A SMALL AREA NORTH-EAST OF BEINN NA CAILLICH, SKYE, SHOWING THE RELATION OF GABBRO AND GRANITE INTRUSIONS TO THE CAMBRIAN LIMESTONES. SCALE, 6 INCHES TO A MILE.

where vertical, notwithstanding their intricate course in ground-plan. Enclosed patches of the limestone, down to 20 yards in diameter, stand up like pillars in the midst of the igneous rock, and these give no evidence of any mechanical disturbance attendant on the intrusion. Nevertheless, analyses of the gabbro and granite prove that their magmas have not dissolved the limestone to any extent that can be detected.

[1] *Tertiary Igneous Rocks of Skye* (1904), pp. 97-100, 131-135.

CHAPTER IV

PETROGRAPHICAL PROVINCES

Petrographical provinces.—Atlantic and Pacific branches of igneous rocks.—Geographical distribution of the two branches.—Evolution of petrographical provinces.—Petrographical provinces in earlier geological periods.

Petrographical Provinces. — Hitherto we have treated igneous rocks without regard to their varied mineralogical and chemical composition. It is now our business to observe that there is a significant distribution in space and time, not only of igneous rocks as a whole, but also of different kinds of igneous rocks. It has been shown that igneous action in any one great cycle has been localised within certain prescribed areas, which were in relation to the principal crust-movements of that age, and, where those movements were of the mountain-building type, were often divided from one another by important orographic lines. It is now to be remarked that the igneous rocks of one such area possess certain common characteristics of the petrographical kind, and sometimes contrast sharply with the rocks of a neighbouring area belonging to the same great system.

Judd's conception of distinct *petrographical provinces*[1] has proved a singularly fertile one. It leads at once to the hypothesis that the igneous rocks of one province stand to one another in a real genetic relationship, community of

[1] *Quart. Journ. Geol. Soc.*, vol. xlii. (1886), p. 54. To the term itself Washington has made objections, which seem to the present writer to have little force: see *The Roman Comagmatic Region* (1906), p. v.

characters being attributable to community of origin. For the presumed relationship implied in such resemblances among associated rock-types, Iddings[1] first employed the convenient term *consanguinity*. This implied community of origin is the starting-point of speculations which have much engaged the attention of petrologists during the last twenty years, the aim being to explain the origin of the diverse associated rock-types by processes of 'differentiation' from a common stock-magma. The doctrine of petrographical provinces, supplemented by that which is implied in the use of the term 'consanguinity' in this connection, has thus come to occupy a prominent place in modern petrology, and it is found to be capable of extended application.

An examination of the rocks belonging to one great period of igneous activity, and of their actual distribution, enables us to distinguish *areas of greater or less extent*, within which the rocks present *a less or greater degree of consanguinity*, the law being that more marked specialisation goes with narrower localisation. A petrographical province, in the original sense of Judd, is found to be part of a larger region, throughout which the rocks exhibit certain points of resemblance, less marked than those which unite rocks belonging to the same province. On the other hand, a province may include, especially in the later stages of its history, several districts, each distinguished by closer petrographical resemblances than those which are common to the whole province. More complete knowledge concerning the geographical distribution of igneous rock-types than has yet been systematized will perhaps lead to an appropriate terminology for these different degrees of consanguinity corresponding with different extents of distribution. At present any such nomenclature can be only provisional.

The kind of resemblance which we have predicated of the igneous rocks of one petrographical province, and accepted as evidence of consanguinity, is compatible with a wide range of diversity among the rocks thus embraced. Moreover, a classification of igneous rocks on this basis would

[1] *Bull. Phil. Soc. Washington*, vol. xii. (1892), pp. 128-130.

traverse all the artificial schemes of systematic petrography, which have been constructed without reference to the genetic principle. Such a classification is a task for the future. At present we shall be content with indicating a broad twofold division of igneous rocks as a whole, and showing that it enables us to map out the great primary regions with some precision.

Atlantic and Pacific Branches of Igneous Rocks.—Such terms as 'acid,' 'intermediate,' and 'basic' are often used as loosely indicative of the systematic position of an igneous rock-type, the reference being primarily to its silica-content, though other characters are in some measure connoted. The implication is that, with some rough degree of approximation at least, the whole assemblage of rock-types can be pictured as ranging along one line of variation. The actual composition of the rocks indicates a far greater complexity. We shall make a decidedly closer approach to the facts if we assume, not one, but two main lines of variation. Each line may be conceived as spanning the interval between the basic and acid extremes, the two diverging most widely near the middle of their course, so that they may be figured diagrammatically by two arcs meeting at both ends. For reasons which will appear, we shall speak of them as the *Atlantic and Pacific branches*. Their chemical characteristics will be considered later; but we may note here as especially distinctive that, comparing types of like acidity, the rocks of the former branch are richer in alkalies, and those of the latter in lime and magnesia. Iddings[1] accordingly used the expressions 'alkali group' and 'subalkali group'; and a similar distinction of 'alkalische' and 'alkalisch-erdige' occupies a more or less definite place in the schemes of some other petrologists.[2]

The more salient mineralogical characteristics of the two branches are summarily as follows:

[1] 'The Origin of Igneous Rocks,' *Bull. Phil. Soc. Washington*, vol. xii. (1892), pp. 89-214 (pp. 183, 184).

[2] See Löwinson-Lessing, 'Studien über die Eruptivgesteine,' *Compte-Rendu VII. Congr. Géol. Internat.* (1897), pp. 193-464 (p. 239)

Atlantic.	*Pacific.*
Alkali-felspars form a large proportion of the more acid and intermediate rocks, and occur in many rocks of low acidity.	Alkali-felspars not abundant except in the more acid rocks, and wanting in the basic; soda-lime-felspars abundant.
Microperthitic and cryptoperthitic intergrowths frequent.	Zonary banding of felspars frequent.
Felspathoid minerals often found (leucite, nepheline, sodalite, primary analcime; also melilite).	Felspathoid minerals not found.
Quartz confined to the more acid rocks.	Quartz not only in acid rocks, but also in many intermediate ones.
Pyroxenes and amphiboles often include soda-bearing kinds.	Pyroxenes represented by augite, diopside, and the rhombic group; amphiboles by common hornblende.
Micas and garnets of common occurrence.	Micas not common except in the more acid rocks.

Among plutonic rocks, for example, the Atlantic branch includes the alkali-granites, syenites and nepheline-syenites, essexites, theralites, picrites, etc.; while in the Pacific branch are found granites, quartz-diorites and diorites, gabbros and norites, and peridotites. It would, of course, be easy to cite rocks, such as some of the monzonites, which do not fall very near to either of the two ideal lines of variation. That the rough two-fold division indicated does nevertheless correspond with a real and important distinction is sufficiently proved by the significant geographical distribution of the two branches. In tracing this from a study of the geological literature of particular districts, it is necessary to remark that lax and comprehensive classificatory names, such as diabase, porphyrite, basalt, and the like, are often insufficient for assigning the rocks to one or other branch. Even in classical European districts, such as the Harz and the Siebengebirge, a revision by more accurate methods of study is discovering that some of the 'diabases' of older writers are essexites, teschenites, or theralites; some of the 'basalts' are monchiquites; etc.

In 1880 Judd[1] drew attention to the marked contrast

[1] *Volcanoes* (1880), p. 202.

between the Tertiary volcanic rocks of Bohemia and Hungary, situated on opposite sides of the Carpathian range. Fuller knowledge of the rocks, intrusive as well as extrusive, of the two areas serves to emphasize the contrast. The volcanic rocks of Bohemia are, as a whole, rich in alkalies, and a large proportion of them contain felspathoid minerals; while the associated intrusive rocks include such types as sodalite-syenite, essexite, bostonite, tinguaite, etc., all characteristically of Atlantic facies. In Hungary, on the other hand, occurs a great development of dacites, andesites, and allied rocks, besides intrusions of banatite, diorite, etc., marking the Pacific affinities of the whole assemblage. Bohemia and Hungary, in respect of this broad petrographical distinction, belong to different regions.[1] That the Bohemian rocks are, as a whole, of much more basic composition than the Hungarian is a characteristic of the second order of importance, and belongs to the particular petrographical provinces in question.

Iddings (*op. cit.*, 1892), while considering the distinction of alkali and sub-alkali groups too indefinite to be made a basis of classification, noted the general law of their distribution in the Western Hemisphere. He pointed out that rocks relatively rich in alkalies are found along the eastern side of the Rocky Mountains and Andes, and farther east in Canada, the United States, and Brazil; while the sub-alkali group is developed throughout the Great Basin of the Western States, on the Pacific coast into Mexico and Central America, and along the whole extent of the Andes. No account is taken here of the geological age of the rocks, but the greater part of those referred to are Tertiary or late Cretaceous.

In 1896 the present writer[2] attempted to show that the two-fold division emphasized by Iddings is of very wide

[1] By averaging a number of chemical analyses, it appears that the mean alkali-percentage is 7·30 for the Bohemian rocks and 5·90 for the Hungarian, despite the fact that the mean silica-content is much lower in the former than in the latter.

[2] 'The Natural History of Igneous Rocks: I., Their Geographical and Chronological Distribution,' *Science Progress*, vol. vi. (1896), pp. 12-33.

application, and pointed out "the very general correspondence of the areas of the alkali and sub-alkali groups respectively with the areas of the Atlantic and Pacific types of coast-line as defined by Suess. The one type is found around the Atlantic and part of the Indian Ocean and in the Polar basins, the other, generally speaking, around the Pacific." It was accordingly suggested that we may distinguish "an Atlantic and a Pacific facies of eruptive rocks, corresponding with distinct phases in crust-movements of a large order." A somewhat similar proposal has since been made by Becke,[1] starting from a comparison of the igneous rocks of the Bohemian Mittelgebirge with those of the Andes. The choice of this small and peculiar Bohemian province, with its great preponderance of basic rocks, as representative of the whole Atlantic branch is misleading; and it has caused Becke to lay stress upon the distinction of denser and lighter rocks as characteristic of his Atlantic and Pacific tribes ('Sippe') respectively. Comparing types of similar acidity, it will be found that the alkali-rocks are not denser but lighter than those of the other branch.

We shall see that the two great branches of igneous rocks, as characterized above, have a well-marked areal distribution, and define two petrographical regions of the first order of magnitude, which stand in relation to the grandest tectonic features of the globe.

Geographical Distribution of the Two Branches.—In tracing briefly and generally the actual distribution of the Atlantic and Pacific branches, we shall confine our attention to the latest and best-known of the great systems of igneous rocks—viz., that of *Tertiary age*, including also, on the one hand, late Cretaceous, and on the other hand Recent.

In the Western Hemisphere the two branches characterize over large areas, as remarked by Iddings,[2] the Atlantic and

[1] 'Die Eruptivgebiete des böhmischen Mittelgebirge und der amerikanischen Andes. Atlantische und pazifische Sippe der Eruptivgesteine,' *Tscherm. Min. Petr. Mitth.* (2), vol. xxii. (1903), pp. 209-265. Compare also Prior, *Min. Mag.*, vol. xiii. (1903), pp. 261, 262.

[2] *Bull. Phil. Soc. Washington*, vol. xii. (1892), p. 184; *Journ. Geol.*, vol. i. (1893), pp. 839, 840.

Pacific slopes respectively; but this broad statement needs some amplification. In South America the belt of the Andes, with its andesites, quartz-diorites, etc., has a most typical Pacific facies; but the dividing-line must be drawn not far east of this mountain-belt, for in some places, as in Mendoza and Salta (Argentina), the alkali rocks approach quite near to the Andes. Where the various chains diverge in Columbia, the line runs north-east and east, following the border of the Sierra de Bogota, the Sierra de Merida, and the range of northern Venezuela, and thence by Trinidad round the outer border of the Antillean chain, where the igneous rocks are of the same types as in the Andes and Central America. The serial arrangement of the West Indian volcanic centres emphasizes the fact that, tectonically and petrographically, these islands belong to the Pacific, not to the Atlantic. The igneous rocks of the Mexican plateau are likewise of Pacific types, and the line may be drawn provisionally along the Sierra Madre in Eastern Mexico, leaving the nepheline-syenites, etc., of the San José district of Tamaulipas[1] on the east side. Farther north the alkali rocks of the Apache Mts. in Western Texas afford another guide, and from New Mexico the boundary between the two great regions follows the eastern border of the Rocky Mts. system. On one side are the Rosita and Cripple Creek districts of Colorado, the Black Hills, and the various areas of alkali rocks in Montana; on the other the extensive tracts of sub-alkali rocks which are so prominent a feature of the Great Basin and the Pacific slope.

To obtain some idea of the degree of sharpness of division between the two regions, we select the western part of Montana, with the adjacent Yellowstone Park, for which a large number of chemical analyses of rocks have been made by the United States Geological Survey. The percentage of alkalies given by these analyses affords a rough test, sufficient for our present purpose. Dividing the area into eight districts, and taking the average for each, we obtain the results given below. Laying down the figures on a map

[1] Finlay, *Ann. N.Y. Acad. Sci.*, vol. xiv. (1904), pp. 247-318.

(Fig. 21), we see clearly how the amount of alkalies increases as we recede from the Rocky Mts.:

	Alkalies per Cent.
Yellowstone Park (96 analyses)	6·78
Madison and Gallatin Valleys (17)	4·17
Butte, Boulder, Elkhorn, and Marysville (25)	6·58
Castle Mt. district (14)	7·46
Crazy Mts. (25)	8·16
Little Belt Mts. (17)	8·09
Highwood Mts. (20)	9·75
Bear-paw Mts. (7)...	11·93

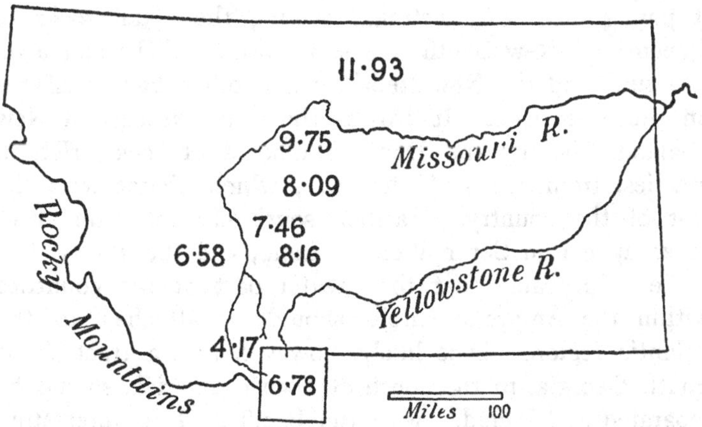

FIG. 21.—SKETCH-MAP OF MONTANA, WITH THE YELLOWSTONE PARK, SHOWING THE AVERAGE ALKALI-CONTENT OF THE IGNEOUS ROCKS IN DIFFERENT DISTRICTS.

The western border of the Pacific Ocean, from the Aleutian Isles to the equator, presents everywhere an assemblage of andesites and other sub-alkali rocks; but the limits of the great Pacific region are less clearly defined on the Asiatic than on the American side. This is due, not only to lack of information and the absence over large areas of any known Tertiary igneous rocks, but also to more complex tectonic conditions—the intervention of great plateaux and the coming in of the Eurasian mountain-system, with its general east to west trend, contrasting with the Cordilleran system of the New World. In the south the old Continental plateau of the Deccan, unaffected by Tertiary mountain-folding, presents a series of 'younger' igneous rocks of Pacific facies. In the East Indian Archipelago we recover the simpler relations, and the coincidence of the Pacific petrographical region with the Pacific tectonic region is very evident. The boundary follows the outer border of the 'great Malay arc' of Suess by Sumatra, Java, and Flores, leaving Timor in the region of alkali rocks. The islands of the Torres Straits, with their andesitic and basaltic lavas, belong to the Pacific region, and the line may be drawn just south of them. Curving more south-eastward, it passes clear of Australia, leaving the alkali rocks of Queensland, New South Wales, Victoria, and Tasmania on one side, and the New Hebrides and other Pacific islands on the other side. It crosses the southern part of New Zealand, dividing the Dunedin district of rocks, rich in alkalies, from the andesites, etc., which characterize the rest of the country. Farther south characteristic alkali rocks appear in South Victoria Land, and the scanty data make it probable that the greater part of the countries within the Antarctic Circle should be attached to the Atlantic region. It is likely, however, that a tract about South Georgia, to the south of South America, should be separated and included with the Pacific. It is interesting to note that these conclusions from the petrographical side agree with those based by Reiter[1] on tectonic considerations. He has proposed to include the Antarctic coasts and islands from the Balleny Isles (near South Victoria Land) to the South Orkneys with the Pacific type, relegating the remainder to the Atlantic.

The boundaries, which we have followed with such rough accuracy as is attainable, are shown on the sketch-map in Fig. 22. The numerous volcanic centres scattered through the Central and Southern Pacific Ocean itself present an assemblage of typical sub-alkali rocks. A few isolated occurrences of rocks rich in alkalies are known; but these

[1] *Die Südpolarfrage und ihre Bedeutung für die genetische Gliederung der Erdoberfläche* (1886), *cit.* Suess, vol. ii., p. 260.

exceptions afford no ground for discrediting, as Cross[1] would seem to do, a law of world-wide application. A trachyte and a nepheline-basalt are known in the Sandwich Isles,

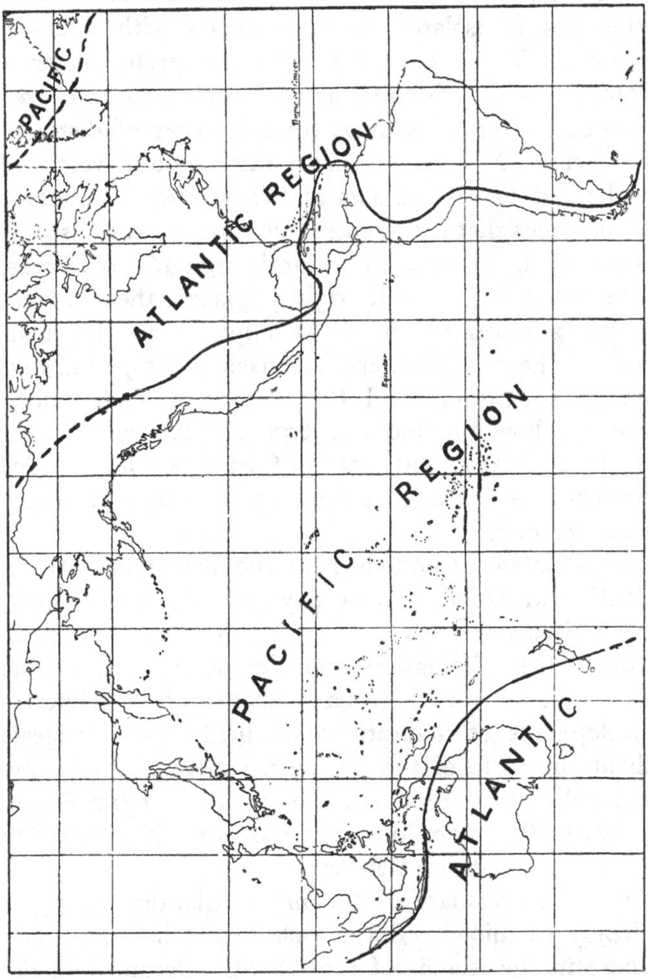

FIG. 22.—LIMITS OF THE PRINCIPAL PART OF THE PACIFIC PETROGRAPHICAL REGION FOR THE TERTIARY AND RECENT IGNEOUS ROCKS.

some limburgites in the Samoa group, and a nepheline-basalt from one of the Caroline Isles. The actual distri-

[1] 'An Occurrence of Trachyte on the Island of Hawaii,' *Journ. Geol.*, vol. xii. (1904), pp. 510-523.

bution of crustal stresses must often be much more complex than has been laid down in any analysis of their effects yet attempted; and, even if the manner in which they control differentiation were known, it would be hazardous to apply the principles to isolated oceanic islands with unknown surroundings. The case of the Galapagos group, however, is instructive. These islands, situated some 800 miles west of the Andes, nevertheless have volcanic rocks of distinctly Atlantic facies. Now Suess[1] makes this group of volcanoes a critical point in his syntaxis of the American mountain-chains, and notes that it presents "the same form of association as occurs elsewhere in the Atlantic region." In another place[2] he remarks: "Wolf rightly places these islands among the grouped volcanoes as opposed to the serial volcanoes of the mainland, and emphasizes the petrological contrast between them and the volcanoes of Ecuador." Here we see how the tectonic type characteristic of the Atlantic is associated with rocks of what we have styled Atlantic facies, even when it appears exceptionally within the Pacific territory.

Certain isolated exceptions near the boundary between the Atlantic and Pacific regions may be interpreted as indicating a limited degree of overlapping of the two, or some oscillation of the dividing-line from time to time. Thus the occurrence of some leucitic rocks in Java,[3] amidst an enormous development of andesitic and allied types, suggests that, about the close of the Tertiary, a magma of Atlantic type encroached for a time upon the frontier of the Pacific region. Leucitic rocks occur among the andesites still farther north, in southern Celebes.

The western boundary of the main Atlantic region we have already examined. Rocks rich in alkalies are found everywhere in the islands of the Atlantic Ocean, from the Azores to Tristan d'Acunha, and along both the western

[1] *Das Antlitz der Erde*, vol. ii. (1888), p. 263; Engl. trans., p. 206.

[2] *Ibid.*, vol. i., p. 691; Engl. trans., p. 539.

[3] Verbeek and Fennema, *Description géologique de Java et Madoura* (1896).

and the eastern side of Africa, besides many districts in the interior of the continent. The limits of this region in the North Atlantic Ocean are indicated by the phonolite of the Wolf Rock, off Land's End, and the rockallite of Rockall, the boundary thus approaching closely the shores of the British Isles. Its extension north-westward can be approximately laid down, if we may assume a late geological age for the fresh nepheline-syenites and sodalite-syenites of Greenland. These are found about Julianehaab, in the south; while the extensive plateaux of basalt on both coasts enable us to attach the remainder of Greenland to the 'Brito-Icelandic' province, with Pacific affinities. This latter includes also Jan Mayen, Spitzbergen, and Franz Josef Land, and may possibly connect directly with the main Pacific region.

There remain Europe, Northern Africa, and Western Asia. The extreme complexity of this part of the world from the tectonic point of view is reflected in the distribution of the igneous rocks, and we shall limit ourselves here to a very general view. Those petrographical provinces which lie beyond the outer border of the Alps, from France to Bohemia, all have a strongly-marked alkaline or Atlantic facies. This is only in a measure true of the western division of the interior belt of igneous activity, including Spain, Western Italy, and part of Algeria, with the islands between. This Iberian-Tyrrhenian province presents probably a greater range of petrographical diversity than any other in the world. Most of the Spanish rocks are of Atlantic facies, but in the districts of Cartagena and Cabo de Gata there is also a considerable development of dacites and andesites. In Sardinia there are typical alkali rocks at Mte. Ferru, but andesites and dacites at Siliqua. In the Æolian Isles most of the lavas belong to Pacific types,[1] but some are of more doubtful affinities, and leucite-basanites have been described from Vulcanello and Stromboli. Only in Western Italy and Algeria is the Atlantic facies consistently maintained, and it is remarkable that some of the

[1] Stark, *Tscherm. Min. Petr. Mitth.* (2), vol. xxiii. (1904), pp. 469-532.

most richly alkaline rocks are found at the Mte. Vulture centre, on the very line of the Apennines. On the eastern side of this orographic boundary-line the intra-Alpine igneous belt has the Pacific facies. This is not very strongly pronounced in the Adriatic province,[1] but in the Hungarian, Balkan, and Ægean there is a great preponderance of andesites, dacites, and other characteristic Pacific types. Similar rocks, of late geological age, are found in the Caucasus, Asia Minor, Persia, Baluchistan, the Kuen-lun range, and the highlands of Northern Tibet; and this probably indicates a continuous tract of sub-alkali rocks from the Eastern Mediterranean to the Pacific itself. It is true that this great arm of the main Pacific region presents in its western part some exceptional occurrences of alkali rocks. In some of these the petrographical peculiarity seems to be directly connected with tectonic features. Thus, the Ditro district, with its nepheline-syenites, etc., though lying within the Carpathian arc, is divided by a minor range from the andesitic districts of Transylvania. Nepheline-bearing lavas are recorded from Fünfkirchen and the Bakony Forest, and from isolated localities in the Balkans. A nepheline-basalt occurs with the rhyolites and andesites of the Troad, and in the same part of Asia Minor are found the kulaites of Washington, while leucitic rocks are known near Trebizond and again about Lake Urumiah, in the north-west of Persia. These, however, are relatively insignificant exceptions to the general Pacific facies of the region.

Evolution of Petrographical Provinces.—We have seen that, for the igneous rocks of the latest great cycle at least, it is possible to map out very extensive petrographical regions, defined with reference to the two great branches of igneous rocks. It would be easy to show, further, that each region comprises a number of petrographical provinces, which, sharing the general characteristics of the region, possess yet a more or less strongly-marked individuality. Hitherto only a few of these provinces have been studied systematically from the genetic point of view, but it is

[1] Stark, *Tscherm. Min. Petr. Mitth.* (2), vol. xxv. (1906), pp. 319-334.

evident that others often have pronounced chemical and mineralogical peculiarities. For example, among provinces of Atlantic facies, the younger igneous rocks of Montana and of Colorado are characterized by approximate equality of the two alkalies or a preponderance of potash, while in Western Texas and in Arkansas soda preponderates. In Bohemia there is a strong preponderance of soda with a low silica-content. Among Pacific provinces there are differences no less marked—*e.g.*, the British Tertiary rocks are collectively much more basic than the Hungarian and Ægean.

Instead of chemical, we might make use of mineralogical peculiarities, as was done in characterizing the two great branches themselves. For example, rhombic pyroxenes, so widely prevalent in many provinces of Pacific facies, are of exceptional occurrence among the Tertiary igneous rocks of Britain, though they are very common in the Ordovician suite. Leucite is a prominent constituent in some areas of alkali rocks, such as Western Italy, but is wanting in others. Ægirine, cossyrite, haüyne, allanite, zircon, and other minerals have often a significant geographical distribution, and the case is strengthened if we regard not only individual minerals, but mineral associations. A review of the igneous rocks of earlier periods would furnish many other examples of special mineralogical characteristics of particular provinces, such as the melanite garnet of Western Sutherland and the perthitic felspars of the Christiania district.

The existence of petrographical regions, provinces, and districts is clearly a fact of the first order of importance, the interpretation of which must hold a fundamental place in any theory of igneous action. Becker[1] has proposed to account for the distribution of different kinds of igneous rocks by assuming a heterogeneous constitution of the primitive globe, which has persisted as a permanent influence. To this hypothesis there are two decisive objections. It ignores the obvious relation of this distribution to the great tectonic features of the Earth's crust; and it would

[1] *Amer. Journ. Sci.* (4), vol. iii. (1897), pp. 34-37; see also Haug, *Traité de géologie*, vol. i. (1907), p. 323.

imply that the existing distribution has remained the same throughout geological time, which we shall show is not the case.

The facts, indeed, seem to admit of only one explanation—viz., that the same causes which controlled the distribution of igneous activity in any one of the great cycles were also fundamentally concerned in the primary distribution of different kinds of igneous rocks. In other words, the *mechanical forces* which determined the situations and habits of the igneous intrusions and extrusions must have intervened, at an earlier stage and at deeper levels in the Earth's crust, in those preparatory processes which gave special characteristics to the rock-magmas in different areas. Accordingly, we attribute the differences between petrographical regions and provinces, not to any original want of uniformity in the composition of the globe, but to real processes of *magmatic differentiation.* That the actual distribution of different kinds of igneous rocks is broadly a geographical one points, then, to extensive differentiation in the horizontal sense. The fact that, where adjacent regions or provinces are sharply divided, the boundary is some important orographic line, clearly indicates the orogenetic forces as in such cases the determining cause of this early differentiation. Where the principal forces concerned have been of the plateau-building or continent-building (epeirogenetic) type, the frontiers of regions and provinces are commonly less sharply delimited.

While the considerations adduced imply that a petrographical province is established as such at a very early stage of the igneous cycle, and a certain stamp of individuality is thenceforth set upon it by the apportioning to it of an initial magma of a certain character, it is clear that the varied assemblage of rock-types which such a province presents must be the final result of later developments. A comparison of different provinces, and in particular of such as we may suppose to be in different stages of their history, seems to show that each one passes through a certain evolutionary process. The history is, from the petrographical point of view, one of *progressive specialisation*, resulting in increasing diversity and sometimes peculiarity of rock-types. With the appearance of highly specialised types in the later stages, there is also very generally a *narrower localisation*. These effects we cannot but ascribe to continued or repeated differentiation, by whatever causes produced. This later differentiation is not always on the same lines as that which distinguished the province from its neighbours; and, as regards particular types, it may either exaggerate or minimise the special characteristic of the province. Exceptionally a rock-type which, from the descriptive point of view, is of Atlantic facies may be evolved from a Pacific stock, or conversely.

Iddings[1] has compared the volcanic rocks of different parts of the great belt which extends along and to the west of the American Cordillera. He points out that among the lavas of the Andes there is an enormous preponderance of andesites and dacites, with a comparative scarcity of basalts and rhyolites, especially of the more basic basalts and the more acid rhyolites. Similar andesitic rocks recur in Central and North America, but they are there accompanied in increasing quantity by the more extreme types, both basic and acid. Iddings infers that the chain of the Andes, with its great active volcanoes, represents an early stage of petrographical evolution, and may be said to be still in its youth; while Mexico is, in this respect, in a more advanced stage of development, and the corresponding tract in the United States, where vulcanicity is almost extinct, represents comparatively old age.

Even in the United States west of the Rocky Mts. the petrographical variation, though of wide amplitude, is along rather narrowly limited lines, and a general simplicity of type characterizes the whole assemblage of igneous rocks, extrusive and intrusive. This is in strong contrast with the rich diversity of types found in Montana, Texas, Arkansas, etc., on the opposite side of the Continental axis. This contrast may be referred partly to difference in the com-

[1] *Journ. Geol.*, vol. i. (1893), pp. 164-175.

position of the initial magmas of the eastern and western regions, partly to differences implied in their respective tectonic features. As regards the former point, the literature of petrography makes it very apparent that there is, on the whole, much *more diversity among the igneous rocks in Atlantic than in Pacific provinces.* In other words, differentiation attains a wider range, and on more varied lines, in alkaline than in sub-alkaline magmas. If, again, we consider the question with reference to tectonic development, it appears that, other conditions being comparable, there is *more diversity among the igneous rocks in mountain than in plateau regions.* In other words, crustal strains of a broad and simple type seem to import a relatively slow progress of differentiation as compared with local disturbances of a more acute kind.

These generalisations are well illustrated by the British Tertiary province. The initial magma here was clearly a basic one not rich in alkalies; and the rocks to which it has given birth show much less range of variety than, *e.g.*, the much smaller province of the Bohemian Mittelgebirge, where the initial magma, also of basic composition, belonged to the alkali branch. The British province has the plateau type of structure, but interspersed in it are certain centres of relative elevation indicating local disturbance in the nature of incipient mountain-building. Throughout the whole cycle we can trace the interaction of the two sets of forces, no less by petrographical than by tectonic evidence. The several groups of igneous rocks which succeeded one another in this province fall into two distinct categories: the *regional*, distributed over the length and breadth of the province, and the *local*, limited to the neighbourhood of the special centres. The former are all of basic nature, with no wide diversity in essential composition, while the latter range from ultrabasic to thoroughly acid types. Here we see that the distinct centres, defined by the local distribution of crustal stresses, were not only special foci at which igneous activity was from time to time localised, but were likewise the principal seat of magmatic differentiation.

Petrographical Provinces in Earlier Geological Periods.—If, as we have been led to conclude, the arrangement of petrographical regions and provinces in any great cycle of igneous activity is closely related to the main orographic features of that period, there will be no necessary coincidence of provinces of different ages. Some degree of correspondence may be looked for, inasmuch as there is a certain tendency for successive systems of crust-movements, even at wide intervals of time, to follow in some measure the same general lines. For instance, Suess has shown that the Alps and the Pyrenees were foreshadowed at an earlier (Hercynian) date by the Variscan and Armorican chains, situated in somewhat more northerly latitudes; and we must expect, accordingly, a certain degree of overlapping of the Hercynian belt of igneous rocks by the later Alpine system. The data concerning petrographical provinces in the earlier geological periods are less complete than might be desired. We shall be content here to summarise the general petrographical characters of the British igneous rocks of different ages, with special reference to the distinction of alkaline and sub-alkaline provinces.

The Lewisian igneous rocks belong very decidedly to what we have termed the Pacific branch, and this is equally true of the younger pre-Cambrian groups of England and Wales. Among these earliest igneous rocks of Britain acid types largely preponderate. Owing, perhaps, to the ampler exposures, the igneous rocks of the Ordovician present a somewhat greater diversity than those of the Uriconian, etc., but in their general facies there is no essential difference. In Western Sutherland and Ross, however, Teall has described a group of plutonic rocks including nordmarkite, syenite, nepheline-syenite, and borolanite, with minor intrusions of allied types—all relatively rich in alkalies. Here for the first time we encounter an assemblage with typically Atlantic characters. It is proved to be younger than the Cambrian strata of Assynt, but older than the great overthrusting. It has a restricted areal distribution, partly

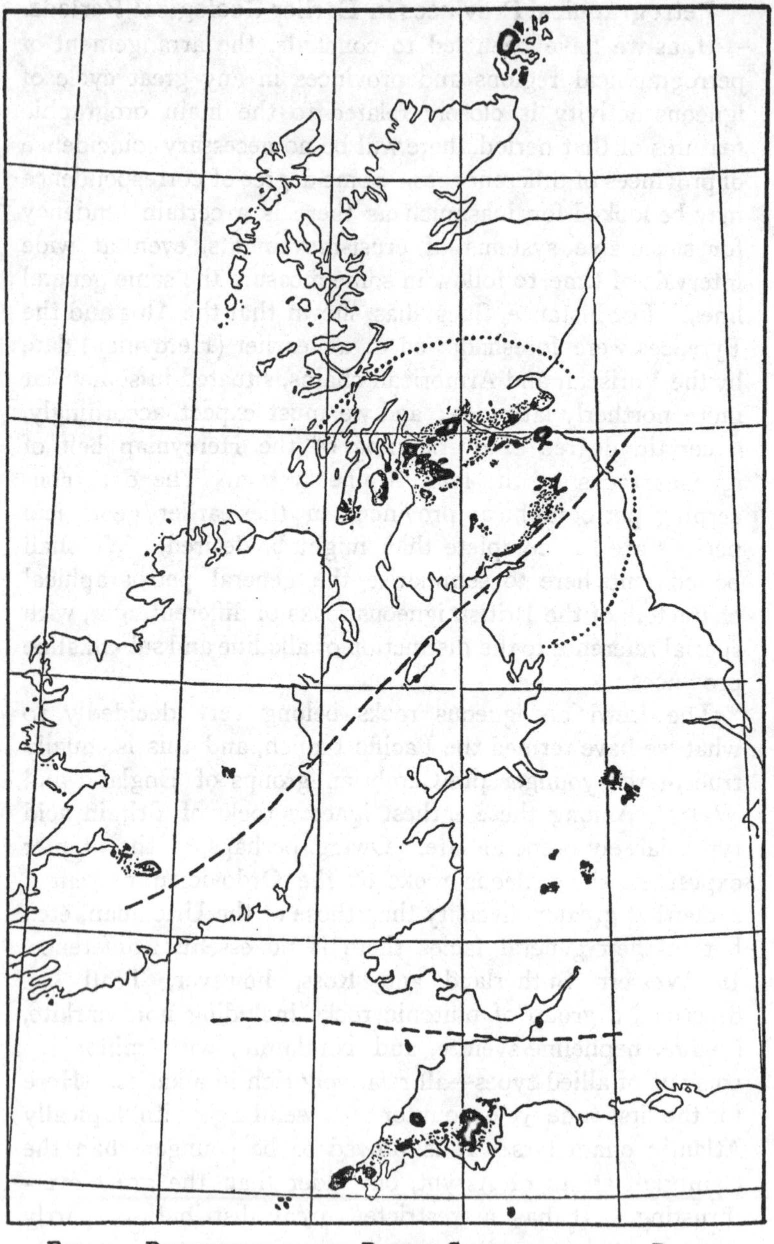

FIG. 23.—DISTRIBUTION OF THE BRITISH CARBONIFEROUS AND PERMIAN IGNEOUS ROCKS.

coinciding with that of the earlier Lewisian system, which belongs to the other great branch.

The igneous rocks of Old Red Sandstone age in Scotland and Ireland all belong to the sub-alkaline branch. Since the rocks of about the same age in the Christiania district embrace a great variety of types characterized by richness in soda, there evidently existed at this period well-marked petrographical provinces, of which the Norwegian and British belonged to opposite branches. Moreover, the Devonian igneous rocks of the south-west of England must be assigned to still another province, apparently of Atlantic affinities, though less strongly characterized. The basic lavas and tuffs are compared by Flett[1] with the spilites and schalsteins of Westphalia, Nassau, and the Harz, the felspar being generally oligoclase or albite. The intrusive dolerites and proterobases are also peculiar, some of them having essexitic characters. It appears, then, that this area must be separated from the Caledonian and attached to another region which extends eastward across Europe.

In the Lower Carboniferous we again have a well-marked areal distribution of different igneous rocks in the British area (Fig. 23). At this period, however, Scotland and Ireland belonged to a province of distinctly Atlantic affinities. The basic lavas seem to be in part comparable with those of the essexite family in the Devonian of Southern Norway, and there are also mugearites,[2] trachytes, and various types carrying primary analcime, nepheline, and soda-pyroxenes. The intrusive rocks are of still more pronounced alkaline character. To this age we may probably refer the dykes described by Flett[3] in the Orkneys. Very different are the normal basalts and basaltic tuffs of the Isle of Man, Derbyshire, and the Bristol district, with their accompanying intrusions, and the intrusive dolerites, etc., referred to a

[1] *The Geology of the Country around Plymouth and Liskeard*, *Mem. Geol. Sur.* (1907).

[2] Flett, *Summary oj Progress Geol. Sur.* for 1907, pp. 119-126.

[3] *Trans. Roy. Soc. Edin.*, vol. xxxix. (1900), pp. 865-905.

Carboniferous age[1] in the Midlands of England. The Scoto-Irish and English provinces may be divided, as shown on the map, by a north-east to south-west line, having the direction of the Caledonian axes of folding. In many districts in the southern half of Scotland there is developed a group of east to west dykes and sills, apparently of late Carboniferous age or later, consisting of quartz-dolerite, a typical sub-alkaline rock. Here, therefore, we have a Pacific province overlapping and partly coinciding with an earlier Atlantic province. Its limits cannot be laid down without more complete data, but the known occurrences in Scotland are roughly circumscribed by the dotted line in Fig. 23, which is prolonged into England to embrace the Whin Sill of Teesdale, evidently belonging to the same group of intrusions.

The Carboniferous and Permian igneous rocks of Cornwall and Devon belong, not to the Caledonian, but to the Hercynian belt of activity, with east to west extension. Some of the volcanic rocks formerly assigned to the Culm seem to belong in fact to the Devonian, and more information is needed concerning the rest. The granites, with their attendant dykes of quartz-porphyry and lamprophyre, are doubtless closely akin to those of Brittany. According to the field evidence their age may be either late Carboniferous or Permian. The igneous rocks interbedded in the Lower New Red Sandstone of the Exeter district, and reappearing farther west at Cawsand and elsewhere, include various alkaline basic types, and are regarded by Teall[2] as the reduced representative of the Permian volcanic series of France and Germany.

The British Tertiary province includes a large part of Scotland and the northern portions of England, Wales, and Ireland. The rocks, remarkable for the strong preponderance of basic types, belong to the Pacific or sub-alkali branch.

It is clear from the foregoing hasty survey that, so far as

[1] These, or some of them, may possibly be Tertiary (Watts, *Proc. Geol. Assoc.*, vol. xv. (1898), pp. 399, 400).

[2] *The Geology of the Country around Exeter*, *Mem. Geol. Sur.* (1902), chap. iv.

concerns the British area, petrographical provinces have not been fixed from one geological period to another, but have undergone frequent rearrangement. There are not wanting indications that the arrangement in a given period of igneous action stands related to the crust-movements of that period, though this could be established only by a much more complete examination of the facts. It is at least manifest that the distribution of different groups of igneous rocks in Britain cannot be explained by any initial want of uniformity in the composition of the Earth's crust in this tract.

CHAPTER V

MUTUAL RELATIONS OF ASSOCIATED IGNEOUS ROCKS

Different orders of relationship.—Chronological sequence of associated rock-types.—The variation-diagram.—Plutonic complexes.—Variation in a single rock-body.—Heterogeneity due to simultaneous injections.—Variation among minor intrusions.

Different Orders of Relationship.—The doctrine of evolution, as applied to igneous rocks, is not of a kind which can be established by deductive reasoning, but rather by examining in the light of this hypothesis the actual facts of petrology. This we have done in the preceding chapter as regards the salient features of the geographical distribution of different kinds of rocks. We have now to consider from the same point of view the paragenesis of particular rock-types, their actual manner of association in the field, and the order in which the associated and presumably consanguineous types have made their appearance.

Even a cursory view of the distribution of different igneous rocks cannot fail to remark the frequent recurrence in widely separated regions of certain associations of two or more rock-types, analogous to those mineral associations to which the term *paragenesis* is applied. Such pairs as basalt and rhyolite, gabbro and granite, essexite and nepheline-syenite, minette and aplite, readily suggest themselves in this connection. The associations so often repeated cannot be fortuitous, and they seem to admit of no explanation other than that of community of origin.

Pursuing the idea of blood-relationship by descent from a

common stock, we see that we must expect *different orders of relationship* among the cognate rocks of one natural assemblage. If two given rock-types be derived from the hypothetical common parent magma by one or two or more removes in different cases, they may belong to the same or different lines of descent, and to the same or different generations. While the drawing up of a formal genealogical tree must be, in the present state of knowledge, largely conjectural, much light is thrown on the relationships of different rocks by their mutual association and the order in which they have been intruded or extruded. Thus, a plutonic rock often has associated with it various groups of dyke-rocks which may be regarded as its *satellites* ('Ganggefolgschaft' of some German writers). They have been subsequently intruded in the vicinity of the plutonic rock, and their petrographical characters connect them more closely with it than with other plutonic rocks of the same series. Though not necessarily derived through that type, they belong presumably to the same line of descent. If directly derived, they may be derived with or without differentiation. Thus, a granite may be attended by a group of granite-porphyry dykes not essentially different from it in composition; but, on the other hand, many granites have as satellites groups of dykes, such as minettes and aplites, which differ widely from the plutonic type and from one another.[1] Brögger[2] marks this distinction by calling the non-differentiated dykes *aschistic* and the differentiated ones *diaschistic*.

Again, it is sometimes found that two groups of dyke-rocks, satellites of the same plutonic rock and differing more or less notably from it in composition, differ from it in opposite directions, and in such a manner that the two, mixed in certain proportions, would reproduce approximately the composition of the plutonic rock-type. The presumption is that the magmas of the two satellite groups have arisen from the splitting up (by whatever physical processes) of a magma

[1] Rosenbusch, *Tscherm. Min. Petr. Mitth.* (2), vol. xii. (1891), pp. 386-388.

[2] *Eruptivgesteine des Kristianiagebietes*, I. (1894), p. 125.

like that of the plutonic rock; and, to express the special relation which the two groups bear to one another, Brögger[1] uses the term *complementary*. Thus, he finds that, in the neighbourhood of Gran and elsewhere, plutonic intrusions of essexite are accompanied by dykes and sheets of two types, camptonite and mænaite (lime-bostonite); the former a melanocratic type—*i.e.* with predominance of the dark minerals—and the latter leucocratic—*i.e.*, with light minerals predominant. Analyses show that a mixture of 9 parts of camptonite with 2 of mænaite would have nearly the composition of the essexite. It is, of course, conceivable that a parent-magma may give rise, not to two, but to three or more co-ordinate derivatives, and Brögger[2] has elsewhere cited examples which he interprets in this manner.

It is often found that a number of rock-types, associated at one centre, or at least in one petrographical province, and having like geological relations and a general community of characters, vary widely in composition, but in such a manner that the variation is along certain defined lines. From the genetic point of view, we must suppose that the various rocks have resulted from differentiation of the same kind in different degrees. Such a set of related rock-types may conveniently be called a *series*. A series in this sense may be, for instance, a number of different lavas erupted successively at one volcanic centre; or a set of plutonic intrusions in the same district; or, again, dyke-rocks which are satellites of different members of a plutonic series, and each related in the same way to its own principal. Brögger[3] has used the term 'Gesteinsserie' in a more limited and artificial sense, and it does not appear to the present writer that the conception so defined corresponds with reality.

Chronological Sequence of Associated Rock-Types.—When the association of different kinds of igneous rocks in one petrographical province, or at one centre, is considered with reference to their derivation from a common stock by

[1] *Quart. Journ. Geol. Soc.*, vol. l. (1894), p. 31.
[2] *Eruptivgesteine des Kristianiagebietes*, III. (1898).
[3] *Ibid.*, I. (1894), pp. 169-181.

magmatic differentiation, special significance attaches to the order in which the several rock-types have been intruded or extruded. The phenomena of modern volcanoes throw little light on this question. Notwithstanding the very peculiar nature of the lavas of Vesuvius, Fuchs (1870) concluded that the chemical composition of flows erupted in historic times shows little variation; and von Lasaulx (1880) arrived at a like conclusion in the case of Etna. Lang[1] subsequently endeavoured to trace some law of variation in both cases; but the variation actually shown by chemical analyses is of small amplitude, and, if significant at all, is periodic, not secular. We may infer that the causes which bring about any important change in the volcanic rocks of a given centre operate with extreme slowness as measured by human events.

Turning to the geological record, we find that different writers have arrived at widely different generalisations concerning the order of succession of presumably cognate rock-types in one petrographical province. Bertrand,[2] in his survey of igneous action in Europe, dealing with the question necessarily in a very summary manner, seemed to find that the latest rocks of each great system are the most basic; though he also notes a tendency in the latest stages to peculiar and highly specialised types. Sir A. Geikie,[3] on the other hand, referring in particular to the British area, concludes that, on the whole, "each eruptive period witnessed the same sequence of change from basic to acid lavas." Brögger,[4] while recognising that different laws may hold good for the volcanic rocks and the plutonic, lays stress on the latter as indicating the "order of primary differentiation." He finds for this case an order of increasing acidity, with in many instances a final reversion to basic types. Iddings,[5] on the other hand, claims a greater importance

[1] *Zeits. für Naturw.* (Halle), vol. lxv. (1892), pp. 1-30.
[2] *Bull. soc. géol. Fra.* (3), vol. xvi. (1888), p. 593, etc.
[3] *Quart. Journ. Geol. Soc.*, vol. xlviii. (1892), Proc., p. 178.
[4] *Eruptivgesteine des Kristianiagebietes*, II. (1895), pp. 165-181.
[5] *Quart. Journ. Geol. Soc.*, vol. lii. (1896), pp. 606-617.

for the extrusive rocks, and enunciates the rule "that in a region of eruptive activity the succession of eruptions commences in general with magmas representing a mean composition and ends with those of extreme composition."

From our point of view a discussion of the relative volume and importance of extrusive as compared with intrusive rocks seems irrelevant. If volcanic and plutonic rocks follow different laws, it is necessary to inquire what is the actual law of sequence for each. We shall see some grounds for believing that in each of the three phases of igneous activity which we have distinguished—volcanic, plutonic, and minor intrusions—some general law of succession is more or less widely applicable, and that the law is different for each of the three phases. If we merely arrange all the igneous rocks of one province in chronological order, without regard to the distinction of the three different phases, no intelligible law of succession is apparent.

As regards the *volcanic phase*, von Richthofen,[1] forty years ago, laid down a general order of succession, which he had established primarily in Hungary and Transylvania and in parts of the Western States of America, but found to hold also in other regions. His scheme, applied in the first instance to volcanic rocks of Tertiary and post-Tertiary age, recognised five natural orders, normally following one another in succession: (*a*) propylite [hornblende-andesite], (*b*) andesite [pyroxene-andesite], (*c*) trachyte [mica-hornblende-andesite], (*d*) rhyolite, (*e*) basalt. We supply the modern designation of the rocks in brackets, where von Richthofen's names might be misleading. The sequence seems at first glance to have no simple relation to the relative acidity of the several rocks; but, regarded more closely, it is seen to represent an *increasing divergence from the initial type*. Thus *a*, *b*, *e* are in order of increasing basicity, and *a*, *c*, *d* in order of increasing acidity. There is concurrent variation in these two opposite directions, and members

[1] 'The Natural System of Volcanic Rocks,' *Mem. Calif. Acad. Sci.*, vol. i. (1868), pp. 39-133; also *Zeits. deuts. geol. Ges.*, vol. xx. (1868), pp. 663-726, and vol. xxi. (1869), pp. 1-80.

belonging to the two diverging lines partly alternate with one another. That this is the true interpretation is confirmed by the fact that, although many provinces of sub-alkaline lavas show an order differing in some points from that given above, they still fall under the same law of divergence. For instance, rhyolite may follow basalt instead of preceding it. For the Eureka district of Nevada the order, as given by Hague, may be summarised: (*a*) hornblende-andesite, (*b*) hornblende-mica-andesite, (*c*) dacite, (*d*) rhyolite, (*e*) pyroxene-andesite, (*f*) basalt. This differs from the order of von Richthofen, but it still shows one line of increasing acidity, *a*, *b*, *c*, *d*, and another of increasing basicity, *a*, *e*, *f*. In Mexico[1] the order is: (*a*) pyroxene-andesite, (*b*) hornblende-andesite, (*c*) hornblende-mica-andesite, (*d*) hornblende-hypersthene-andesite, (*e*) hypersthene-andesite, (*f*) augite-andesite and labradorite, (*g*) basalt, (*x*) rhyolite. The place of the rhyolite is not accurately ascertained, but it is younger than the hornblende-andesite at least. Here again we have *a*, *b*, *c*, *x* showing increasing acidity, and *b*, *d*, *e*, *f*, *g* increasing basicity.

The law of increasing divergence, as maintained by Iddings, certainly has a very wide application, though the initial type is different in different cases. It is often a trachyte among rocks of the Atlantic branch. For example, in the Siebengebirge[2] we have: (*a*) trachyte-tuff, (*b*) rhyolite, (*c*) sanidine-oligoclase-trachyte, (*d*) hornblende-andesite, (*e*) basalt. Exceptionally one of the two divergent lines may be suppressed, as at Monte Ferru, Sardinia,[3] where there is merely a sequence of increasing basicity from sanidine-plagioclase-trachyte to leucite-basalt. Again, if we attribute the divergence to progressive differentiation in an intercrustal magma-reservoir, it is not difficult to conceive that fresh accessions of magma to the reservoir may some-

[1] Aguilera and Ordoñez, *Datos para la geología de México* (1893), p. 47.

[2] Mangold, *Ueber die Altersfolge der vulkanische Gesteine . . . im Siebengebirge*, Inaug. Diss. (1888), Kiel.

[3] Doelter, *Denkschr. k. Akad. Wiss. Wien, Math.-nat. Cl.*, vol. xxxix. (1878).

times bring about a repetition of the same sequence. Instances of this might be cited: an interesting case is the volcanic succession in the Berkeley Hills, near San Francisco.[1] Writing *a* for andesite, *b* for basalt, and *r* for rhyolite-tuff, the order is as follows:

Lower Berkeleyan	*a, b, r; a, b, r.*
Upper Berkeleyan	*a, b, r; a, b* [? hiatus].
Campan	*a, b, r, b, r.*

This shows five, or perhaps six, recurrences.

In the *plutonic phase* it is very generally found that the most basic rock has been intruded first, then less basic types in order, and the most acid last. This law of *decreasing basicity*, or increasing acidity, is of such wide application that we cannot doubt its significance. As an example, illustrating at once the general law and the way in which it may be qualified by partial exceptions, we take the sequence of the plutonic intrusions of the Christiania district, as given in Brögger's[2] latest contribution:

1. Oldest products of differentiation of the stem-magma: lime-alkali-rocks—(*a*) essexites (basic and ultra-basic), (*b*) akerites (intermediate).

1-2. Passage-rocks between the first and second classes: (*c*) larvikite-monzonites.

2. Middle products of differentiation of the stem-magma: intermediate alkali-rocks—(*d*) larvikites, (*e*) lardalites (nepheline-syenites).

3. Youngest products of differentiation of the stem-magma: acid alkali-rocks—(*f*) pulaskites and nordmarkites, (*g*) ekerites (soda-granites), (*h*) biotite-granites, (*k*) rapakiwi-granites.

The general law of decreasing basicity is here sufficiently evident, though it is not without exceptions. Thus the lardalites, which succeeded the larvikites, are poorer in silica, though richer in alkali; and the rapakiwi rocks, which

[1] Lawson and Palache, *Bull. Dep. Geol. Univ. Calif.*, vol. ii. (1902), p. 438.

[2] *Nyt Mag. for Naturvid.*, vol. xliv. (1906), pp. 113-144.

close the succession, are somewhat less acid than the biotite-granites. Seeing in the general law the operation of a differentiation of the first order, we may suppose the exceptions to result from a subsidiary differentiation on different lines. The similar series of plutonic rocks in Massachusetts, the Montreal district, etc., show a like order of succession; and Brögger[1] has pointed out that the Monzoni rocks, originating from an initial stock less notably rich in alkali, present a parallel sequence, although the actual types are different. In provinces belonging to the Pacific branch, the plutonic rocks show the same law of decreasing basicity in the order of their intrusion. Those of Old Red Sandstone age in Scotland afford a good illustration.

In the *phase of minor intrusions* the succession of different rock-types sometimes appears more complicated than in the two preceding phases. This may be attributed to a progressive elaboration of the processes of magmatic differentiation, which in this late phase may follow concurrently different lines. Taking first the simplest case, where minor intrusions related to a particular centre fall into several distinct groups with a *serial* relationship, we find that a simple rule holds good in a large number of cases. The order is one of *decreasing acidity*, or increasing basicity, and is therefore the reverse of the normal order in the plutonic phase. For instance, the plutonic rocks of the British Tertiary cycle show the regular sequence:

(*a*) ultrabasic, (*b*) basic, (*c*) acid;

but for the local groups of minor intrusions connected with the plutonic centres the order is:

(*a*) acid, (*b*) basic, (*c*) ultrabasic.

Where two groups of minor intrusions hold a *complementary* relationship to one another, diverging from the presumed common source, it may be supposed that, as in the two diverging lines of a typical volcanic succession, the relative order of intrusion is not necessarily the same in every case. Nevertheless, the order of decreasing acidity holds in most

[1] *Eruptivgesteine des Kristianiagebietes*, II. (1895), p. 163.

cases for which we have data. Rosenbusch has remarked that, of associated aplites and minettes, the former have been the earlier intruded, and this generalisation is verified in many different regions. Among the minor intrusions of Old Red Sandstone age in Scotland, the various lamprophyric dykes and sheets intersect, not only the non-differentiated or little-differentiated types (porphyrites, etc.), but also in general the quartz-porphyries which are their proper complements. On the other hand, bostonites intersect camptonites, both in Norway and the Montreal district, and other exceptions to the usual sequence might be cited.

When two or more pairs of complementary rocks have been evolved at a given centre by differentiation on the same general lines, we should expect them to show an order of *increasing divergence*. This is illustrated at the Piz Giuf centre in the eastern Aar-massif.[1] The initial magma here seems to have had the composition of a quartz-syenite, and the plutonic intrusions are of syenite and basic granite, followed by more acid granite. The sequence of the associated dykes is: (*a* and *b*) kersantite and granite-porphyry, (*c* and *d*) spessartite and aplite. Each pair is complementary, and the latter pair represents a further degree of differentiation than the former.

The Variation-Diagram.—When a number of cognate igneous rocks have a *serial relationship* to one another, the variation being along certain definite lines, and presumably brought about by differentiation of the same kind, the character of that variation may be made apparent by a *graphical comparison* of the chemical analyses. This brings out at a glance relations which, lying buried in columns of figures, might pass unnoticed. For this purpose we shall make use of a simple diagram, in which the silica-percentages of the several rocks are taken as abscissæ, and the percentages of the other constituents (which, generally speaking, are the bases) as ordinates.[2] Suppose each side of the square in

[1] Weber, *Beitr. z. geol. Karte Schweiz* (N.S.), part xiv. (1904).

[2] Iddings has used percentages by molecules instead of percentages by weight: *Bull. Phil. Soc. Washington*, vol. xii. (1892), pp. 89-214; *Prof. Paper No.* 18 *U.S. Geol. Sur.* (1903).

Fig. 24 divided into 100 parts. Then, if *am* represent the silica-percentage of one of the rocks, and *mn* and *mq* its percentages of alumina and magnesia, and if the figures for the other rocks of the series be plotted in like manner, we can draw through the summits of the ordinates curves *np* for alumina and *qr* for magnesia, which will represent the

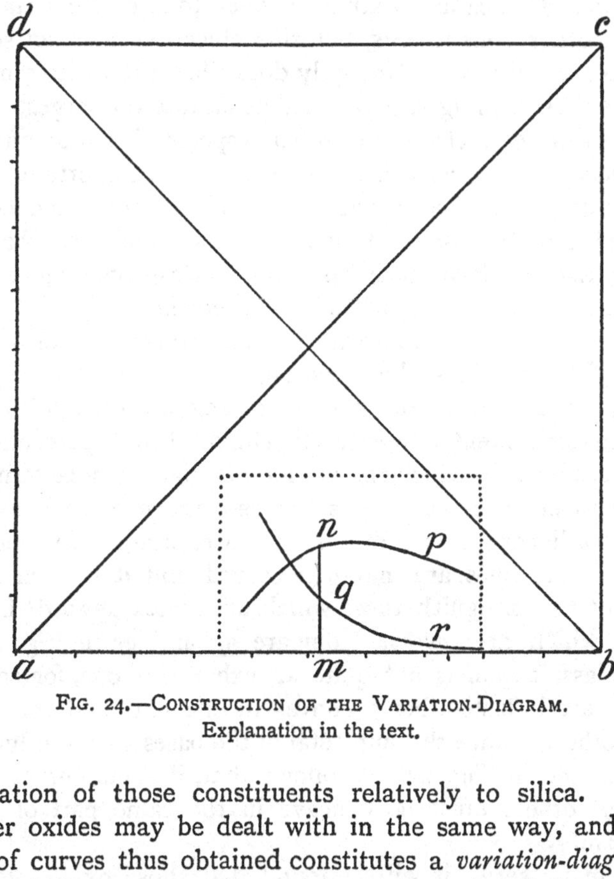

FIG. 24.—CONSTRUCTION OF THE VARIATION-DIAGRAM.
Explanation in the text.

variation of those constituents relatively to silica. The other oxides may be dealt with in the same way, and the set of curves thus obtained constitutes a *variation-diagram*. Regarded as representing in some sense the derivation of the rocks from a common source, it is also a picture of the kind of differentiation ('Differentiationsbild,' Vogt).

The variation of the silica itself is represented by the straight line *ac*, and that of the sum of all the other constituents by *db*, from which it follows that all the curves must be contained within the triangle *dab*. In any actual case the curves fall within a small part of that triangle, and only this part of the diagram need be drawn, as shown by the dotted rectangle.

Certain writers have taken exception to the arbitrary selection of silica as a datum to which to refer the variation of the other constituents, but this objection does not seem to have much force. Not only does silica enter into almost all the rock-forming minerals and constitute the largest part of most of the rocks, but in other respects also it stands on a special footing, as being usually the only important acid constituent appearing in the analyses. Its pre-eminence is conceded in the usage of such terms as 'acid' and 'basic' rocks, and we have seen also that silica-percentage often stands in relation with chronological sequence.

Some of the curves in a variation-diagram may be approximately straight lines; but even approximate rectilinearity is rarely realised in all the lines of a diagram, though it may hold to some extent between certain limits of silica-percentage. The ideal case in which the lines are straight may be termed *linear variation*, inasmuch as the percentage of each oxide is then a linear function of the silica-percentage. In general most of the lines are markedly curved, and it is then important to distinguish those which are *convex* upward[1] from those which are *concave* (compare *np* and *qr* in Fig. 24). The classification is not quite an exhaustive one, for some curves are inflected, being convex in one part and concave in another. Since the sum-total of the bases varies in linear fashion (*db* in Fig. 24), it follows that, if some curves are convex, others must be concave in the same part of the diagram.

In many series it will be found that those oxides which belong chiefly to the light-coloured minerals (alumina, alkalies, lime) have convex curves, while those which belong

[1] Harker, *Journ. Geol.*, vol. viii. (1900), pp. 389-399. Washington has used the terms 'convex' and 'concave' in a reversed sense, *ibid.*, vol. ix. (1901), p. 652.

chiefly to the dark minerals (magnesia, ferrous oxide, titanic oxide) have concave curves. A closer degree of correspondence often becomes apparent between particular curves in a given diagram, indicating a *sympathetic variation* of the constituents to which the curves belong. In other cases we may see evidence of *antipathetic variation*, one curve rising as another declines, and conversely. These mutual relations are too frequently observed to be dismissed as accidental, and must be supposed significant. That particular oxides vary in the same way, and this in many different series of rocks, may be taken as indicating that they are actually combined together in the magmas which undergo differentiation. In other words, the variation is in reality one, not of independent oxides, but of compounds more or less similar to known minerals. To this point we shall return in a later chapter. It is not often practicable to make a mineralogical, instead of a chemical, variation-diagram, because the percentage mineral composition of a rock is not easily determined with any precision, and many of the rock-forming minerals are themselves of variable composition. We shall be content, therefore, to construct a diagram from the chemical analyses, but to regard it as giving at the same time a picture, if not always a very distinct one, of the variation in mineralogical composition.[1]

The manner of applying a mineralogical interpretation to the variation-diagram will readily suggest itself. For instance, in many series of rocks of the sub-alkali branch the alumina is contained principally in the felspars; and the rocks carry alkali-felspars in the more acid types, felspars increasingly rich in lime in the intermediate and basic types, and little or no felspar in the ultrabasic. The alumina-curve, therefore, starting from near the base-line on the left, rises at first

[1] Schroeder van der Kolk has made an elaborate investigation of sympathies and antipathies of the chief elements for igneous rocks as a whole; but the fact that the elements are often differently combined in different rocks obscures the relationships in such an indiscriminate survey. See *Over de Sympathieën en Antipathieën der elementen in de Stollingsgesteenten* (1903, Amsterdam).

very sharply, owing to the coming in of a felspar very rich in alumina (37 *per cent.* in anorthite). It then rises more gently, and finally declines, partly because the alkali-felspars are poorer in alumina (orthoclase 18½ *per cent.*), partly because the coming in of quartz reduces the total proportion of felspar in the rock. In a series belonging to the alkali-branch the summit of the alumina-curve may be higher and farther to the left, owing to the presence of basic silicates of the felspathoid group (nepheline with about 33 *per cent.* of

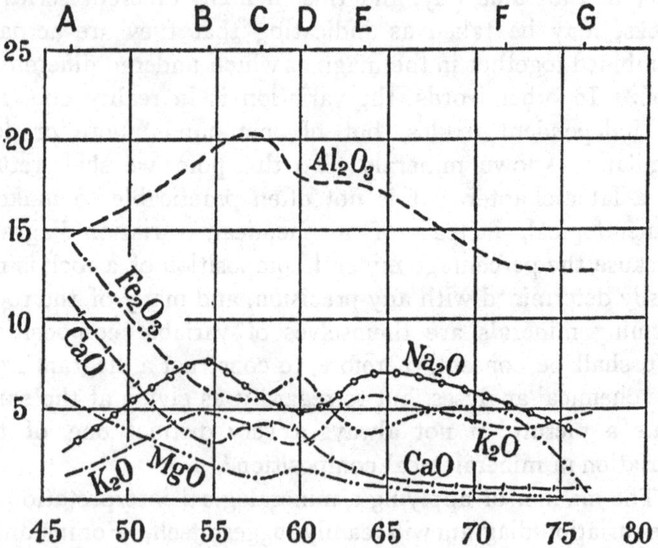

FIG. 25.—MUTUAL RELATIONS OF THE PLUTONIC ROCKS OF THE CHRISTIANIA DISTRICT, FROM ANALYSES GIVEN BY BRÖGGER. TOTAL IRON (WITH MANGANESE) RECKONED AS FERRIC OXIDE.

A, Essexite, average for Gran district; B, lardalite, normal type; C, larvikite (augite-syenite), mean of three; D, akerite (more acid augite-syenite), mean of eight; E, nordmarkite (quartz-syenite), average; F, ekerite (soda-granite); G, biotite-granite, mean of three.

alumina). Again, in most series of rocks, the magnesian minerals are most abundant in the basic, and especially the ultrabasic types, and the magnesia-curve therefore declines from left to right. It declines, however, more steeply at first than afterwards, because in the ultrabasic rocks the magnesia is chiefly orthosilicate, and in the basic rocks chiefly metasilicate; hence the concave shape of the curve, contrasting with the convex shape of the alumina-curve.

We have premised that the analyses plotted are those of rocks connected by a real serial relationship. If the analyses are numerous, and some of them fall near together, it may be found that the summits of the ordinates, for a given constituent, do not fall exactly on a flowing curve. This is a necessary consequence of imperfection in the analyses, and a certain degree of 'smoothing' is legitimate.

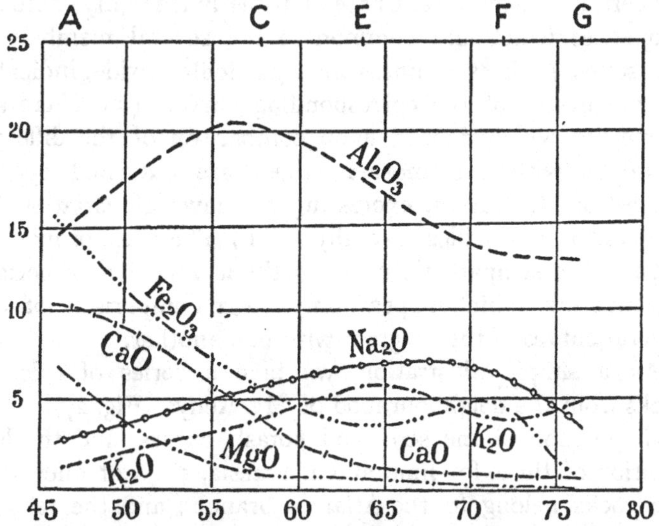

FIG. 26.—VARIATION-DIAGRAM FOR THE PRINCIPAL PLUTONIC ROCKS OF THE CHRISTIANIA DISTRICT, OMITTING THE LARDALITE AND AKERITE OF THE PRECEDING DIAGRAM.

More considerable discrepancies must be taken to indicate that the variation is not in fact wholly on one definite line, but is complicated by the effects of a subsidiary differentiation. In this way *the variation-diagram becomes a criterion* of the simplicity or complexity of the variation which it represents. Among the plutonic rocks of Christiania, Brögger at first ranked lardalite and akerite as co-ordinate with essexite, larvikite, and the rest; but he has since come to the conclusion that the first is a derivative from the larvikite magma, and the second holds a like subordinate relation to essexite. This might be inferred from the variation-diagram, where the inclusion of lardalite and akerite would necessitate some very unnatural flexures of the curves (compare Figs. 25 and 26).

If the curves justify themselves by their smoothness and regularity, they stand as a graphical expression of the laws of variation in the given series of rocks, and *the variation-diagram becomes a characteristic* of that series for comparison with other series. From this point of view the distinctive properties of such a series are of three kinds: (i.) There are properties which are common to the several members of the series, such as richness in a particular oxide, indicated by the height of the corresponding curve. (ii.) There are properties which depend upon comparison of the different members with one another; these are indicated by the shapes of the curves, expressing the laws of variation for the constituent oxides severally. (iii.) There are properties, such as the sympathetic or antipathetic variation of certain constituents, which appear only on a comparison of the several curves of the diagram with one another.

As a simple illustration, we take a series of volcanic rocks from the neighbourhood of Mt. Kenya (Fig. 27). The high position of the soda and potash curves, and the low position of those for lime and magnesia, show at once that the rocks belong to the Atlantic branch; and the relative position of the two alkali-curves shows that this province is one of those in which soda dominates over potash. The nearly level course of these two curves is a special characteristic of this series, and so also is the peculiar behaviour of the iron-curve. This, instead of declining steadily, rises to a summit at the comparatively high silica-percentage 65. At the same time the alumina-curve falls sharply, the antipathetic relation between the two constituents being connected with the replacement of alumina by ferric oxide in the soda-bearing pyroxenes and amphiboles. The Christiania rocks, like those of East Africa, belong to the alkali-branch, with a predominance of soda, and there are, accordingly, points of resemblance between Figs. 26 and

27; but there are also significant differences. The most remarkable feature of the Christiania plutonic rocks is the close parallelism between the soda and potash curves, connected with the fact that a large part of the alkalies is contained in perthitic felspars of approximately constant composition. This sympathy is lost, with the perthitic character, at the acid end of the series.

For comparison with the thoroughly alkaline lavas of

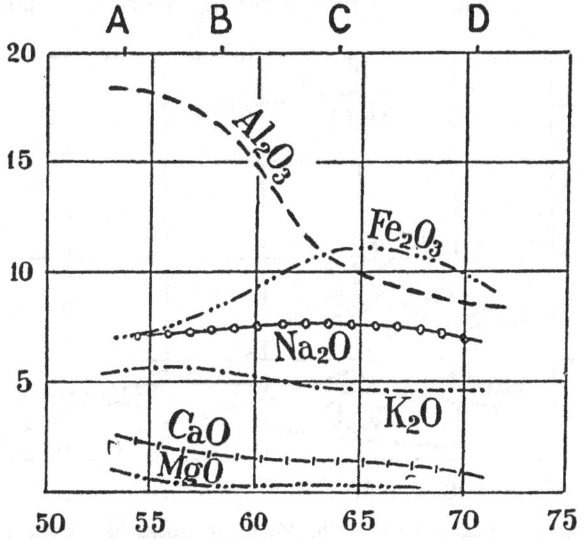

FIG. 27.—VARIATION-DIAGRAM OF VOLCANIC ROCKS FROM BRITISH EAST AFRICA, FROM ANALYSES BY PRIOR.[1] TOTAL IRON RECKONED AS FERRIC OXIDE.

A, Kenyte, Mt. Kenya; B, phonolite, Mt. Kenya; C, phonolitic obsidian, Lake Nakaru; D, soda-rhyolite, Lake Naivasha.

Mt. Kenya, we take the volcanic rocks of Lassen Peak, in California, which are of typical Pacific facies. They are basalts, andesites, dacites, and rhyolites of ordinary types, and the variation-diagram (Fig. 28), though constructed from a large number of analyses, requires very little 'smoothing.'

Plutonic Complexes.—It is the plutonic rocks that afford the simplest illustrations of the kind of relationship met with

[1] *Min. Mag.*, vol. xiii. (1903), p. 247.

among a series of cognate rock-types. In the plutonic phase it often happens that there is a somewhat narrow localisation of activity at certain centres; and here successive intrusions of different magmas may follow one another, apparently at short intervals, in such a manner that the later intruded masses impinge upon or invade the earlier. For such a series of rocks, intimately associated and beyond doubt genetically related, we shall use the term *plutonic complex*. The several types may be divided by sharp boundaries, or

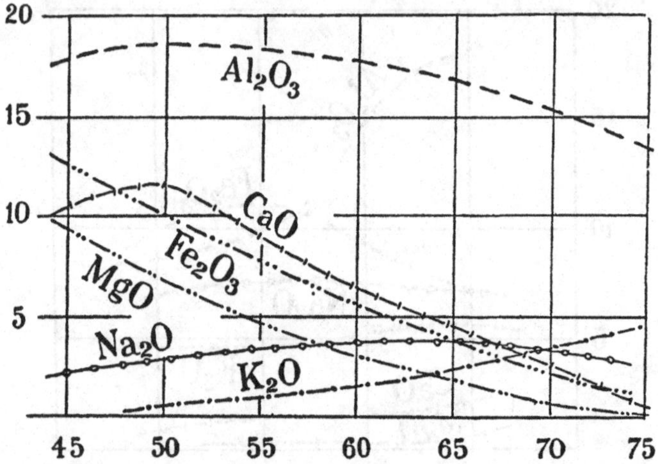

FIG. 28.—VARIATION-DIAGRAM OF VOLCANIC ROCKS OF LASSEN PEAK, CALIFORNIA, FROM THIRTY-FOUR ANALYSES.[1]

The quartz-basalts are excluded as being of abnormal origin (see below, Chapter XIV.).

may graduate one into another; the distinction between the two cases being probably less fundamental than may at first sight appear.

Plutonic complexes with discontinuous junctions are well illustrated among the Tertiary intrusions of the Inner Hebrides and other British districts. The order of intrusion is always that of decreasing basicity (peridotite and allivalite, eucrite, gabbro, granite). In the laccolitic complexes a newer rock is always intruded partly beneath an older, and

[1] *Bull. No.* 148 *U.S. Geol. Sur.* (1897), pp. 192-200.

encroaching upon it; and where the boss-habit is found, a newer rock is always intruded beside an older, and partly through it. Between ultrabasic and basic rocks the junctions are often extremely intricate, with a broad belt of mechanical intermixture, so that we may pass quite gradually from peridotite traversed by gabbro veins to gabbro enclosing blocks and fragments of peridotite. Between basic and acid rocks such mechanical intermixture is often complicated by chemical reactions, with the production of hybrid varieties. In these places the basic intrusion must have been followed by the acid after no long interval, for we cannot suppose that the later magma was sufficiently superheated to produce any important refusion of solid rock already cooled. When the acid rock gives evidence of marginal chilling against the basic one, no sign of such chemical reactions is found.

A deceptive effect of gradual transition—though easily detected—may be brought about locally by reactions of the kind just mentioned. But, apart from such special phenomena, there are many cases in which the variation between the different members of a plutonic complex is continuous in one place and discontinuous in another. As typical examples, we will examine the complexes belonging to the group of 'newer granites' in Scotland. This group is represented by many intrusions, large and small, from Caithness to Galloway, and from Peterhead to the Ross of Mull, and they are described in the publications of the Geological Survey. The epoch of intrusion is later than that of the great Mid-Palæozoic crust-movements. The rocks, in different cases, break through all formations up to the volcanic series of the Lower Old Red Sandstone, while pebbles of many of them occur in the conglomerates of the Upper Old Red Assuming that all belong to one natural group, their date is thus fixed with some precision. The rock-types represented range from ultrabasic to highly acid, and the chief facts of their variation may be summarised as follows:

(i.) Rocks of acid composition preponderate, especially in the largest complexes, such as those of the Cairngorms (principally biotite-granite) and the Beinn Cruachan district (hornblende-granite and tonalite).

(ii.) The several types in one complex are usually sharply divided at a definite junction, but sometimes pass gradually into one another, and this difference may even be seen in different parts of one boundary-line. This proves that the distinction is not necessarily very significant. It possibly depends on the depth to which the complex has been exposed by erosion, so that a sharp boundary might, if we could follow it downward, give place to a gradual transition. Clearly no long interval separated the intrusions of successive rocks, and this is sometimes confirmed by effects due to local fusion of an earlier rock by a later magma.

(iii.) Wherever an order of succession is apparent, it is that of decreasing basicity.

(iv.) As a rule, though not without exceptions, the younger and more acid rocks, making usually the principal part of the complex, have been intruded in such a manner as to leave the older and more basic rocks on their margin as an inconstant border. This is especially so when one type passes gradually into another. The most acid granites occur as veins traversing the other members of the complex.

(v.) The general nature of the variation in these complexes may be gathered from the most usual rock-types. These are, primarily, granites, usually with biotite or hornblende; various rocks classed as tonalites and quartz-diorites, also with biotite or hornblende, and sometimes augite; basic diorites, rich in ferro-magnesian elements; and, of less frequent occurrence, peridotites of more than one type. This may be taken as the main line of variation for the group as a whole. In other cases, as in parts of Aberdeenshire, a rhombic pyroxene becomes an important constituent in the more basic rocks. Again, some of the rocks would fall into the monzonite family, as understood by Brögger. This family includes plutonic rocks, ranging from basic to acid, in which orthoclase and plagioclase felspars occur in roughly equal proportions. Quartz-monzonites, in this sense, are found on Beinn Nevis and Criffel, and the kentallenites of Ballachulish and Glen Orchy represent a basic type of the same family. The variation shown by the principal types in any one district is, however, of a serial kind.

(vi.) The actual variation of chemical composition in one typical series is illustrated in Fig. 29, which represents the

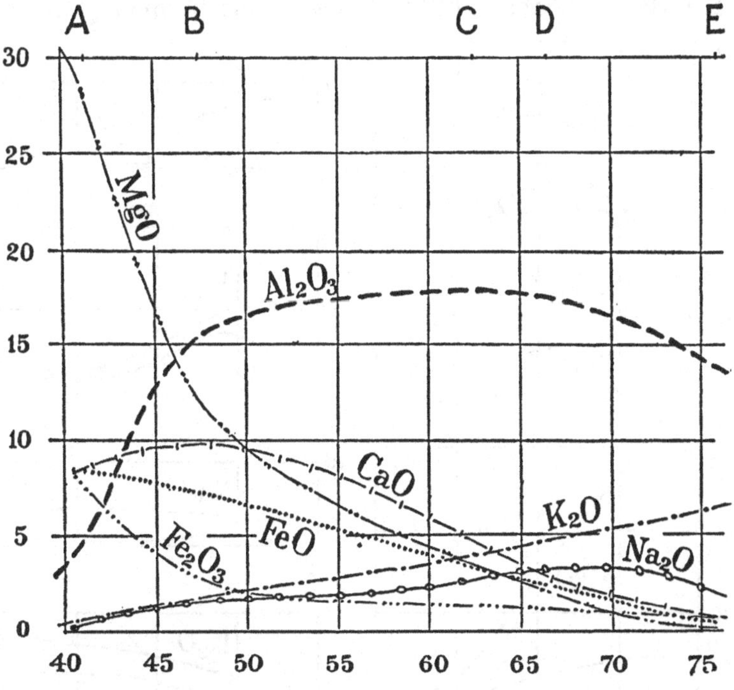

FIG. 29.—VARIATION-DIAGRAM OF PLUTONIC COMPLEX OF GARABAL HILL AND MEALL BREAC, NEAR ARDLUI, LOCH LOMOND, FROM ANALYSES BY PLAYER.[1]

The first rock, partly serpentinised, is recalculated to 100 *per cent.*, omitting water. One exceptional type, not found in place, is omitted.
A, olivine-diallage-rock; B, biotite-diorite; C, hornblende-biotite-granite; D, porphyritic-biotite-granite; E, eurite vein.

complex described by Dakyns and Teall at Garabal Hill and Meall Breac, west of the head of Loch Lomond. This variation-diagram is a very characteristic one for a common association of sub-alkaline rocks. The flat convex shape

[1] *Quart. Journ. Geol. Soc.*, vol. xlviii. (1892), p. 215.

of the alumina-curve, the declining concave curves of the magnesia and ferric oxide, the lime with its maximum near the basic end, the soda with its maximum near the acid end, and the steady rise of the potash-line are all characteristic features.

(vii.) The range of variation in chemical composition for the whole assemblage of rock-types in this group of plutonic

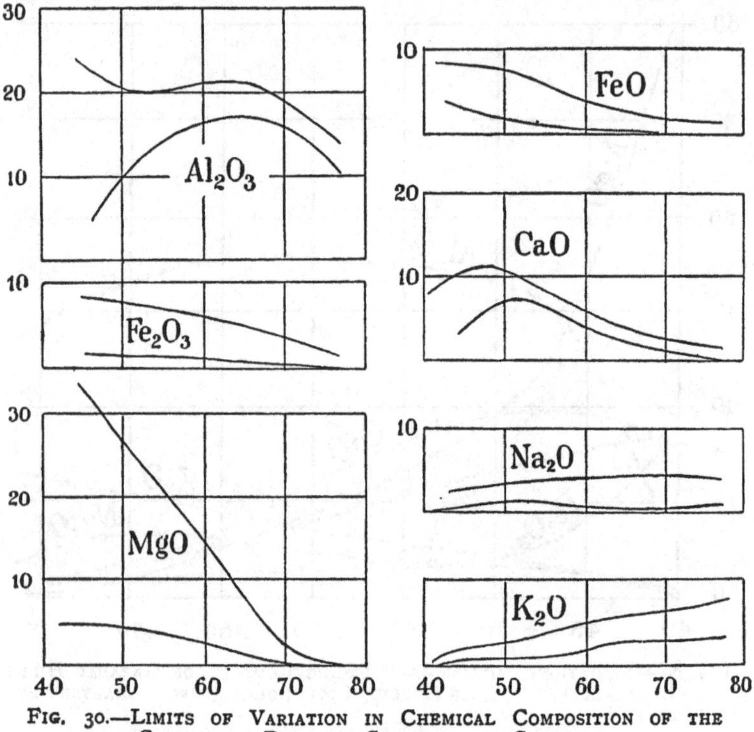

FIG. 30.—LIMITS OF VARIATION IN CHEMICAL COMPOSITION OF THE CALEDONIAN PLUTONIC COMPLEXES OF SCOTLAND.

The percentages of the several oxides vary between limits indicated by the upper and lower curves in each case.

complexes is, of course, much wider; but it is nevertheless confined within certain limits. Some idea of these limits is given by Fig. 30, compiled from a large number of analyses. It will be seen that the limiting curves in this generalised diagram resemble in shape the actual curves of variation for the particular series just considered.

(viii.) Turning to the mineralogical variation, and to the Garabal Hill complex as a typical case, we have to note a significant point on which Dakyns and Teall have laid stress. This is the *parallelism between the order of intrusion of the several rocks and the order of crystallization of the constituent minerals.* In the table appended both the rock-types and the minerals are given in their proper order. Disregarding the iron-ore, which is of very minor importance, it is seen that, as the rocks succeed each other, the several minerals appear and disappear in the order in which they stand in the list. The earlier rocks are richer in the earlier minerals, and the later rocks in the later minerals. The normal order of decreasing basicity in a series of plutonic rocks thus connects itself with Rosenbusch's normal order of decreasing basicity in the crystallization of minerals from a rock-magma.

	1. Iron Ore.	2. Olivine.	3. Augite (diallage).	4. Brown Hornblende.	5. Green Hornblende.	6. Biotite.	7. Plagioclase.	8. Orthoclase.	9. Quartz.
A. Olivine-diallage-rock	+	+	+	+	—	—	—	—	—
B. Biotite-diorite	+	—	—	—	+	+	+	—	—
C. Hornblende-biotite-granite ...	+	—	—	—	+	+	+	+	+
D. Porphyritic biotite-granite ...	+	—	—	—	—	+	+	+	+
E. Eurite vein	+	—	—	—	—	—	+	+	+

In plutonic complexes belonging to the Atlantic branch it seems that the variation is more often complicated by the effects of subsidiary differentiation, and then ceases to have the serial relation. Thus, the Magnet Cove complex, in Arkansas,[1] includes some very extreme melanocratic types, two of the analyses (biotite-ijolite and jacupirangite of Washington) having nearly the same silica-percentage, but

[1] J. F Williams, *Ann. Rep. Geol. Sur. Ark.* for 1890, vol. ii., pp. 163-295; Washington, *Bull. Geol. Soc. Amer.*, vol. xi. (1900), pp. 389-416, and *Journ. Geol.*, vol. ix. (1901), pp. 607-622, 645-670; compare *Geol. Mag.* (1902), pp. 177-180.

differing widely in other particulars. Taking the mean of these two, and excluding the 'covite' (a nepheline-bearing shonkinite), which has admittedly arisen by subsidiary differentiation, we obtain the variation-diagram of Fig. 31. A special feature of this is the sharp rise of the lime-curve on the left, connected with the abundance of melanite in the ultrabasic types.

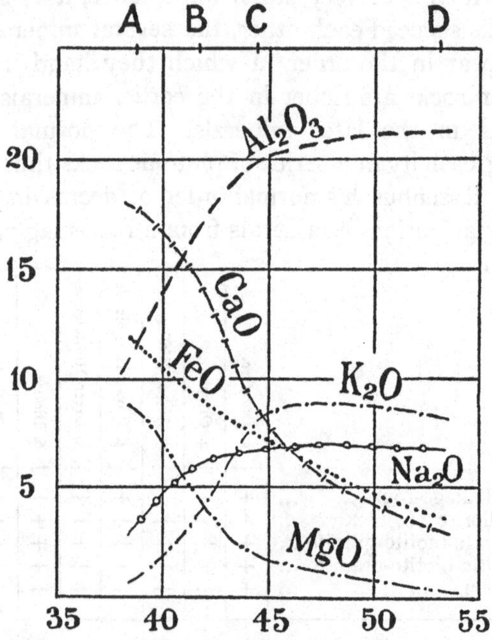

FIG. 31.—GENERALISED VARIATION-DIAGRAM FOR THE PLUTONIC COMPLEX OF MAGNET COVE, ARKANSAS, FROM ANALYSES GIVEN BY WASHINGTON. THE TOTAL IRON RECKONED AS FERROUS OXIDE.

A, mean of jacupirangite and biotite-ijolite; B, ijolite; C, arkite (leucite-nepheline-syenite); D, foyaite.

Variation in a Single Rock-Body.—The variation which we have hitherto considered is that apparent in differences of composition between rocks which, however intimately associated, must be regarded as representing distinct intrusions (or extrusions) of magma. But we have to recognise, in addition, that in many cases a rock-body which is clearly a geological unit, as representing a single intrusion of magma, shows, nevertheless, considerable variation in different parts of its mass. While the former kind of variation is, at least in general, discontinuous, the latter is essentially continuous. From the genetic point of view the former must be ascribed to differentiation, under unknown conditions, prior to intrusion (or extrusion); while the latter is, at least in general, attributable to differentiation subsequent to intrusion, and effected 'in place'—*i.e.*, in the observed environment.[1]

The variation in different parts of a single rock-body is usually, if not always, of the *serial* kind,[2] and can therefore be graphically exhibited in a variation-diagram. There is in most cases a well-marked *orderly arrangement* of the associated varieties with reference to the boundary of the rock-body, taking the form, therefore, of bilateral symmetry in the case of a sheet or dyke and concentric zones in the case of a laccolite or boss. This is sufficient to suggest that differentiation in place is connected, in such cases, with the cooling of the intruded body of magma.

The commonest case is that in which the mass shows a *relatively basic margin*. Sometimes the greater part of the rock presents no great differences of composition, and the modification at the margin comes on rather rapidly, though still in a continuous manner. In other cases there is a steady change in composition from the centre to the boundary. A good example is that of the gabbro of Carrock Fell, in Cumberland, an intrusive mass of generally laccolitic habit exposed over an elongated area about half a mile in width (Fig. 32). In the centre the rock is a quartz-gabbro of relatively acid nature (silica-percentage 59½). This graduates into an ordinary gabbro, which in turn becomes more basic outwards, and especially rich in iron-ores. At the actual border is a very dense ultrabasic rock with as much as 27 *per cent.*

[1] Brögger terms the one primary, or 'deep-magmatic,' and the other secondary, or 'laccolitic,' differentiation; but the latter term, at least, does not seem well chosen: *Eruptivgesteine des Kristianiagebietes*, I. (1894), pp. 178, 179.

[2] Brögger uses the term 'Faciessuite' as contrasted with 'Gesteinsserie.'

of titaniferous iron-ore (silica-percentage 32½). There is throughout a gradual transition from one variety to another, and, for the most part, the constituent minerals are the same, but associated in different proportions.[1]

An analogous instance is that described by Pirsson[2] at Yogo Peak in Montana. Here is an elongated boss-like mass some 5 miles long, the western end of which makes the Peak. The dominant rock is a porphyritic granite, but this

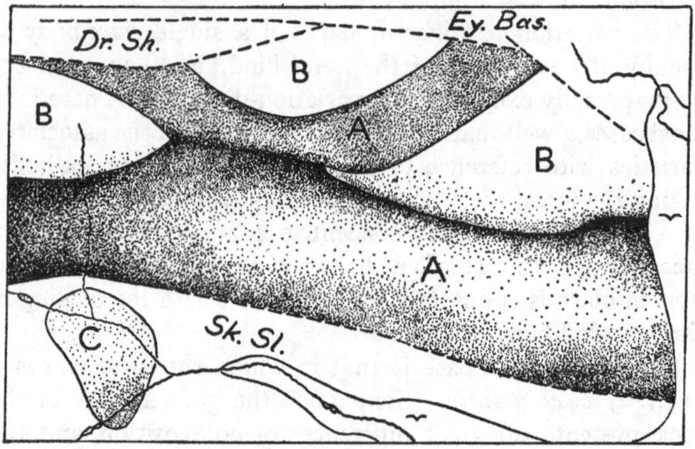

FIG. 32.—SKETCH-MAP OF CARROCK FELL, CUMBERLAND. SCALE, 1½ INCH TO A MILE.

This district illustrates three different kinds of variation in igneous rocks, due to different causes:

(i.) The gabbro (A) becomes increasingly basic towards the margin, as indicated by the shading.

(ii.) The granophyre (B), intruded subsequently to the gabbro, becomes relatively basic along its southern border owing to absorption of gabbro material.

(iii.) The greisen of Grainsgill (C) is a modified pegmatite, thrown out as a fringe on the northern border of the Skiddaw granite.

The surrounding rocks are Skiddaw slates, Eycott basalts, and Drygill shales.

passes gradually through banatite and monzonite into a thoroughly basic rock of the shonkinite type. The last makes the actual border of the mass at both ends—possibly also along the sides, though this was not ascertained

[1] Harker, *Quart. Journ. Geol. Soc.*, vol. l. (1894), pp. 311-336.

[2] *20th Ann. Rep. U.S. Geol. Sur.*, III. (1900), pp. 563-568.

(Fig. 33). The analyses give the variation-diagram shown in Fig. 34. Another interesting example in Montana is the Shonkin Sag laccolite in the Highwood Mts.,[1] where the concentric arrangement is very clearly exhibited (Fig. 35). The central part of the laccolite is a white syenite, which graduates into the melanocratic shonkinite, forming the chief bulk of the intrusion. The shonkinite becomes denser outwards, and passes at the border into a peculiar basic rock which is described as leucite-basalt-porphyry.

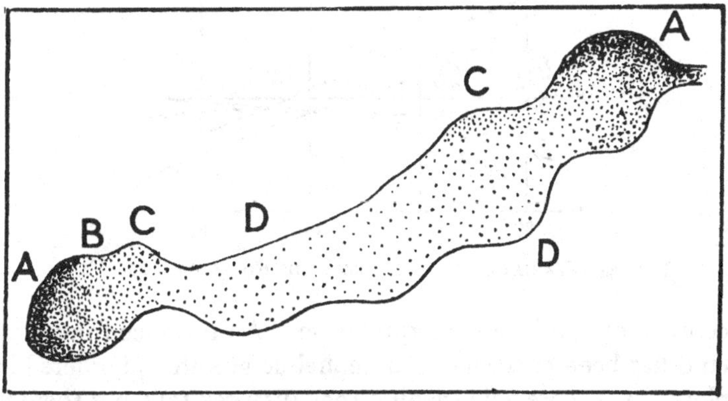

FIG. 33.—GROUND-PLAN OF INTRUSIVE BOSS AT YOGO PEAK, LITTLE BELT MTS., MONTANA. (AFTER PIRSSON.)

D, porphyritic granite; C, banatite; B, monzonite; A, shonkinite.

Excluding pegmatite fringes, which will be considered later, a *relatively acid margin*, with gradual transition, is certainly a rare case. An instance has been recorded by Brögger[2] in the laccolite of Ramnäs, near Christiania. The principal rock here is an akerite with granitoid structure; but towards the border it becomes finer-grained and porphyritic, and at the same time more acid in composition, finally passing into a quartz-porphyry. A series of analyses illustrates this variation, the silica-percentage rising by degrees from $58\frac{1}{2}$ to $71\frac{1}{2}$. The diorite stock of Castle Mt., in Montana,[3]

[1] Weed and Pirsson, *Amer. Journ. Sci.* (4), vol. xii. (1901), pp. 1-17.

[2] *Zeits. Kryst.*, vol. xvi. (1890), pp. 45, 46.

[3] Weed and Pirsson, *Bull. No.* 139 *U.S. Geol. Sur.* (1896), p. 61.

passing at the edge into a quartz-diorite-porphyrite, seems to be comparable with the Ramnäs occurrence; but other alleged cases of an intrusive body becoming gradually more acid towards the margin are either doubtful or clearly sus-

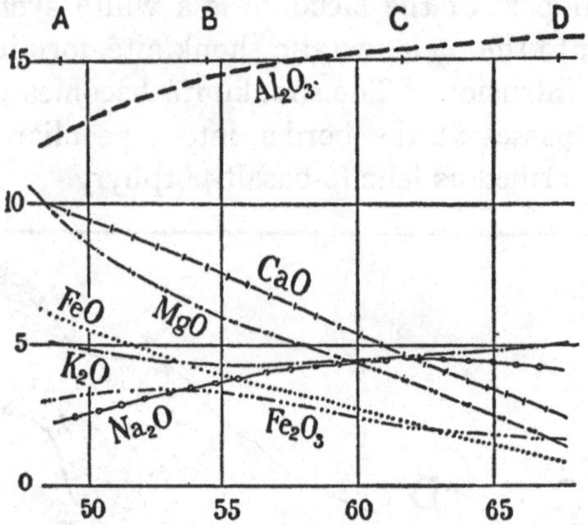

FIG. 34.—VARIATION-DIAGRAM FOR THE ROCKS OF YOGO PEAK.

ceptible of a different interpretation. An instance from Finland has been cited, where a nepheline-syenite is bordered in places by a more acid syenitic rock (umptekite); but there is nothing in Ramsay's account[1] to suggest a gradual transition

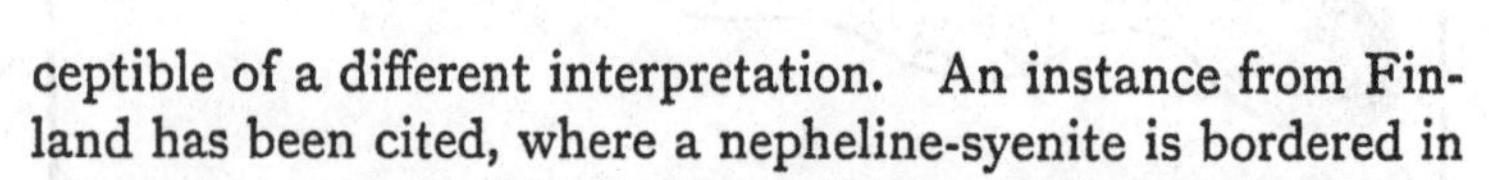

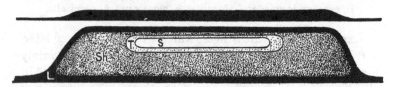

FIG. 35.—SECTION OF SHONKIN SAG LACCOLITE, MONTANA. (AFTER WEED AND PIRSSON.)

The upper figure is drawn to true scale. In the lower one the vertical dimensions are exaggerated six times to show the structure of the interior.
S, syenite; T, transitional rock; Sh, shonkinite; L, leucite-basalt-porphyry.
The several types are stated to graduate into one another.

between the two rocks. Washington regards the Magnet Cove complex (p. 131, above) as an example of an intrusion differentiated in place, with a basic centre and a more acid

[1] *Das Nephelinsyenitgebiet auf der Halbinsel Kola* (1894), pp. 75, 205.

margin; but it is clear from the description by Williams and by Washington himself that the several rocks are in general sharply divided. Indeed, according to the map, the leucocratic rocks are separated from the melanocratic by intervening sediments, having apparently been intruded at a somewhat higher horizon in the same anticlinal dome.

An interesting occurrence is that at Taberg, in Småland, Sweden, where a large mass of titaniferous magnetite-olivine-rock forms the central part, and an olivine-hyperite the marginal part, of what is apparently a single intrusive body (Fig. 36). According to Törnebohm, the one rock passes

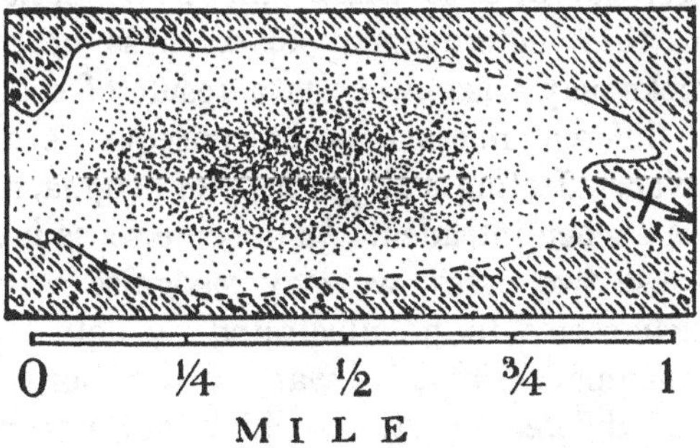

FIG. 36.—GROUND-PLAN OF INTRUSIVE BODY OF OLIVINE-HYPERITE WITH CENTRAL MASS OF MAGNETITE-OLIVINE ROCK, TABERG, SWEDEN. (AFTER VOGT.)

The shading is to illustrate a gradual transition between the rocks.

gradually into the other. This case is one of those discussed by Vogt,[1] who has made a study of the various basic segregations (Ausscheidungen) in plutonic rocks, with special reference to the origin in this way of workable ores of iron. In his view many important masses of iron-ore, the geological relations of which prove them to be primary igneous products, have been formed in place by the 'concentration' in particular places of the metalliferous constituents which were originally distributed through the magma. He distinguishes especially segregations of titaniferous iron oxides and segregations of nickeliferous iron sulphides, both found in

[1] *Zeits. prakt. Geol.*, 1893, pp. 4-11, 125-143, 257-284; 1900, pp. 233-242, 370-382; and 1901, pp. 9-19, 180-186, 289-296, 327-340.

association with rocks of the gabbro and norite family, but the former having a central and the latter a marginal situation. It is to be remarked, however, that, in all those occurrences for which Vogt gives details and diagrams, excepting only that of Taberg, the variation is not of the continuous kind, but the ore-body, whether 'oxidic' or 'sulphidic,' is more or less sharply bounded against the silicate-rock. This would seem to indicate that the two, although closely cognate, have not arisen from differentiation in place, but have been intruded as two distinct magmas, either simultaneously or successively.

Heterogeneity due to Simultaneous Injections.—We have seen that heterogeneity in a mass or complex of plutonic rocks may arise either as a result of successive intrusions of different magmas or in consequence of differentiation in place affecting a single body of magma, presumably homogeneous when intruded. As in some sense intermediate between these two cases, we may recognise a third possibility, which seems to be illustrated by actual examples. Here the magma as intruded was not a uniform mass, but consisted of different portions which did not mingle, or mingled only slowly or partially, during and after the intrusion. The resulting rock-mass may have an irregularly patchy character, on a larger or smaller scale; but more commonly the different portions have been drawn out in consequence of flowing movement, giving rise to a regular *banded structure*, or, again, a parallel arrangement of inconstant lenticles and streaks ('Schlieren').

Banding largely attributable to this cause is well exhibited in the basic and ultrabasic plutonic rocks of the Inner Hebrides. In the gabbros of Skye it is not often strongly marked, but in one locality there are beautiful examples which have been described by Sir A. Geikie and Dr. Teall.[1] Here the rock presents alternating light (felspathic) and dark (pyroxenic) bands, varying in width from a foot down-

[1] *Quart. Journ. Geol. Soc.*, vol. l. (1894), pp. 645-659, Pl. XIII., XXVI.; see also Harker, *Tertiary Igneous Rocks of Skye*, *Mem. Geol. Sur.* (1904), pp. 90-92, 117, 121, Pl. V., VI.

ward, and including thin black seams, very rich in titaniferous iron-ore. Adjoining bands sometimes shade into one another, and are sometimes sharply bounded; but even in the latter case the interlocking of the crystals at the junction forbids the supposition that the bands represent entirely distinct intrusions. On the other hand, the manifest relation of the banding to fluxion, in some places with a sinuous course, is opposed to the explanation by differentiation in

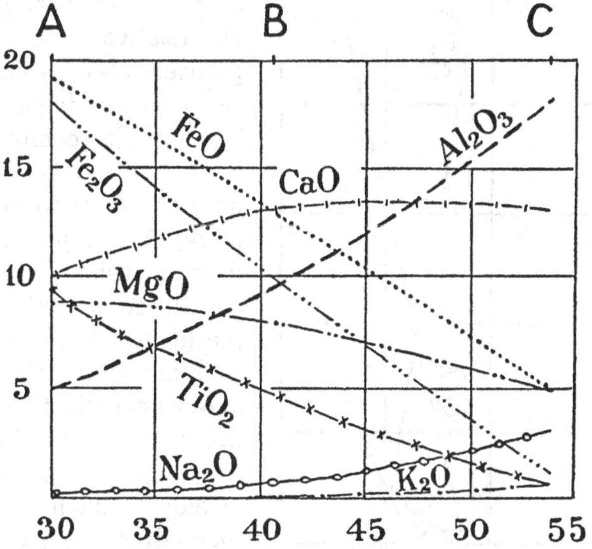

FIG. 37.—VARIATION-DIAGRAM OF THE BANDED GABBROS OF DRUIM AN EIDHNE, SKYE, FROM ANALYSES BY PLAYER.

A, thin ultrabasic seam, mainly of augite and titaniferous iron-ore; B, dark augitic band; C, pale felspathic band.

place, though a certain amount of segregation subsequent to intrusion is not precluded. The authors accordingly ascribe the phenomena to the intrusion of a heterogeneous magma. The variation among the different bands, being connected chiefly with different proportions of felspar and augite, approximates to the linear relation (Fig. 37). Banded structures of like nature are more generally found in the eucrites of Rum, and they are exhibited in a still more striking manner in the rocks of the ultrabasic group.

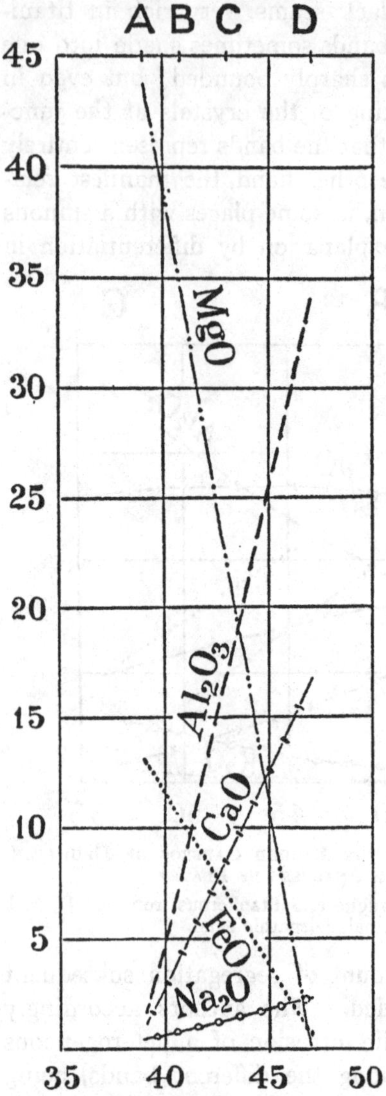

FIG. 38.—VARIATION-DIAGRAM OF THE ULTRABASIC ROCKS OF RUM, FROM ANALYSES BY POLLARD.

A, dunite; B, harrisite; C, allivalite; D, felspar-rock (a narrow band).

The ultrabasic complex of Rum,[1] best exposed on the flanks of Askival and Allival, illustrates heterogeneity arising from three distinct causes: (i.) The rocks consist principally of olivine and a basic felspar, near anorthite, in all relative proportions, pyroxenes being usually of subordinate importance. The whole complex is built up of parallel sheets, usually from 50 to 150 feet thick, which represent distinct intrusions, probably introduced in order from the highest to the lowest. They are alternately richer in olivine (peridotites) and richer in felspar (allivalites). (ii.) The several intruded magmas were themselves heterogeneous, consisting of more peridotic and more felspathic portions, which did not mingle freely, but were drawn out to produce a conspicuously banded arrangement within the several sheets. (iii.) After intrusion there was a further segregation of parts richer in olivine and in felspar respect-

[1] Harker, *Geology of the Small Isles*, *Mem. Geol. Sur.* (1908), pp. 69-77.

ively. Flowing movement having ceased, this did not usually take the form of banding, but gave rise to structures of a 'concretionary' kind, traversing the various bands. Since the variation throughout the complex depends essentially on the different relative proportions of two minerals of fairly constant composition, the variation-diagram is almost accurately of the linear kind (Fig. 38). The steepness of the lines, due to the fact that the principal minerals do not differ greatly in silica percentage, is very characteristic of ultrabasic rocks.

Some of the titaniferous iron-ore masses already referred to seem to be more or less closely comparable with the

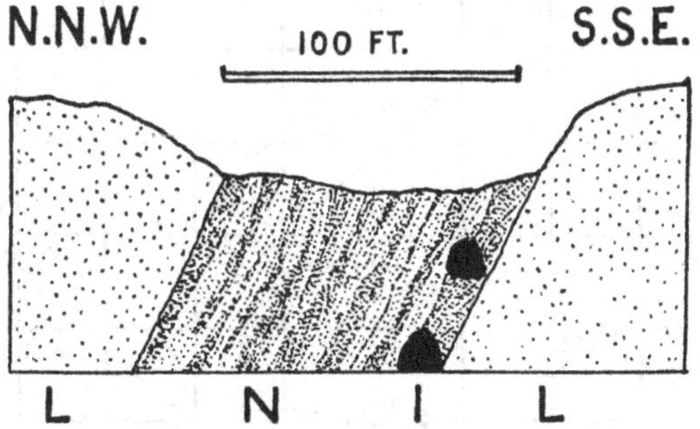

FIG. 39.—SECTION ACROSS THE STORGANGEN, NEAR SOGGENDAL, NORWAY. (AFTER VOGT.)
L, labradorite-rock; N, ilmenite-norite; I, ilmenite-rock.

banded gabbros of Skye, etc. Such are the occurrences in the Ekersund and Soggendal district on the south-west coast of Norway. Here an area some 40 miles long and 25 miles broad consists chiefly of labradorite-rock, which, however, passes into norite on the one hand and enstatite-granite on the other. It is traversed by dykes of various kinds, including some of norite very rich in ilmenite, and others, less regular, of nearly pure ilmenite-rock. One of the former, named Storgangen ('great dyke'), has a width of 100 to 230 feet, and shows a strongly-banded structure (Fig. 39). It contains ilmenite to the amount of about 40 *per cent.* on the average, rising in some bands or streaks to 70 or 80 *per cent.*, and there are masses of almost pure titaniferous iron-ore. These are always sharply bounded, and the appearances suggest that magmas of such peculiar composition are not freely miscible with ordinary silicate-magmas. Nevertheless, the consanguinity of the whole assemblage of rocks is sufficiently evident, and their mutual relations as regards chemical composition are shown in the variation-diagram appended (Fig. 40). The sympathetic

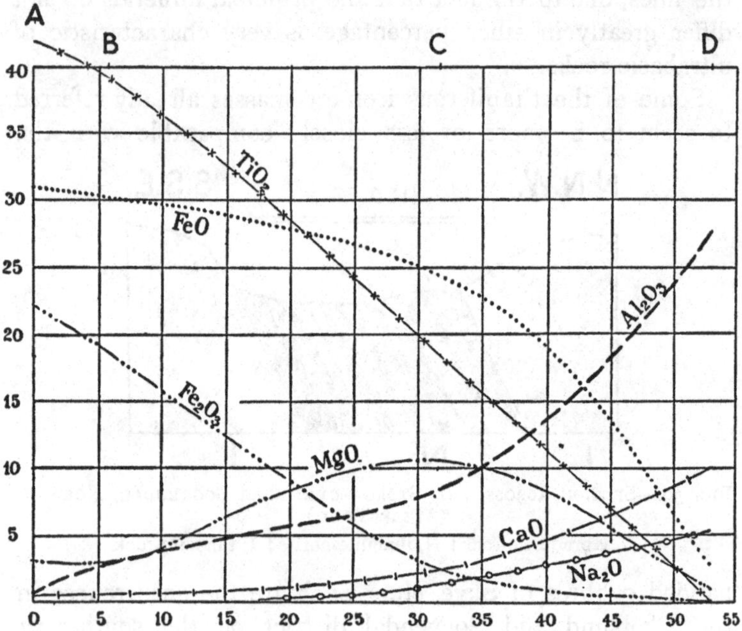

Fig. 40.—Variation-Diagram of the Plutonic Rocks (including Iron-Ore Bodies) of the Ekersund and Soggendal District, from Analyses given by Vogt.

relation of the titanic and iron oxides on the left side of the diagram is characteristic.

Variation among Minor Intrusions.—The dykes and sheets, which at so many centres of igneous activity have followed larger intrusions of plutonic habit, show in general a wider range of petrographical diversity. This is in accord with the doctrine of evolution by magmatic differentiation, and is one of the facts which may guide hypotheses giving shape to that conception. In many districts, where a complete record can be studied, we find among the minor intrusions types which are practically identical in composition with plutonic types in the same district, differing only in textural and structural characters; but we find also types which differ more or less markedly, in composition as well as in structure, from the plutonic rocks to which they appear to be related. It is at least a plausible working hypothesis that all the satellites of a given plutonic rock have been derived from a magma either identical with, or closely similar to, that which produced the plutonic rock itself, this magma having been intruded at the later epochs sometimes without any substantial alteration, sometimes modified by further differentiation, not on the lines of the primary differentiation which gave birth to a series of plutonic types. This hypothesis is equally applicable to the most usual case of a distinct phase of minor intrusions, and to the case (illustrated, apparently, by the Christiania district) in which each plutonic rock has been immediately followed by its own satellites.

Several significant points may be urged in support of this view. We have already remarked that the various groups of minor intrusions often fall into pairs having a complementary relation with regard to the presumed parent-magma. It is evident that, if from a given magma A a portion a be abstracted, of different composition, the remaining portion (A') must differ from A in the opposite sense. Indeed, we may write:

$$A = ma + nA';$$

the equation representing the actual partition of A into two different parts, a and A', in fractional amounts m and n respectively. The change produced in the residual magma by the abstraction of a depends both on the degree of difference between A and a and on the fraction m. Thus, the drafting off of a relatively small amount of a partial magma during the plutonic phase, as in the case of aplite veins, does not modify materially the composition of the main body. It is easy to conceive that the conditions may be different at a later stage, when differentiation may operate in a reduced remnant of the magma—*i.e.*, the part not intruded and consolidated as plutonic rock. Here the circumstances will be favourable for the partition of the magma into portions differing considerably from it, and still more from one another. Again, it is a striking fact that the more extreme rock-types never occur in great force. This is easily understood from the foregoing simple considerations; for it is clear that, if a differs greatly from A, m must be a small fraction.

Whatever be the nature of this later differentiation—a question which is not at present under discussion—we may contemplate as a probable case that in which like processes operate upon the several magmas corresponding with a series of plutonic rocks, A, B, C, etc. There will thus arise a series of derivatives—a, b, c, etc.—each related in a definite manner to its own principal. If the differentiation of the several magmas has proceeded, not only on the same lines, but to the same extent, then to a smooth series of abyssal rock-types will correspond an equally smooth series of hypabyssal types:

A ... ... a
B ... ... b
C c

We must expect, however, that the differentiation, while following the same lines in each case, has not always been carried to the same extent. The relations will then be of this kind:

A ... ... a
B ... ... b
C ... ... c;

and the hypabyssal series will be less regular than the abyssal one. This becomes very apparent when we examine actual assemblages of cognate rocks.

Brögger's conception of a 'Gesteinsserie' was first put forward in connection with the grorudite series among the rocks of the Christiania district, and the chief members of this appear to be related as follows:

Abyssal.	*Hypabyssal.*
Essexite.	(Sussexite.)[1]
Larvikite.	Sölvsbergite (without quartz).
Lardalite.	Tinguaite.
Nordmarkite.	Quartz-sölvsbergite.
Ekerite.	Grorudite.

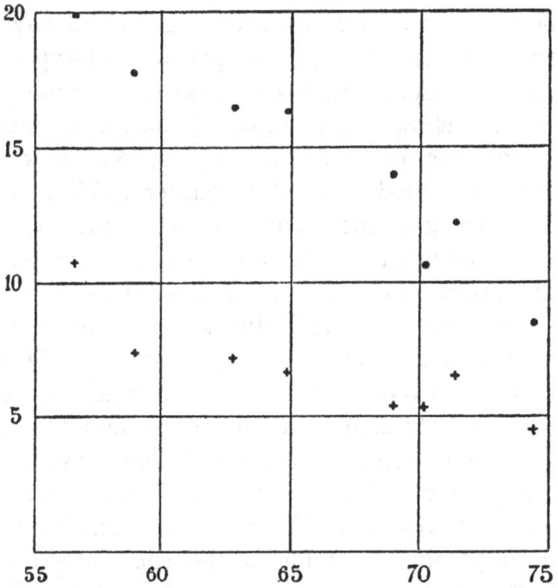

FIG. 41.—DIAGRAM SHOWING THE VARIATION OF ALUMINA AND SODA IN THE GRORUDITE SERIES OF THE CHRISTIANIA DISTRICT, FROM ANALYSES GIVEN BY BRÖGGER.

The dots belong to alumina and the small crosses to soda. Other constituents would show a like degree of regularity or irregularity.

The grorudites, sölvsbergites, etc., of which analyses have been made, clearly fall into a natural series, but a considerable degree of smoothing is necessary in constructing a variation-diagram (see Fig. 41). Within limits, the diagram so obtained would be approximately of the linear kind; but it is impossible to admit even approximate linearity as a

[1] This type has not been found in the district, but is added for the sake of completeness.

fundamental characteristic of such a series. Brögger[1] in one place insists on this view, when he lays it down that "every mean of a number of members of the series corresponds approximately with a possible member of the series"; but in another place abandons it, when he calculates the composition of the hypothetical sussexite on the assumption that, while some constituents vary in arithmetical, others vary in geometrical progression.

In the plutonic complexes already described any orderly arrangement of the different component rock-types is only of a very rough kind. Analogous associations among minor intrusions may show a much more regular arrangement. In this way arise composite sills and dykes, some of which will be further considered in a later chapter (XIV.). Again, a dyke representing a single intrusion has sometimes a relatively basic marginal modification comparable with that found in many plutonic rock-bodies, though with less extreme variation. In the Rainy Lake region of Canada, Lawson[2] has noted some striking examples. One dyke, 150 feet wide, showed a difference of 10 in silica-percentage between the quartz-diabase in the centre and the ordinary diabase at the margin. Brögger and Vogt have described similar cases in the Christiania district. One dyke, about 30 feet wide, at Huk on the island Bygdö, consists of mica-nordmarkite-porphyry, passing at the edge into kersantite, the difference in silica-percentage being 3.[3] Brögger[4] records a similar difference of acidity between the interior and margin of a dyke of grorudite.

[1] *Eruptivgesteine des Kristianiagebietes*, I. (1894), p. 175; compare p. 172.

[2] *Amer. Geologist*, vol. vii. (1891), pp. 153-164.

[3] Vogt, *Zeits. prakt. Geol.*, 1893, p. 4.

[4] *Eruptivgesteine des Kristianiagebietes*, I. (1894), p. 56.

CHAPTER VI

IGNEOUS ROCKS AND THEIR CONSTITUENTS

Chemical composition of igneous rocks.—Melting-points of the rock-forming minerals.—Other physical properties.

Chemical Composition of Igneous Rocks.—Of the chemical composition of rock-magmas we have no direct knowledge, although there exist a large and increasing number of chemical analyses of igneous rocks formed from the consolidation of such magmas. Doubtless most of these analyses represent with some degree of closeness the composition of the magmas which gave birth to them, but there are considerations which forbid us to accept the one as necessarily the equivalent of the other. The volatile constituents of a magma may be in great part eliminated from the final products of crystallization, and we shall see that another complication is introduced in some cases by differentiation in place.

Igneous rocks collectively vary greatly in chemical composition, and there is often a wide range of variation among rocks associated in the same district. Averages for different districts show differences which are smaller in degree, but, as already pointed out, are nevertheless significant. By calculating the mean of a large number of analyses of rocks from many localities, so that different petrographical provinces are fairly represented, we obtain a result which may be used as a standard to which analyses of various rocks may be referred for comparison. We quote such a calculation made by Clarke,[1] and perhaps representing with some degree of

[1] *Bull. Phil. Soc. Washington*, vol. xi. (1889), pp. 131-142. For later averages, computed on a somewhat different basis, see *Bull. No.* 148 *U.S.*

approximation the mean composition of the accessible parts of the Earth's crust.

SiO_2	58·59
Al_2O_3	15·04
Fe_2O_3	3·94
FeO	3·48
MgO	4·49
CaO	5·29
Na_2O	3·20
K_2O	2·90
H_2O	1·96
	98·89

The individual analyses are not symmetrically grouped about the mean. This is shown, as regards silica-percentage, in Fig. 42, where the curve of distribution is clearly not a

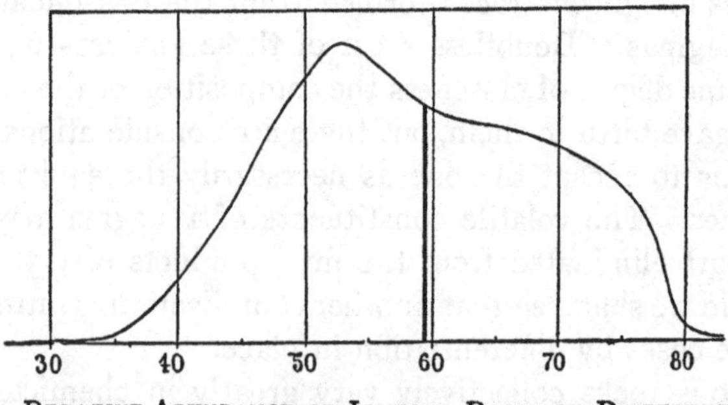

FIG. 42.—RELATIVE ABUNDANCE OF IGNEOUS ROCKS OF DIFFERENT SILICA-PERCENTAGES.

The abscissæ represent silica-percentages and the ordinates the relative abundance of the corresponding rocks, as deduced from the 'superior' analyses in Washington's collection. The ordinate strongly drawn in the figure corresponds with Clarke's mean analysis, and should therefore bisect the area enclosed by the curve.

'curve of error.' It would indicate that the most abundant rocks are those with about 52·5 *per cent.* of silica, but that, though moderately basic rocks are more abundant than

---

Geol. Sur. (1897), pp. 12, 13; *Bull. No.* 228 (1904), pp. 14, 15; *Proc. Amer. Phys. Soc.*, vol. xlv. (1906), pp. 14-32; *Bull. No.* 330 (1908), pp. 13-37; also Washington, *Prof. Paper No.* 14 *U.S. Geol. Sur.* (1903), pp. 106-115.

moderately acid or sub-acid, thoroughly acid rocks are more abundant than thoroughly basic and ultrabasic. It is very likely that fuller information, taking into account the bulk of the rock-masses which the analyses represent, would considerably modify the form of the curve, and throw the mean farther to the right. The abrupt declination of the curve at the ultra-acid end, contrasting with its gradual dying down at the opposite end, is, however, significant.

A glance over any collection of analyses of igneous rocks shows that, while the amount of each constituent varies between wide limits, the variations for different constituents are not of an arbitrary kind, but are controlled by certain general rules. For example, rocks rich in silica are poor in magnesia, and conversely. For a given silica-percentage the magnesia is not, of course, constant; but in the great majority of rocks it does not depart greatly from a mean value, which depends on the silica-percentage. It is possible, therefore, to construct a *generalised variation-diagram*, which will embody the general rules referred to; and, when this is done, it is easy to see that the curves are susceptible of a mineralogical interpretation. In the generalised variation-diagrams given in Figs. 43, 44, we have, however, discriminated between the Atlantic and Pacific branches, the broken curves belonging to the former and the continuous curves to the latter. The analyses employed are those ranked by Washington[1] as 'superior.'

The percentage of *silica*, which is taken as abscissa in the diagrams, varies in unaltered rocks from about 35 (or very exceptionally as low as 25) to about 80. Lower figures—down to zero—are given by segregations of iron-ores, etc.

The percentage of total *alkalies* ranges from zero to about 15, or exceptionally as much as 17½, being highest in rocks of medium acidity belonging to the Atlantic branch: this last remark is true of potash and soda severally. For both alkalies the broken curve lies well above the continuous

[1] *Chemical Analyses of Igneous Rocks . . . , Prof. Paper No.* 14 *U.S. Geol. Sur.* (1903).

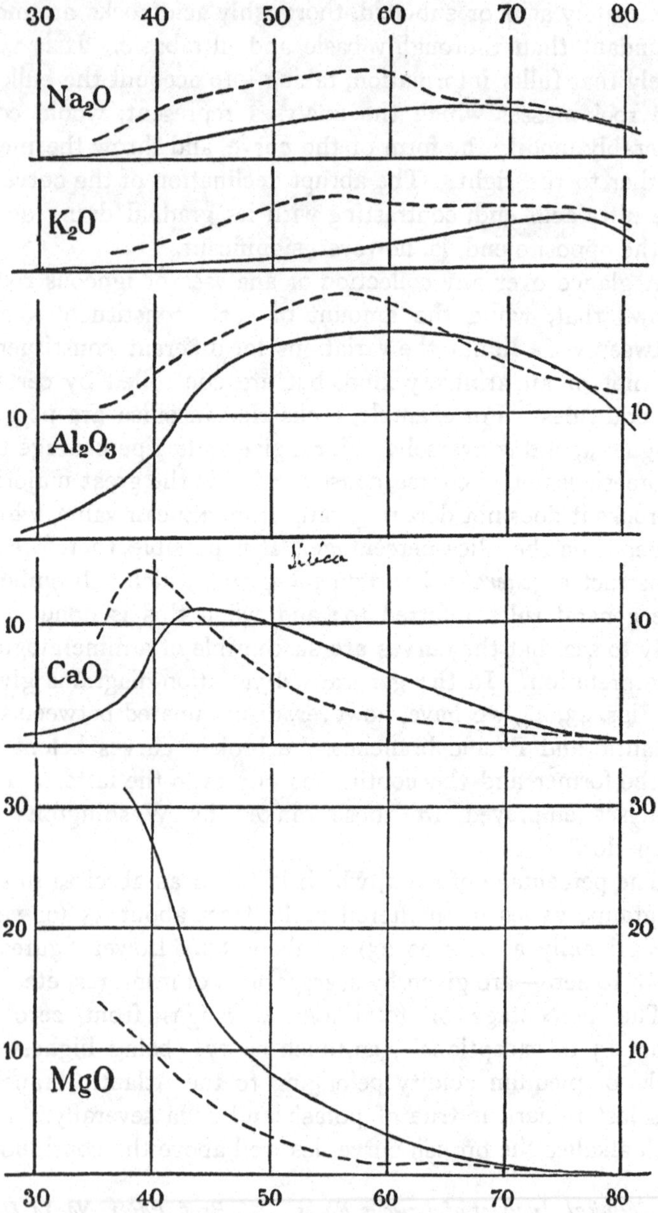

FIG. 43.—GENERALISED VARIATION-DIAGRAM FOR IGNEOUS ROCKS (ATLANTIC AND PACIFIC BRANCHES).

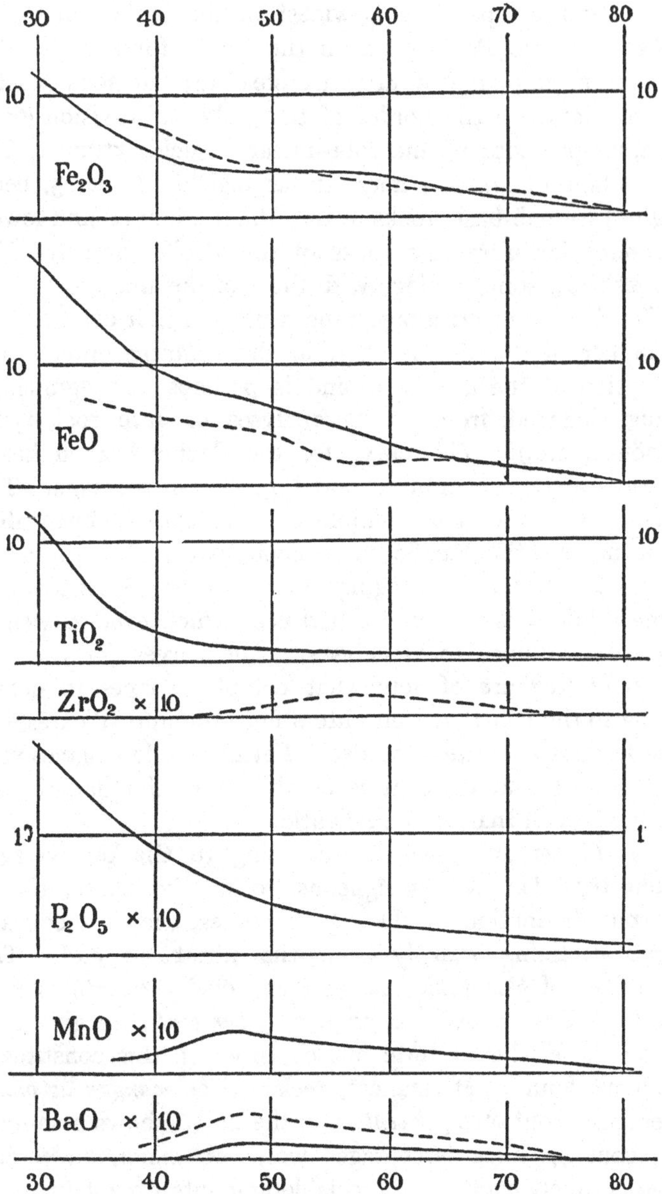

FIG. 44.—GENERALISED VARIATION-DIAGRAM FOR IGNEOUS ROCKS (ATLANTIC AND PACIFIC BRANCHES).

one, and the separation is widest in the basic half of the range. In the Pacific branch the curves for both alkalies start from zero, and at first rise steadily; but soda reaches its maximum on the border of the truly acid division (66 to 67 silica-percentage), and potash near the acid extreme. The percentage of *alumina* may rise as high as 24 or 25, being highest for sub-basic rocks of the Atlantic branch and lowest (or zero) for ultrabasic rocks of the Pacific branch. The curves show some analogy with those of the alkalies.

For *lime* the percentage ranges up to 15, or even 17. It is higher in the Pacific than in the Atlantic branch: the exception at the ultrabasic end is perhaps not significant, being deduced from a few analyses of rare rock-types (alnöites, etc.). The curve for the Pacific branch has a strongly marked summit at about 43 silica-percentage. The curves for *magnesia* are simple concave curves, much alike in form, but the Pacific curve about twice as high as the Atlantic. Magnesia, ranging up to 47 or 48 *per cent.* in some dunites, is the only constituent which ever outweighs silica in ordinary rock-analyses. The curves for the two *oxides of iron* are of somewhat complex shapes in detail, owing to the fact that iron enters not only into silicates, but into ferrates and titanates also. Ferrous oxide behaves very like magnesia, and, like it, is notably more abundant in the Pacific branch than in the Atlantic.

For *titanic oxide*, which, according to Clarke, averages about 0·73 *per cent.* in igneous rocks, the curve, drawn without distinction of the two branches, shows a concave curve declining sharply from the ultrabasic end. The behaviour of *phosphoric acid* (average 0·26 *per cent.*) is very similar. For *zirconia* (average 0·03 *per cent.*) we give only the curve for the Atlantic branch, in which this constituent is chiefly found. Manganese, reckoned as *manganous oxide*, averages about 0·10 *per cent.*, and the curve shows a summit at about 47 silica-percentage; while *chromium*, *nickel* and *cobalt* scarcely occur in appreciable amounts except in ultrabasic rocks. For *barium* and *strontium*, reckoned as oxides, Clarke gives average percentages 0·11 and 0·04 respectively. Approximate curves for baryta are shown on the diagram.

Since in igneous rocks, and indeed in rock-magmas also, these various constituents occur combined to form minerals, mostly silicates, it will be convenient to glance at some of the properties of these minerals before proceeding to discuss the crystallization of rock-magmas.

Melting-Points of the Rock-Forming Minerals.—The trustworthy data relative to the thermal and other physical properties (other than optical) of the rock-forming minerals are at present very scanty. Owing to their high melting-points and other peculiarities, the experimental examination of these minerals presents very considerable practical difficulties; and, in studying general laws, physicists have commonly employed more tractable materials, such as the halides and especially organic bodies. Technological investigations on slags have, however, furnished valuable information; and in recent years researches have been undertaken with express design to a geological application of the results. These suffice to prove that the practical difficulties are, with the aid of the electric furnace and other modern methods, not insuperable; and we may hope that the petrologist will soon be provided with data enabling him to introduce into the discussion of rock-genesis the quantitative element, which hitherto has been almost wholly lacking. The most important contributions have come from the Geophysical Laboratory of the Carnegie Institution at Washington.[1]

While providing us with some accurate determinations, made by methods of precision on pure material, these researches have incidentally served to discredit most of the

[1] Day and Allen, *The Isomorphism and Thermal Properties of the Felspars* (1905); see also *Amer. Journ. Sci.* (4), vol. xix. (1905), pp. 93-142. Allen and White, On 'Wollastonite and Pseudo-Wollastonite, *Amer. Journ. Sci.* (4), vol. xxi., pp. 89-108. Day and Shepherd, 'The Lime-Silica Series of Minerals,' *ibid.*, vol. xxii. (1906), pp. 265-302; also 'Die Kalkkieselreihe der Minerale,' *Tscherm. Min. Petr. Mitth.* (2), vol. xxvi. (1907), pp. 169-232. Allen, Wright, and Clement, 'Minerals of the Composition $MgSiO_3$,' *Amer. Journ. Sci.* (4), vol. xxii. (1906), pp. 385-438.

rough estimates formerly arrived at by less refined methods. It is shown that earlier determinations of melting-points may be in error to the extent of hundreds of degrees.[1] This is due mainly to the fact that the instant of melting was judged merely by observing the appearance of the body tested. Now the collapse and loss of shape which overtakes a crystal at or about some roughly definable temperature is quite naturally called 'melting' in popular language; but the melting with which the physicist and petrologist are concerned is *a discontinuous change of state,* of which the only proper criterion is discontinuity of physical properties; and the melting-point is the temperature at which such discontinuity occurs. In many of the rock-forming minerals the change of state is not accompanied by any alteration, such as sudden softening, which is obvious to the eye. Day and Allen found that for the alkali-felspars the viscosity of the liquid near the melting-point is of the same order as the rigidity of the crystalline body. The only way, therefore, of determining the melting-point of minerals with any precision is to observe the temperature at which 'latent heat' is evolved in crystallizing or absorbed in melting. For the alkali-felspars even this method failed.

Since the words 'solid' and 'liquid' are not always used with a constant signification, it is better to follow Tamman[2] in taking as the most fundamental distinction that between *isotropic* and *anisotropic.* In an isotropic body there are only *scalar* properties—*i.e.*, such as have no relation to direction. An anisotropic body has scalar properties, such as density, but also *vector* properties, such as coefficients of linear expansion, which are related to direction. This implies an ordered geometrical arrangement of the molecules, which is wanting in the former case. Under isotropic are included gaseous, liquid, and glassy; under anisotropic only crystalline, though sometimes with different (polymorphous) modes of crystallization. Thus, the isotropic state includes both liquid

[1] Doelter's successive estimates of the melting-point of anorthite range from 1124° to 1290°, while the true figure is 1532°.

[2] *Kristallisieren und Schmelzen* (1903).

and solid modes of consistence (in the ordinary sense of the words); and the same is true of the anisotropic, for liquid crystals find their place here. The change from liquid to glass, or the reverse, is a continuous change, and the expression 'melting-point' has therefore no meaning as applied a glass (Fig 45, B). The change from isotropic to anisotropic is necessarily discontinuous. Thus the ideal curve of cooling of a simple body which crystallizes has the broken shape shown in Fig. 45, A, with a horizontal portion at the altitude corresponding with the melting-point.

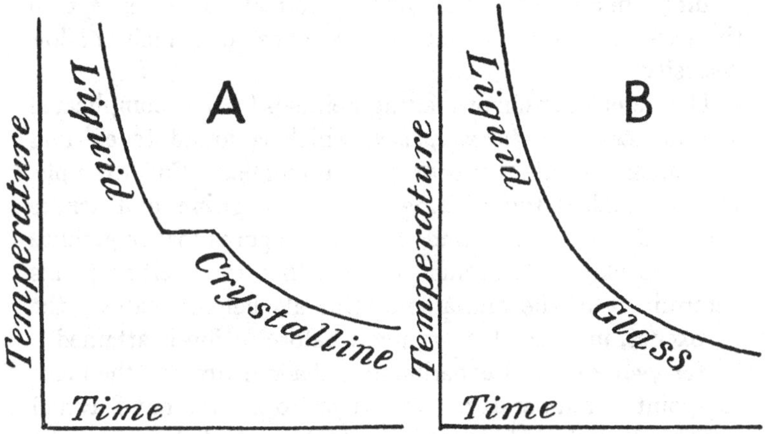

FIG. 45.—IDEAL CURVES OF COOLING.
A, for a body which crystallizes at a certain temperature (the melting-point); B, for a body which consolidates as a glass.

These principles are recapitulated here, because disregard of them has sometimes led to serious misconception. Thus, Barus[1] carried out a series of careful experiments to determine the thermal constants of diabase, including the volume-change and latent heat of fusion. In these experiments the fused rock was allowed to cool to a glass, or to a mass chiefly of glass—for it appears that some crystallization did occur during the process. Consequently such abrupt, or at least rapid, change of volume and evolution of heat as were observed represent only a fraction of the proper volume-

[1] *Bull. No.* 103 *U.S. Geol. Sur.* (1893).

change and latent heat of fusion.[1] We shall see later that a complex rock like diabase has no melting-point proper, and that the actual volume-change and evolution of heat during the crystallization of such a magma are not simple quantities.

Experiments for the determination of melting-points of rock-forming minerals have been made by Joly,[2] Cusack,[3] Doelter,[4] Brun,[5] and others. All these experiments, besides being made on natural (and therefore impure) crystals, involve the subjective element already referred to. The results, therefore, are of little practical value, except in the case of minerals which fuse at once to a 'melt' of low viscosity.

The determination of melting-points is further complicated in some cases by *polymorphism*, which is found to exist in a number of the rock-forming minerals. For example, Doelter makes the melting-points of ægirine and acmite 940° and 945°, and he finds the same figures for ænigmatite and riebeckite. It seems probable that the melting-points determined for the amphibole minerals belong really to the pyroxenes, into which they invert before fusion is attained.

For *quartz* several experimenters have estimated the melting-point at temperatures from 1670° to 1780°; but Day and Shepherd find that quartz is stable only below 800°, at which temperature it inverts to tridymite, with a melting-point near 1600°.

The melting-points of the less alkaline *felspars* have been

[1] Vogt considers that the latent heat found is only one-fifth of the true value, and concludes that Barus' estimate of the effect of pressure on the melting-point is therefore five times too great. But it is to be observed that Barus also underestimates the discontinuous volume-change, and the two errors will partly counterbalance one another in this calculation. Vogt, *Silikatschmelzlösungen*, II. (1904), pp. 209, 210.

[2] *Proc. Roy. Ir. Acad.* (3), vol. ii. (1891), pp. 38-64.

[3] *Ibid.*, vol. iv. (1897), pp. 399-413.

[4] *Tscherm. Min. Petr. Mitth.* (2), vol. xx. (1901), pp. 210-232; vol. xxi. (1902), pp. 23-30; vol. xxii. (1903), pp. 297-321; *Physikalisch-chemische Mineralogie* (1905), pp. 99, 100.

[5] *Arch. sci. phys. et nat. Genève* (4), vol. xiii. (1902), pp. 352-375.

closely determined by Day and Allen, anorthite giving 1532° and the other plagioclases in order lower figures (see below, Fig. 77). Extrapolation suggests a melting-point of about 1230° for albite, and orthoclase is probably not very different. For the *felspathoids* we have only roughly approximate estimates. Brun gives for nepheline 1270°, sodalite 1310°, haüyne 1450°, and leucite 1420°. The last figure belongs properly to meta-leucite (regular), into which leucite (rhombic) inverts at some temperature above 450°.

In the *pyroxene and amphibole groups* dimorphism, or rather polymorphism, introduces considerable complication. According to Allen, Wright, and Clement, the magnesium metasilicate is tetramorphous; the four forms being monoclinic pyroxene (clino-enstatite), rhombic pyroxene (enstatite), monoclinic amphibole, and rhombic amphibole. Here the relation is of the irreversible 'monotropic' kind, the highest form (of relatively rare occurrence in nature) being theoretically stable at all temperatures, with a melting-point of 1521°. In the case of wollastonite and its dimorphous form (pseudo-hexagonal but really monoclinic) the relation is 'enantiotropic,' with inversion-point at 1180°, and the higher form (pseudo-wollastonite) melts at 1512°. For diopside Vogt found a melting-point of 1225°, but Day has recently determined that of the pure artificial substance as 1375°. The alkali-pyroxenes have quite low melting-points—940°-945°, according to Doelter. The amphiboles as a group are low-temperature minerals, and probably pass at higher temperatures into the more stable pyroxene form.

As regards other minerals, Brun gives for olivine (poor in iron) 1750°, magnetite 1260°, hæmatite 1300°, sphene 1210°, apatite 1550°, spinel and zircon nearly 1900°. Day has recently determined the melting-point of fluor at 1387°.

Other Physical Properties.—For the rock-forming minerals, so far as is known, the volume-change with rise of temperature is always an expansion. The *coefficient of dilatation* ranges from 0·000016 in orthoclase to 0·000035 in quartz, the latter agreeing nearly with iron. These are mean values between 0° and 100° C.; for a range of 0° to 1000° the mean coefficients would probably be nearly twice as great. The *coefficient of linear expansion* is, of course, in crystals a vector quantity; it may differ considerably in different directions, and be negative for some. It follows that, when an igneous rock cools down from the temperature at which its consolidation was completed, the crystals of different minerals contract in different degrees, and also change their shape as well as their volume, setting up a complex system of strains and stresses within the rock. These may conceivably be relieved in various ways, as by the opening of cleavage-cracks, bending of flexible crystals, and shearing along gliding-planes (secondary twin-lamellation); and such phenomena, therefore, do not necessarily prove crushing of the rocks by external forces.

For rock-magmas the coefficient of dilatation is doubtless somewhat greater than for solid rocks. Barus found 0·000047 for molten diabase between 1100° and 1500°.

The *volume-change in fusion* is almost certainly an expansion in the case of any of the rock-forming minerals at ordinary pressures, but we are virtually without experimental data concerning the amount of this discontinuous change. A rough estimate may be reached by indirect reasoning, as follows: In Fig. 46 are shown the volume-curves for the isotropic (glass-liquid) and anisotropic (crystalline) states of a given substance. The vertical distance between them in any place represents the difference of specific volume between the two states at the corresponding temperature. Although it is difficult to measure this quantity experimentally at the melting-point, the measurement is easily made at a low temperature, being calculated simply from the specific gravities of crystal and glass.[1] For different minerals this quantity ($v_2—v_1$), at atmospheric temperature, ranges from 0·0087 in anorthite to 0·0734 in quartz. These figures correspond with a percentage contraction in passing from glass to crystal of 2·35 to 16·35. Now, experiment shows that the glassy substance expands at first somewhat less rapidly than

[1] A number of data, for minerals and rocks, are cited by Roth, *Allgemeine und chemische Geologie* (1883), vol. ii., pp. 52, 53.

the crystalline, but at higher temperatures somewhat more rapidly; so that the two curves of Fig. 46 tend first to separate slightly, and then to converge again. Probably, therefore, the volume-change $v_2—v_1$ is nearly the same at the melting-point as at low temperatures.

While we may safely assume that all the rock-forming minerals severally expand in melting and contract in crystallizing, we cannot at once extend the statement to igneous rocks. Specific volume is an additive property for a crystalline aggregate of different minerals, but not for a fused magma of the same. We shall see later that such a magma must be regarded as a solution, in which, therefore, we must be prepared to find some condensation of volume. When a mineral crystallizes out from a mixed rock-magma, the concomitant volume-change therefore consists of two parts—(*a*) that consequent upon the abstraction of the constituent from the solution, and (*b*) that proper to the change of state. The former is an expansion, the latter a contraction. In general the net result is a contraction, as may be tested by a comparison of specific gravities at atmospheric temperature of crystalline rocks and the glasses obtained from their fusion. Experiments on various rocks by Delesse and others show a contraction ranging in different cases from 6 to 14 *per cent.* There are, however, exceptions. In many of the Tertiary basic dykes in the western isles of Scotland there is a selvage of black glass, which is from 1½ to 4 *per cent.* denser than the crystalline interior; although, as Delesse[1] proved for one example, the composition is the same. Since the component minerals are known to be severally denser in the crystalline than in the glassy state, the vitreous rock has a density considerably higher than that which might be calculated from its several (glassy) components. We infer that the mixed glass is of the nature of a solution involving a very notable degree of condensation. It is at least highly probable that like relations hold good at high temperatures; so that a basic rock of this type crystallizes with expansion

[1] *Ann. des mines* (5), vol. xiii. (1858), p. 369.

and melts with contraction. Experimental researches[1] seem to indicate this behaviour in certain slags and basic magmas, but observations of this kind are affected by many sources of error. In spite of exceptions, it is at least a very general rule that, under low pressures, rock-magmas contract in crystallization; and Stübel's theory of vulcanicity, based on a presumed expansion, is not well grounded.

The *specific heat* at low temperatures has been determined for most of the common rock-forming minerals. The mean value between 0° and 100° ranges mostly from 0·18 to 0·20, which is high as compared with most of the common metals. The property being an additive one, crystalline rocks have a similar specific heat, though it is to be remarked that any notable amount of uncombined water in a rock must raise the value accordingly. The specific heat increases with rise of temperature up to about 0·30 at 400°–600°, and then remains nearly constant. A rough estimate for the mean specific heat of crystalline rocks from their consolidation down to atmospheric temperatures is about 0·27. The specific heat at low temperatures is not very different for the crystalline and isotropic states; but at high temperatures the isotropic has a considerably greater specific heat—in the case of microcline, between 1000° and 1250°, about 40 *per cent.* greater.

The *thermal conductivity* of minerals and rocks is very low as compared with that of metals. In centimetre-gramme-second units the conductivity (k) varies in different igneous rocks from 0·0036 to 0·0060, the higher figure belonging to the granites. Quartz and quartzites have a still higher conductivity, 0·0095. The coefficient of diffusivity of temperature ($k \div cs$, where c is the specific heat and s the density) ranges from 0·0067 to 0·0113, or for quartz 0·0206. Acid rocks seem to be, in general, better conductors than basic, and coarse-grained better than fine-grained. In crystals conductivity is a vector quantity, and this is also the case in cleaved or

[1] See, *e.g.*, Fleischer, *Zeits. deuts. geol. Ges.*, vol. lvii. (1905), pp. 201-214, and vol. lix. (1907), pp. 122-131.

foliated rocks, which conduct better along than across the structural planes.

The isotropic state constantly possesses greater energy than the crystalline; and at the melting-point the energy-difference between the two states (represented by the vertical distance between the two curves in Fig. 47) measures the *latent heat of fusion* (L). No direct measurement of this quantity has yet been made for any rock-forming mineral.

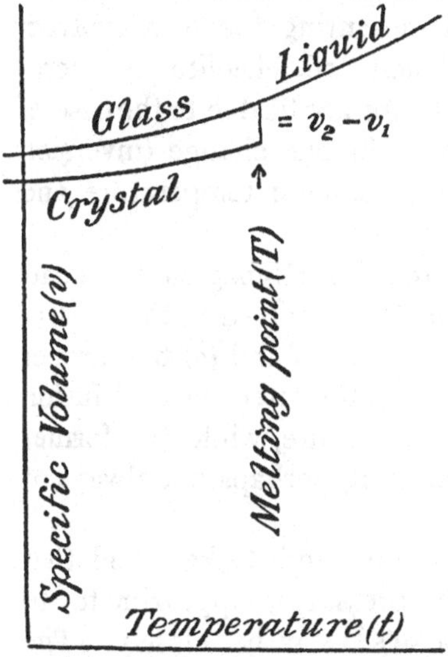

FIG. 46.—VOLUME-ISOPIESTICS: CURVES EXPRESSING THE VARIATION OF SPECIFIC VOLUME WITH TEMPERATURE AT CONSTANT PRESSURE.

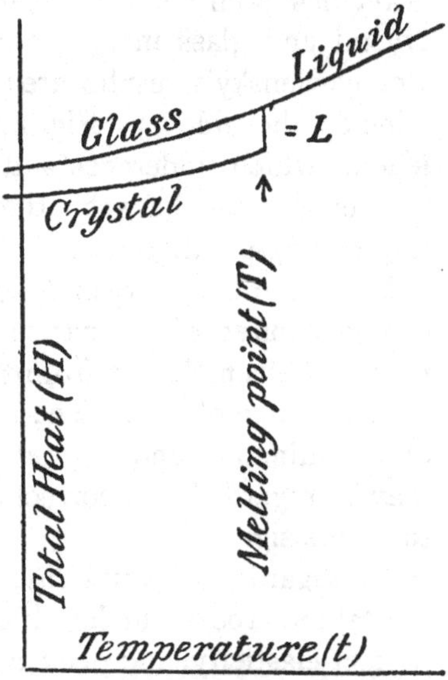

FIG. 47.—ENERGY-ISOPIESTICS: CURVES EXPRESSING THE VARIATION OF ENERGY-CONTENT WITH TEMPERATURE, AT CONSTANT PRESSURE.

Vogt, however, has made some approximate estimates, mostly calculated from Åkerman's determination of the 'total heat of fusion' of slags. The results, reckoned in calories, are: melilite 90, anorthite 100, diopside 102, enstatite 125, olivine 130. We may apply here the indirect reasoning already used for the discontinuous volume-change in fusion (compare Figs. 46 and 47). The energy-difference at the melting-point (L) could be determined from that at atmospheric temperature if we knew accurately the forms of the two curves in Fig. 47. This would involve a complete knowledge of the two specific heats; for the specific heat at any temperature is the rate at which the energy-curve rises.[1] It appears (p. 160) that in general the two curves are at first nearly parallel, but diverge somewhat at high temperatures. Consequently the latent heat of fusion will be somewhat greater than the energy-difference at low temperatures. The latter has been determined for some minerals by dissolving crystal and glass in acid and measuring the heat evolved. Bogojawlensky's results are: leucite 26, elæolite 74, microcline 83, diopside 93. Fig. 47 is not applicable to the case of leucite, which undergoes a discontinuous change (inversion to meta-leucite) between the atmospheric temperature and the melting-point.

When a mineral crystallizes from a rock-magma, the heat-change consists of two parts: (*a*) that involved in the separation of the constituent from the solution, and (*b*) that proper to the change of state. The latter is the latent heat of fusion of the mineral itself at that temperature, while the former may be regarded as a correction of it, perhaps not always of the same sign.

As regards *compressibility*, Adams and Coker[2] find that crystalline rocks, under high pressures, approximate to perfect elasticity, and obey Hooke's law rather closely. The following figures are to be multiplied by 100,000,000 to give the bulk-modulus of elasticity in C.G.S. units: Quincy granite, 2,750; Peterhead granite, 3,300; Montreal nepheline-syenite, 4,290; Canadian anorthosite, 5,760; Sudbury diabase, 7,329. Of these, only the last is less compressible than cast-iron (6,897). The corresponding figure for quartz (calculated from Amagat's results) is about 4,212; and we may infer that

[1] If c_1 and c_2 be the specific heats for crystal and glass, and t the temperature,

$$\frac{dL}{dt} = c_2 - c_1.$$

[2] *An Investigation into the Elastic Constants of Rocks* . . ., Washington (1906); also *Amer. Journ. Sci.* (4), vol. xxii. (1906), pp. 95-123.

the alkali-felspars are more compressible than quartz, but the lime-felspars and ferro-magnesian minerals less compressible. The compressibility of crystalline rocks is not greatly influenced by temperature. No experiments on fused minerals or rock-magmas are recorded, but we may assume with confidence that these are more compressible than the crystallized minerals and rocks.

An important question is that of the *dependence of melting-point on pressure.* James Thomson's equation may be conveniently written:

$$\frac{dT}{dp}=0{\cdot}02416\ \frac{T}{L}(v_2-v_1);$$

where T is the temperature of fusion reckoned from absolute zero, p the pressure in atmospheres, L the latent heat of fusion in calories, and v_2-v_1 the increment of specific volume in fusion measured in ccm. per gm.[1] To obtain some idea of the actual effect of increased pressure in raising the melting-point, suppose that a given mineral melts under atmospheric pressure at 1200° C., with a volume-change in fusion 0·05, and a latent heat of 100 calories. Then the elevation of melting-point per atmosphere will be—

$$0{\cdot}02416\times 1{,}473\times 0{\cdot}05\div 100\ ;$$

that is 0·0178. The increase of pressure required to raise the melting-point by 1° will be about 56 atmospheres, and a pressure equivalent to a column of rock 10,000 feet deep will raise the melting-point about 15°.

A calculation of this kind cannot, however, be extended to very great pressures, for the reason that the quantities v_2-v_1 and L, which appear in the equation, are themselves functions of the pressure. Pressure raises the melting-point only provided v_2-v_1 is positive—*i.e.*, provided the body melts with expansion. This is true of the rock-forming minerals at moderate pressures; but, since a liquid is more compressible in general than a solid, it seems probable that with increasing pressure v_2-v_1 will diminish, and may eventually pass through

[1] The numerical factor is obtained thus: 1 atmo. = 1,033 gm. per sq. cm.; 1 gm. cal. = 42,750 gm. cm.; 1,033 ÷ 42,750 = 0·02416.

zero and become negative. This would imply that the melting-point rises to a maximum value at a certain pressure, and is lowered by further increase of pressure. Such behaviour has been verified for numerous organic bodies, and Tamman[1] supposes the property to be a general one. The 'curve of fusion' (plotted with pressure as abscissa and melting-temperature as ordinate) will then have a distinct summit, and the seemingly anomalous behaviour of a body

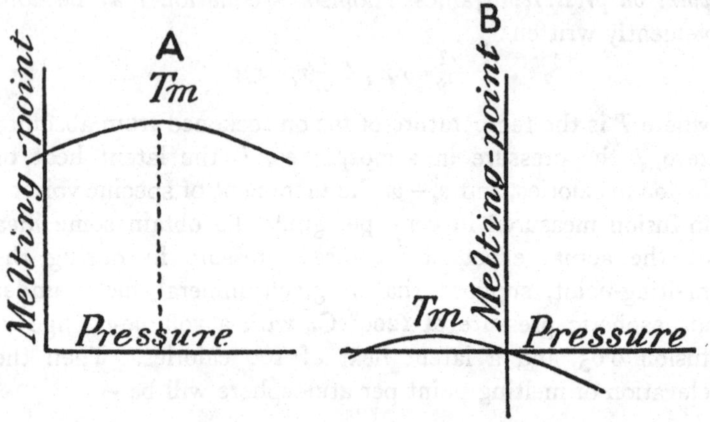

FIG. 48.—IDEAL CURVES OF FUSION.

A, For the common case; B, for a body like water or bismuth, which at ordinary pressures melts with contraction. *Tm* is the maximum melting-point.

like water will be reconciled (Fig. 48). For silicate-minerals the pressure requisite to bring about the reversal of behaviour —from melting with expansion to melting with contraction— is doubtless a very high one;[2] but, as Arrhenius has pointed out, the principle is of fundamental importance in any speculation concerning the deep interior of the globe.

[1] *Kristallisieren und Schmelzen* (1903).

[2] We have seen reason for believing, however, that in some exceptional cases a mixed rock-magma crystallizes with *net* expansion, even at ordinary pressures (p. 159). In this case pressure will lower the melting-points of the minerals in the magma.

CHAPTER VII

ROCK-MAGMAS

Constitution of rock-magmas.—Rock-magmas as solutions.—Vogt's researches on slags.—Order of crystallization.—Actual temperatures of rock-magmas.

Constitution of Rock-Magmas.—A fundamental question, as preliminary to any discussion of the crystallization of igneous rocks, is that of the chemical constitution of the rock-magmas from which they are formed. The chemist presents his analysis of an igneous rock in the form of a column of percentages of silica, alumina, and other oxides (perhaps with some sulphides or other unimportant constituents). For certain purposes it is sometimes convenient to translate the result into percentages of the elements: oxygen, silicon, aluminium, etc. On the other hand, we may be able, either by calculation from the chemist's data or by direct mechanical analysis, to present the result in terms of the component minerals: orthoclase, quartz, biotite, etc. For a crystalline rock any of these alternatives represents the composition, but only the last represents the actual constitution. We have to inquire which manner of presentation corresponds with the true constitution in the case of a molten rock-magma.

Some petrologists have supposed, and others seem implicitly to assume; that the constituents exist in the magma in the form of free oxides (with sulphides, etc.).[1] This view implies a suspension of chemical affinities which

[1] See, *e.g.*, Iddings, 'The Origin of Igneous Rocks,' *Bull. Phil. Soc. Wash.*, vol. xii. (1892), pp. 154-156; Löwinson-Lessing, 'Studien über die Eruptivgesteine,' *Compte Rendu VII Congr. géol. intern.* (1897), pp. 327-331.

is not easily understood. It implies, also, that the compounds known to us as rock-forming minerals are formed—with liberation of a large quantity of heat—in the act of crystallization; although, from analogy, we should expect such nascent compounds to appear as amorphous precipitates. The objections from the petrographical side are, to the geologist, more weighty, and seem to be conclusive. The wide variation met with among igneous rocks certainly results from processes of differentiation carried out, for the most part, in fluid rock-magmas; and, quite as certainly, it is a variation, not of independent oxides, but of silicates (and some other compounds) more or less analogous to the known rock-forming minerals. The common constituent minerals of igneous rocks are not numerous, and a given mineral may occur in very different kinds of rocks. Very significant is the fact that a number of simple minerals (periclase, wollastonite, monticellite, gehlenite, sillimanite, cordierite, etc.) which are readily formed, *e.g.*, in the thermal metamorphism of sediments, are absent from igneous rocks, or are present only when there is reason for suspecting the contamination of the magma by dissolved sedimentary material. We have already seen that variation, whether in particular series of rocks or in igneous rocks as a whole, is controlled by laws, which admit of no other than a mineralogical interpretation. The sympathetic variation, *e.g.*, of soda and alumina can be explained only on the supposition that a large part, at least, of the soda was actually combined with alumina (in such molecules as albite, nepheline, and acmite) throughout the differentiation. We must conclude that a rock-magma is, in the main, *a mixture of definite silicate-compounds*, the only free oxides present—at least, in a magma approaching the point of crystallization—being those parts, if any, of the silica, alumina, ferric oxide, and water which after crystallization appear as quartz, corundum, hæmatite, and free water.

The point comes out very clearly if we have regard to the order in which the minerals crystallize. Thus, the bulk-analysis of an ordinary granite shows more alumina than is required to make felspars with the alkalies and lime present, and this excess is contained in micas or aluminous hornblende, minerals of variable composition. But the last-named minerals, with others of minor importance, have crystallized before the felspar. By their abstraction the composition of the remaining magma was accurately adjusted, so that the molecules of alumina equalled the sum of the molecules of potash, soda, and (remaining) lime. In other words, it was reduced accurately to the composition of a mixture of felspars and quartz; which is inexplicable, except on the supposition that it was actually a mixture of felspars and quartz. The experiments of Morozewicz on artificial magmas illustrate the same point. Here the 'excess' of alumina made such minerals as corundum, sillimanite, and cordierite, but these always crystallized out first, leaving a magma of felspars and quartz.

We cannot conclude, however, that the several compounds which exist in a rock-magma at a high temperature are necessarily identical with those which eventually crystallize out from it at lower temperatures. We must suppose that, in general, there is adjustment of *chemical equilibrium*; and this is governed by the conditions of temperature and pressure, the former being the more important. Accordingly, reactions may take place between the several compounds present as the magma cools. It seems certain that such complex molecules as, *e.g.*, those of the hornblendes and micas can exist only at relatively low temperatures, and will not form at all (from a magma containing their constituents) unless the requisite conditions are realised. Much light is thrown on this subject by experiment in the laboratory. Thus Fouqué and Michel-Lévy,[1] by fusing together microcline and biotite, obtained instead an aggregate of leucite and olivine, with some magnetite. The reaction may be roughly represented by such an equation as:

$$2KAlSi_3O_8 + K_2Mg_2Al_2Si_3O_{12} = 4KAlSi_2O_6 + Mg_2SiO_4;$$
$$\text{microcline} + \text{biotite} = \text{leucite} + \text{olivine}.$$

[1] *Synthèse des minéraux et des roches* (1882), p. 77.

The orthoclase-biotite association, so characteristic of plutonic rocks, is stable at low temperatures (and high pressures), but the leucite-olivine association, so frequently found in volcanic rocks, at high temperatures (and low pressures).[1] Again, Doelter[2] has shown that the equation—

$$\underset{\text{augite}}{CaMgSi_2O_6, MgAl_2SiO_6} = \underset{\text{olivine}}{Mg_2SiO_4} + \underset{\text{anorthite}}{CaAl_2Si_2O_8}$$

—represents a real reaction, which is reversible, and doubtless has an important bearing on the occurrence and relative proportions of the minerals in rocks crystallized under different conditions.

A question of some importance relates to the part which water plays in the constitution of rock-magmas at high temperatures. At atmospheric temperatures its behaviour is almost neutral, the 'heat of neutralisation' being, according to Thomsen, only about one-hundredth that of the feeble silicic acid, and it can thus compete with silica only in a negligible degree. But between 0° and 50° the relative avidity of water increases rapidly; and Arrhenius[3] points out that if this continues, water must be at magmatic temperatures a much stronger acid than silica. He concludes that at high temperatures the water in a rock-magma must exist as hydrates and basic silicates, and part of the silica as silicic acid (H_2SiO_3) and acid silicates.

Again, we have seen that some of the compounds which figure as rock-forming minerals are dimorphous or polymorphous. We may suppose, for example, that any free silica in a magma above 800° C. is in the form of tridymite, and becomes quartz only when that temperature is passed in the process of cooling.[4] Where the relation is of the monotropic kind, there is no fixed inversion-point. If we may apply Ostwald's law of 'successive reactions,' a higher

[1] Iddings has pointed out that some minettes and some leucite-basalts give almost identical analyses.

[2] *Physikalisch-chemische Mineralogie* (1905), p. 121.

[3] *Geol. Fören. Förh. Stockholm*, vol. xxii. (1900), pp. 395-419.

[4] The inversion-point will be displaced by pressure, but probably to no important extent.

form reaches the equilibrium form by passing through those intermediate, and it would appear that the form which actually crystallizes will depend on the rate of cooling and crystallization. Thus the monoclinic pyroxene of composition $MgSiO_3$ is formed in rapidly crystallized slags, but only very exceptionally under natural conditions. It seems probable that polymorphism is in general connected with polymerization, the lower forms having the larger molecules. Thus the four forms of magnesium metasilicate may be conjecturally written: Monoclinic pyroxene, $MgSiO_3$; rhombic pyroxene, $Mg_2Si_2O_6$; monoclinic amphibole, $Mg_4Si_4O_{12}$; and rhombic amphibole some higher multiple of the simple empirical formula. There seems to be no good reason for supposing that polymerization, any more than combination, takes place in the act of crystallization. Vogt[1] has attempted to determine the actual state as regards polymerization of minerals as they exist in the magma, using Van 't Hoff's formula for the lowering of melting-point; but it is very doubtful whether the data yet obtainable have sufficient precision to warrant such an application.

The case of isomorphous series of minerals does not constitute any exception to the general principle. Labradorite, for example, is not a compound, but a mixture of two compounds. These compounds, anorthite and albite, exist as such in the magma, and only the mixture of them to make labradorite takes place in the act of crystallization.

Rock-Magmas as Solutions.—We have seen that the constituents of a molten rock-magma exist there in the form of definite compounds, mostly silicates, and in general identical with those compounds which are familiar in the less complex rock-forming minerals. We shall next give some reasons for believing that the relation of the several compounds in the magma is one of *mutual solution*.

(i.) The specific properties of a given mineral are greatly modified by the presence of other minerals in the magma. It has long been recognised that the order in which the minerals crystallize is not determined by their relative

[1] *Silikatschmelzlosungen*, I. (1903), pp. 40-44.

fusibility as separately tested. For instance, the large leucite crystals in the lavas of Vesuvius enclose crystals of augite, which are clearly of earlier formation, though more fusible. Take the melting-point of leucite at 1420°, and that of the Vesuvian augite at 1220°. Then, since the augite cannot have crystallized above its melting-point, the leucite, which followed it, must have crystallized below the melting-point of augite, and therefore at least 200° below its own melting-point. We infer that, in presence of the augite and other constituents of the mixed magma, the leucite has had its melting-point (or freezing-point) lowered by at least 200°.

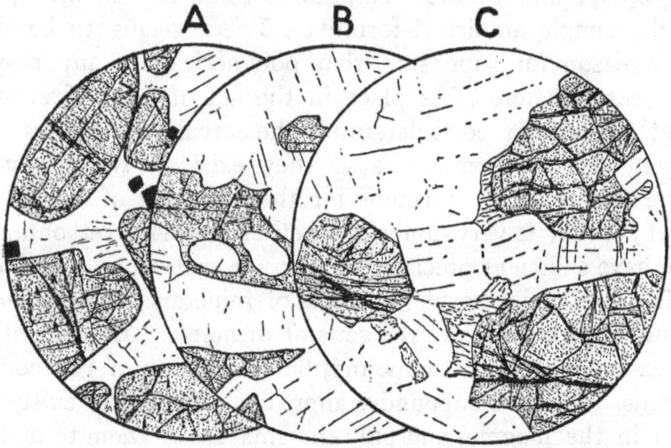

FIG. 49.—OLIVINE-ANORTHITE-ROCKS (ALLIVALITES), ISLE OF RUM. ×20.
A, olivine in excess, and crystallized first; B, anorthite in excess, and crystallized first; C, eutectic proportions, simultaneous crystallization.

This *lowering of freezing-point* is the most characteristic property of solutions.

(ii.) The argument just employed does not depend upon the particular melting-points assigned, for it is well known that leucite crystallizes sometimes before and sometimes after augite, and a like observation applies to other minerals. The actual order of crystallization is dependent, as we shall see later, upon a number of factors; but, in proportion as the conditions are simplified in different cases, the analogy with crystallization from an ordinary saline solution becomes more apparent. The allivalities of Rum, already mentioned (p. 140), afford a good example (Fig. 49). These rocks consist essentially of olivine and anorthite in varying relative proportions. In the more peridotic varieties the olivine has crystallized first; in the more felspathic varieties the anorthite; and when the two minerals were present in certain proportions they have crystallized simultaneously. Here we have another characteristic property of solutions—viz., that *the first mineral to crystallize is that which was in excess as compared with certain standard proportions.*

(iii.) Considering for the sake of simplicity a mixture of two minerals only, the abstraction of the one which was in excess will eventually reduce the residual magma to the standard proportions premised. If the analogy with saline solutions still holds, the two minerals will then proceed to crystallize simultaneously in the same definite (eutectic) proportions. For reasons to be considered later, this is not always the case in rock-magmas, especially with rapidly crystallizing minerals like olivine and anorthite; but *eutectic aggregates* do, nevertheless, represent the later products of crystallization in very many igneous rocks. Teall[1] first suggested, in 1888, that the interstitial micropegmatite of acid rocks is of this nature. Its occurrence is that of a residual product; the intimate intergrowth of the felspar and quartz is proof of simultaneous crystallization; and, as Vogt has since shown, the relative proportions of the constituents at least approximate to constancy. Eutectic or approximately eutectic mixtures, not necessarily with graphic or other special structures, are of very widespread occurrence in igneous rocks.

(iv.) In a general survey of the characters of igneous rocks, it would be easy to point out other features explicable only by the known properties of solutions. For instance, we have already seen (p. 159) that in the mixture of a number of different constituents in a magma there is a *condensation* or contraction, sometimes even greater than

[1] *British Petrography* (1888), p. 401. See also Presid. Address, *Quart. Journ. Geol. Soc.*, vol. lvii. (1901), pp. lxxv, lxxvi.

that which occurs in crystallization. It is not necessary to discuss the matter further at this stage, for the solution theory rests on general inductive grounds—*i.e.*, on its ability to explain the observed facts of petrology as they will be successively brought out in the following chapters.

The solution theory of igneous rock-magmas, first clearly enunciated by Bunsen[1] in 1861, has been developed by Lagorio, Teall, Morozewicz, and others, and especially in recent years by Vogt. Some of these petrologists have discussed the question, whether any particular constituent of a magma can be regarded as a general solvent, analogous to water in aqueous solutions. Lagorio assigned this part to a wholly hypothetical silicate, which he designated 'Normalglas.' Its assumed composition is $R_2O,2SiO_2$, where R=K, Na, and the equally hypothetical compounds $CaO, 2SiO_2$ and $Al_2O_3,6SiO_2$ were supposed to play a like rôle. Barus and Iddings have expressed the opinion that silica plays in rock-magmas the part of the water in aqueous solutions; but this idea rests on the assumption that the silica exists in the magma in a free state. Löwinson-Lessing supposes, more vaguely, that "those constituents of the magma which, at the given moment, predominate quantitatively over the others" may be regarded as the solvent. Becker makes the eutectic mixture the solvent, the solute being the excess of one or more constituents over the eutectic proportions. In a mutual solution of a number of constituents the actual relations, as regards partition of solutes among solvents, are presumably of a complex kind; but the question, if, indeed, it is other than a question of terminology, is rather a theoretical one. In one respect the water in an aqueous solution does appear to play a special part, as distinct from other constituents which are said to be dissolved in it: it serves in some sort as a medium in which they may become dissociated into ions, while the water itself suffers little, if any, dissociation. Although there is undoubted dissociation in rock-magmas, we can-

[1] *Zeits. deuts. geol. Ges.*, vol. xiii. (1861), pp. 61-63.

not point to any particular constituent as playing a part analogous to that of the water.

The experiments of Barus and Iddings[1] may be taken as proving (*a*) that rock-magmas are conductors of electricity; (*b*) that, at a given temperature, an acid magma has a higher conductivity than a basic one; (*c*) that, for a given magma, the conductivity increases with rise of temperature. We may infer that there is in rock-magmas a noteworthy degree of *ionization* or electrolytic dissociation; and it is probable that there is more in an acid than in a basic magma at a like temperature. It is perhaps not proved that ionization increases with rise of temperature, for the increased conductivity may be due to diminished 'ionic viscosity.'

The theory of electrolytic dissociation is based almost wholly upon study of the special case of aqueous solutions. Moreover, most of the laws which have been deduced are true only for very dilute solutions, and may further be vitiated by the presence of 'complex ions.' In the case of rock-magmas we have to do with a mixed solvent; we cannot assume the solutions in general to be dilute; and we have no direct knowledge of the nature of the ions. We are, therefore, reduced to conjecture, guided by analogy. It may probably be assumed that, at a given temperature, a given electrolyte is more completely dissociated in proportion as it is in more dilute solution; but we do not know whether the limit of total dissociation is ever approached. As regards the relative degree of dissociation in different minerals, we may be guided by the close connection between ionization and chemical activity. Thus, the superior conductivity of an acid as compared with a basic magma may possibly be due to the abundance of easily dissociated alkali-felspars in the former. In the simpler types of silicates (such as enstatite) it may be assumed that the metal-atom (Mg) forms the positive or cation, and the acid-radicle (SiO_3) the negative or anion. For more complex silicates we may plausibly follow the analogy of such double salts as $KAg(CN)_2$ and Na_2PtCl_6, which in aqueous solution are dissociated, so that the strong

[1] *Amer. Journ. Sci.* (3), vol. xliv. (1892), pp. 242-249.

alkali-metal alone forms the positive ion: the ions of orthoclase will then be K and $AlSi_3O_8$. Quartz cannot be supposed an electrolyte; but, if at a high temperature part of the silica in a magma is in the form of silicic acid, H_2SiO_3, this may be dissociated into ions—either H and $HSiO_3$ or H_2 and SiO_3.

Vogt's Researches on Slags.—For the first comprehensive attempt to apply the principles of solutions to the crystallization of igneous rock-magmas we are indebted to Professor Vogt,[1] of Christiania. To demonstrate the applicability of the laws of solutions to silicate-magmas, he deals first with artificial slags, on the basis of an extended series of researches carried out by Åkerman and by Vogt himself. The slags in question crystallize as aggregates, in various associations and relative proportions, of olivine and fayalite, monoclinic and rhombic pyroxenes, lime-felspars, melilites, spinels, magnetite, apatite, etc. These compounds, setting aside a few which are rare or unknown in nature, are the ordinary constituents of the more basic igneous rocks; and, excepting in the absence of water, the molten slags are practically identical with basic rock-magmas. Quartz, the alkali-felspars, muscovite, hornblende, and some rock-forming minerals of minor importance, do not occur in slags, and there is good reason for believing that some of them cannot be crystallized from a magma containing no water or other flux (Chapter XII.). Leaving aside this question for the

[1] Of Vogt's earlier works on slags the most important are 'Studier over Slagger,' *Bihang til k. Svenska Vet.-Akad. Handl.*, vol. ix. (1884); 'Om Slaggers . . .,' *Jernkontorets Annaler* (1885); 'Beiträge zur Kenntniss der Gesetze der Mineralbildung in Schmelzmassen,' *Arch. for Math. og Naturvid.*, vols. xiii., xiv. (1888-90). For a general discussion of rock-magmas as solutions see 'Die Silikatschmelzlösungen, I., II.,' *Vidensk.-Selsk. Skrifter, Math.-naturv.-Klasse* (1903) No. 8, and (1904) No. 1; and for a general summary of this, 'Die Theorie der Silikatschmelzlösungen,' *Ber. des V Intern. Kongr. angew. Chemie zu Berlin* (1903), Sekt. III. A, vol. ii., pp. 70-90. Some parts of the subject are more fully developed in 'Physikalisch-chemische Gesetze der Krystallisationsfolge in Eruptivgesteinen,' *Tscherm. Min. Petr. Mitth.* (2), vol. xxiv. (1906), pp. 437-542; vol. xxv. (1906), pp. 361-412.

present, we have still in Vogt's researches on slags the means of arriving at some general results, which, with proper qualifications, must be applicable as well to acid as to basic magmas.

Vogt's principal memoir (1903-04) is in two parts. In the first he inquires what mineral is the first to crystallize from a magma of given mineral composition. The percentages of the several constituent minerals are calculated from a chemical analysis of the crystallized slag, and the actual order of crystallization is determined by a microscopical examination. Selecting slags composed essentially of two minerals in various relative amounts, it is found that the mineral which is first to separate is, with few exceptions and irregularities, the one which is in excess as compared with certain definite (eutectic) proportions. For instance, in different slags consisting essentially of olivine and diopside, the former or the latter crystallizes first according as their ratio (in parts by weight) is greater or less than about 30 : 70. This limiting or eutectic ratio between the orthosilicate and metasilicate corresponds with an acidity[1] 1·6. Fig. 50 embodies the results of observations on a large number of slags having the general composition of Ca-Mg-(Fe-) silicates with little alumina. The ordinates represent acidity, and the abscissæ the relative proportion of Mg (with Fe). Each marked point on the diagram represents in this way the composition of a particular slag, and the sign used to mark it indicates the mineral which crystallizes first in that slag. It will be seen that the points belonging to a given mineral are gathered in a certain field of the diagram. The dividing-lines between the several fields may be called eutectic lines, since they pass through points which represent eutectic proportions of the several minerals. Three such lines meet in a point which represents the composition of the ternary eutectic mixture of the three minerals.

The second part of Vogt's memoir deals with the lowering of freezing-points in these slags. Åkerman had already

[1] By acidity is understood the oxygen-ratio of silica to bases, which is 1 for an orthosilicate and 2 for a metasilicate.

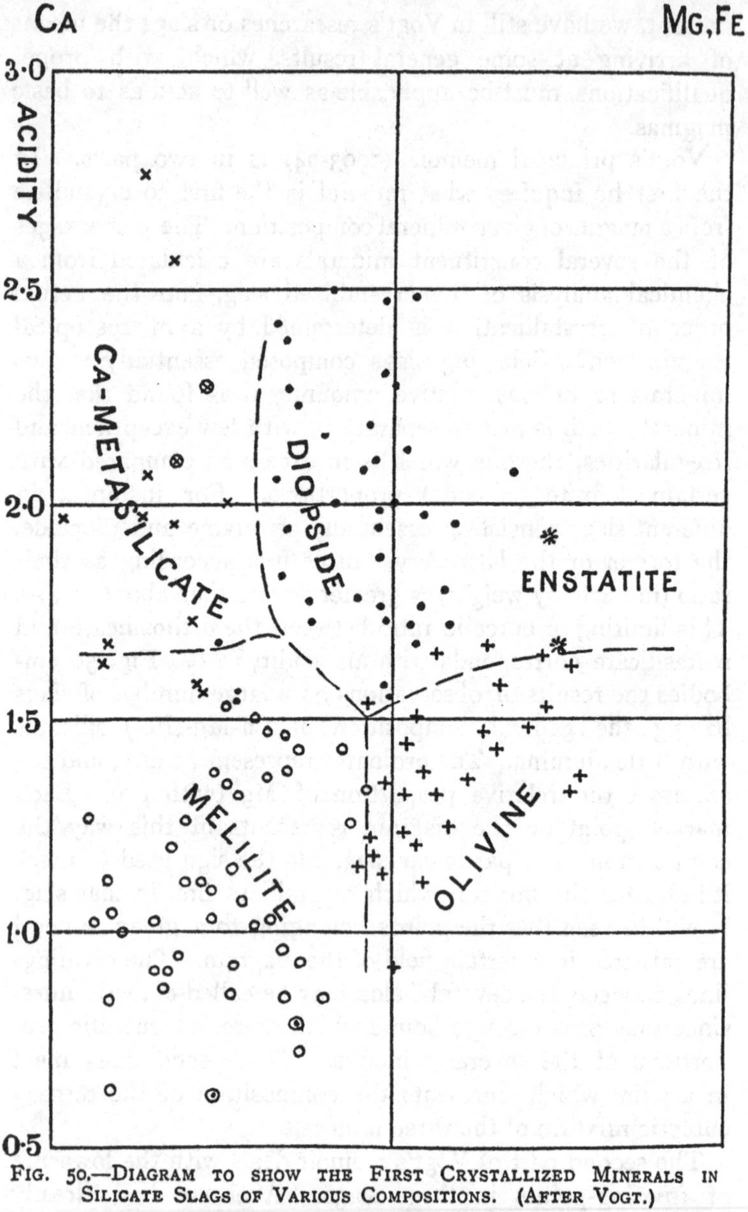

FIG. 50.—DIAGRAM TO SHOW THE FIRST CRYSTALLIZED MINERALS IN SILICATE SLAGS OF VARIOUS COMPOSITIONS. (AFTER VOGT.)

In the calcium metasilicate field the two crosses within small circles represent wollastonite, and the simple crosses pseudo-wollastonite. In the lower part of the diagram the dots within small circles represent spinel, followed by melilite.

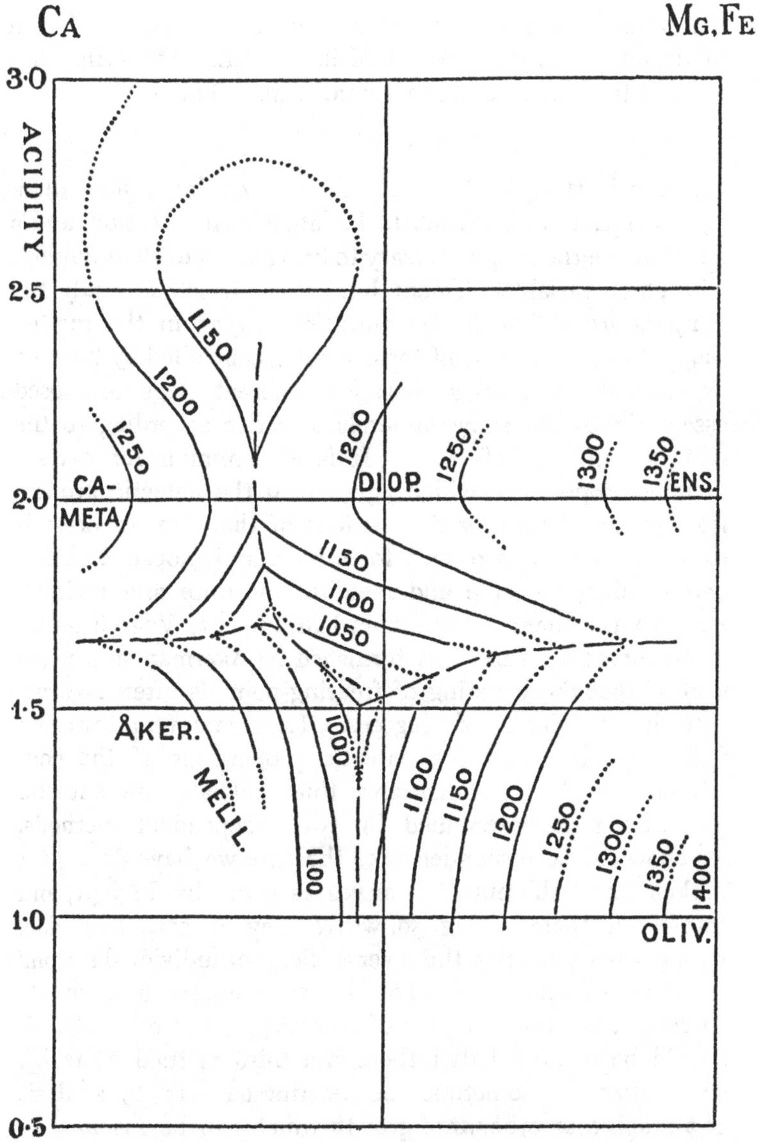

FIG. 51.—DIAGRAM TO SHOW THE TEMPERATURE AT WHICH CRYSTALLIZATION BEGINS IN SILICATE SLAGS OF VARIOUS COMPOSITIONS. (AFTER VOGT.)

The temperature-determinations have a relative rather than an absolute significance.

determined for some hundreds of examples the 'total heat of fusion'—*i.e.*, the amount of heat required to bring unit mass of the slag at 0° C. to a fluid state. This is

$$cT + L + h,$$

where c is the specific heat, T the 'melting-point' (in a certain generalised sense), L the latent heat of fusion, and h a certain overheating necessary to bring about evident fluidity. The chief variable element here is T, which is really the temperature at which crystallization begins in the molten slag. Vogt makes use of these results, controlled by his own experiments, and finds that, for different slags composed essentially of the same minerals, T varies according to the relative amounts of these. It is a minimum for certain definite proportions, which agree with the eutectic proportions already found by observations of the order of crystallization. This was proved for olivine and diopside, olivine and melilite, diopside and melilite, anorthite and melilite, etc. In the diagram reproduced in Fig. 51, Vogt has laid down curves from the data furnished by Åkerman, and made it clear that the lowering of freezing-point is often measurable by hundreds of degrees. The maximum lowering indicates, of course, the eutectic proportions of the constituents. The two diagrams thus indicate the eutectic proportions as determined by two independent methods. To show the correspondence of the two, we have drawn (in broken lines) the eutectic curves as given by Fig. 51, and transferred them to Fig. 50, where they are seen to divide, with few irregularities, the several 'fields of individualisation.'

Vogt has given, as a result of his researches, the approximate eutectic ratios for a number of different pairs of minerals. It should be remarked that there is a third method by which such ratios can sometimes be determined—viz., by analysis of a graphic or other intergrowth which can be assumed to have a eutectic composition. This is applicable especially to quartz and the alkali-felspars. The following figures are taken from Vogt's memoirs:

Orthoclase : quartz	... ...	72·5 : 27·5
Orthoclase : albite	... ...	42 : 58
Anorthite : diopside	... ...	*ca.* 65 : 35
Anorthite : olivine	... ...	*ca.* 70 : 30
Diopside : olivine	... ...	68 : 32
Diopside : enstatite	... ...	45 : 55
Hedenbergite : hypersthene	... ...	*ca.* 40 : 60
Melilite : anorthite	... ...	*ca.* 65 : 35
Melilite : olivine ...	... ...	74 : 26
Melilite : diopside	... ...	*ca.* 60 : 40
Fayalite : magnetite	... ...	67–80 : 33–20

These may be accepted as having a provisional and approximate value. There seems to be no ground for Johansson's[1] supposition that eutectic ratios should correspond with molecular proportions of the constituents.

Order of Crystallization.—Discussion relative to the essential nature of igneous rock-magmas and the process of their consolidation has turned largely upon the order in which the several constituent minerals crystallize as the magma gradually cools. The actual order of crystallization in a given rock is judged from the mutual relations of the crystals as seen in thin slices. If crystals of one primary mineral completely enclose crystals of another, the evidence of their relative age is absolute. Suppose, however, that crystals of two minerals (A and B) are seen in juxtaposition, and that the shape of the dividing-line shows it to be properly a boundary of A, not of B. If we can assume that the two crystals formed successively, this proves that A was the earlier of the two. Crystals, however, are not formed instantaneously at a point of time, and it is not impossible that their periods of growth may overlap. We know that some minerals crystallize more slowly than others, and it is conceivable that a crystal B may be begun earlier and completed later than another crystal A. The relative idiomorphism indicates the order in which the minerals *ceased* to crystallize; while the significant order, from the theoretical point of view, is that in which they *began* to crystallize. It

[1] *Geol. Fören. Förh. Stockholm*, vol. xxvii. (1905), pp. 119-148.

must be remembered, therefore, that the evidence is not always free from ambiguity.

If, discarding the conception of solution, we could picture the several minerals as coëxisting in a rock-magma in perfect independence, their specific properties not modified, we should expect each mineral to crystallize out from the cooling magma at its own proper melting-point, and the several minerals would then appear in the reverse order of fusibility. Fouqué and Michel-Lévy wrote in 1882: "The principle which appears to us to have governed the formation of igneous rocks consists in the fact that the minerals have consolidated according to the order of their respective fusibilities." They found it necessary, however, to qualify this pronouncement by attaching importance likewise to other factors, and especially to the influence of 'agents minéralisateurs.' Summarily, they apply the law of fusibility to basic magmas, in which mineralisers have not played any important rôle. Now, there is no doubt that mineralising agents (water and other fluxes) lower the melting-points of minerals to a greater or less extent, and may affect their relative fusibility in the magma. But the observed order of crystallization compels us to recognise that the several minerals also lower the melting-points of one another, according to the laws of solution. On the theory of the authors quoted, the order of crystallization in basic rocks should be constant, but we have already seen (p. 171) that it depends on the relative proportions of the minerals present.

In 1882 Rosenbusch[1] formulated certain empirical rules as expressing the order of crystallization of the several minerals in a large number of igneous rock-types. He ranges the rock-forming minerals in four groups as follows:

I. Iron-ores and accessory constituents (magnetite, hæmatite, ilmenite, apatite, zircon, spinel, sphene, etc.).

[1] 'Ueber das Wesen der körnigen und porphyrischen Structur bei Massengesteinen,' *Neu. Jahrb.*, 1882, vol. ii., pp. 1-17. See also later editions of *Mikroskopische Physiographie der Massigen Gesteine.*

II. Ferro-magnesian silicates (olivines, pyroxenes, amphiboles, micas, etc.).

III. Felspathic constituents (felspars and felspathoids, including leucite, nepheline, sodalite, melilite, etc.).

IV. Free silica.

Rosenbusch states that in general these four groups of minerals crystallize in order as enumerated. This applies to the non-porphyritic rocks, in which each constituent is supposed to crystallize at one particular stage without recurrence. The porphyritic structure, in which there is a recurrence of one or more constituents in a second generation, imports a modification of the statement, which will be considered later. More explicitly, what is regarded as the normal sequence is laid down in the following rules:

(i.) The separation of crystals in a silicate-magma follows an *order of decreasing basicity*, so that at every stage the residual magma is more acid than the aggregate of the compounds already crystallized out.

(ii.) The relative amounts of the several constituents present in the magma affect the order of crystallization in such a manner that, in general, those present in smaller amount crystallize out earlier.

(iii.) Having regard to the several bases represented in the various constituents, crystallization begins with the separation of iron-oxides and spinellids, proceeds with the formation of magnesium and iron silicates, then silicates of calcium, then those of the alkali-metals, and ends with the crystallization of the remaining free silica.

The first law and the third, which is an amplification of it, inasmuch as they lay down a constant order for the several minerals, appear to be in direct conflict with the laws of solutions. There are, as Rosenbusch recognised, very important exceptions to the order of decreasing basicity. In a large number—probably the majority—of basic rocks the felspar has crystallized before the augite. In many granites orthoclase has crystallized simultaneously with quartz or after it; and when microcline is present, it is usually the latest product of crystallization. It would be easy to cite examples of olivine crystallized after felspar, biotite after felspar and quartz, pyrrhotite after felspar and hornblende, etc. Magnetite has separated out at different stages, and in some cases has remained in the magma to the last.[1] Nevertheless, the general sequence laid down by Rosenbusch as the normal one is found to rule in a large number of cases. Some considerations will be noticed below which help to explain how this empirical law comes to have a certain degree of validity.

Rosenbusch's second law is directly opposed to that which results from the principles of solutions. It is put forward apparently as modifying the operation of the first law, and in particular as applying to the frequent reversal of the 'normal' order in the case of Groups II. and III. (*e.g.*, augite crystallizing after felspar). In this connection, however, the rule given does not seem to accord with the actual facts. The case of some of the minerals of Group I. has probably a different significance. It is true that the minor accessory minerals, such as apatite and zircon, usually crystallize at a very early stage, although their relative amounts in the magma may be very small. But, as we shall point out below, it is by no means clear that these are constituents of the solution on the same footing as other minerals, since there is reason to believe that they have only a limited mutual solubility with the ordinary rock-forming silicates.

Michel-Lévy,[2] writing in 1889, maintains the distinction between "rocks of igneous fusion" and those in which mineralising agents have played an important rôle; and he accepts Rosenbusch's laws, with exceptions, for the former only. On the ground of synthetic experiment, he asserts that most of the minerals of these rocks crystallize in order according to fusibility, but that other constituents, such as spinels and magnetite, are due to "veritable chemical pre-

[1] Compare Teall on basalts from Franz Josef Land, *Geol. Mag.*, 1897, pp. 553-555.

[2] *Structures et Classification des Roches Éruptives* (1889), p. 37.

cipitations, which last throughout the whole time of crystallization."

Lagorio[1] has considered the actual progress of crystallization in rock-magmas, with special reference to the concurrent change in composition of the remaining fluid magma—a point of view which brings the question directly into touch with the solution hypothesis. As data he employed a large series of comparative analyses made for the purpose, analysing the whole rock, the porphyritic elements or spherulites, and the ground-mass or glassy residue. He found that the progressive acidification of the magma demanded by Rosenbusch's law of decreasing basicity is by no means a general principle. In volcanic rocks of intermediate acidity any glassy residue is often notably more acid than the total rock; but in acid and basic rocks the residual 'base' often has about the same silica-percentage as the whole. In very acid rocks it may be less acid, and in very basic rocks more acid.

Morozewicz,[2] approaching the subject from the experimental side, came to the conclusion that the order of crystallization in an igneous rock depends, not on any one factor alone, but on several factors jointly. One of these is the relative amounts of the several compounds present in the magma; but he also lays stress on the "individual capacity of the substance to form supersaturated solutions," or its solubility in the magma. Substances, such as alumina, having a low solubility, will crystallize out early, even when present in small quantity; while others, like plagioclases and pyroxenes, will crystallize earlier or later according to their relative amounts in the magma. It is not easy to follow the conception of solution entertained by Morozewicz, and a like confusion is found in the writings of some others[3] who have experimented on the 'mutual solubility' of

[1] *Tscherm. Min. Petr. Mitth.* (2), vol. viii. (1887), pp. 421-529.

[2] *Ibid.* (2), vol. xviii. (1898), pp. 228, 229.

[3] Doelter, *Centralbl. für Min.* (1902), pp. 199-203; Lenarčič, *ibid.* (1903), pp. 705-722, 743-751; Schweig, *Untersuchungen über die Differentiation der Magmen* (1903), Inaug. Diss., Stuttgart.

minerals as affecting the order of crystallization. It is conceivable that between, *e.g.*, corundum and felspar there is only a limited mutual solubility, in the sense that the two compounds in the molten state are not freely miscible in all proportions. To this possibility we shall return later. But, if this be set aside, the relation between corundum and felspar must be of the same kind as that between augite and felspar, the order of crystallization depending on the relative proportions of the two minerals in the magma as compared with the eutectic proportions. Any such consideration as fusibility or solubility is already included in this, and cannot be reckoned as another determining factor.

If we suppose, with Vogt, that the rock-forming silicates and oxide-minerals are all perfectly miscible in the fluid state, the solution theory leads at once to the simple law that the order of crystallization is determined by the relative proportions of the several minerals present as compared with the eutectic proportions. We shall have to remark later, however, that considerations relative to supersaturation, etc., import very material qualifications of this broad rule, and the actual conditions determining the order of crystallization are often of a complex kind.

Actual Temperatures of Rock-Magmas.—While a definite mineral compound has a precise melting-point, which is also its freezing-point, a mixture of several minerals (unless they happen to be in exact eutectic proportions) has a certain *temperature-range of crystallization.* In general, only a small part of the crystallization is effected in the upper portion of this range, but this depends upon how far the initial composition of the magma departs from eutectic proportions. Rock-magmas as extruded at the surface or intruded among solid rocks have temperatures lying within the range of crystallization for each particular magma. This is proved by petrographical evidence, which shows that in general a magma as extruded or intruded carries crystals already formed; and we have also seen from general considerations that no noteworthy superheating can be supposed in the intercrustal magma-reservoirs from which the supplies are drawn.

This consideration enables us at once to set limits—although somewhat wide limits—to the possible temperature of a magma of given mineralogical composition. An upper limit is prescribed by the melting-points of the less fusible minerals, and a lower limit by the final temperature of consolidation, which, in an anhydrous magma, is theoretically that corresponding with the multiple eutectic of the several essential minerals. In many cases other considerations enable us to narrow these limits, and in particular to show that the upper one is far too high. Thus, assuming the phenocrysts in a lava to be of intratelluric crystallization, it is evident that any lava must have been poured out at a temperature lower than the melting-point of its most fusible phenocrysts.[1] We can go farther than this. Vogt's researches show that, in slags comparable with anhydrous basic magmas, there is a general lowering of melting-points measurable by hundreds of degrees. A very large proportion of the common rock-types probably do not depart very widely from eutectic proportions of their chief constituents; and, in particular, we shall see later that the ground-mass of a porphyritic rock, which makes up the greater part of its bulk, tends to be of approximately eutectic composition. We may infer that the greater part of a rock-magma usually remains fluid down to temperatures not very much higher than the final temperature of consolidation. In the case of such a rock as a basalt this would be at least 200° lower than the melting-point of the most fusible of the minerals. Such data as we possess from direct observation confirm this conclusion, the temperature of emission of the lavas of Etna and Vesuvius being, according to several independent measurements,[2] in the neighbourhood of 1000° C.

While 1000° may be taken as representing with some

[1] Brun in this way determines the temperature of the basalt flows of Stromboli as below 1230°, the melting-point of the enclosed crystals of augite; *Arch. sci. phys. et nat. Genève* (4), vol. xiii. (1902), pp. 85-87.

[2] Bartoli, *cit.* Brauns, *Tscherm. Min. Petr. Mitth.* (2), vol. xvii. (1898), p. 491; Doelter, *Sitz. k. Akad. Wiss. Wien, Math.-Nat. Kl.*, vol. cxii. (1903), pp. 681-705, and *Petrogenesis* (1906), pp. 16, 17.

approximation the ordinary temperature of a basaltic lava, it is certain that most rock-magmas have temperatures considerably lower than this. We have allowed for the fact that the several minerals act as fluxes to one another, as proved by Vogt's experimental researches; but natural rock-magmas, unlike slags, contain also in varying amount more potent fluxes—viz., water and other volatile substances—which occasion a further lowering of melting-points. The effect depends on the amount of these substances present in the magma, and concerning this we have no quantitative knowledge, for only a fraction of the amount remains in the rock when consolidated. It may be remarked incidentally that for this reason the fusion of a rock in a crucible gives no useful information relative to the temperature of its crystallization. Under plutonic conditions the greater part of the volatile fluxes is retained until a late stage of crystallization, and eliminated only when it has done its work; but from an extruded lava the greater part of the water, etc., escapes before crystallization is far advanced. Indeed, the loss of these substances, by raising the melting-points in the magma, may be the immediate cause of crystallization, quite as much as any actual cooling. It appears, therefore, that *intruded magmas crystallize at lower temperatures than extruded magmas of like composition.* Further, there is ample evidence that, as maintained by Fouqué and Michel-Lévy, water and other volatile fluxes are more abundant in acid than in basic magmas, and play there a more important part in crystallization. It follows that, *under like conditions, acid magmas crystallize at lower temperatures than basic magmas.* We must expect, then, the highest temperatures (*ca.* 1000°) in basaltic lavas, and the lowest temperatures in granitic magmas, for which Lehmann's[1] estimate of about 500° is probably by no means too low.

There is abundant experimental evidence that many of the rock-forming minerals have a certain temperature-range of stability for each. For polymorphs of the enantiotropic kind the limits are more or less precise, and have been in

[1] *Alltkrystallinischen Schiefergesteine* (1884), pp. 54, 55.

some cases determined with sufficient accuracy. Thus, the inversion-point for tridymite and quartz is about 800°, and we must conclude that all rocks containing primary[1] quartz have crystallized at lower temperatures than this. For polymorphs with monotropic relation there is no inversion-point, but we may still recognise a relative stability of the different forms, governed by temperature. Thus we may confidently assume from experimental evidence that augite crystallizes at high temperatures, enstatite probably not so high, and hornblende at low temperatures. Minerals of very complex chemical constitution, such as the aluminous hornblendes, micas, tourmaline, etc., can apparently exist only at low temperatures, and none of them have been artificially reproduced without the aid of special fluxes and reagents (see below, Chapter XII.). Knowledge concerning this subject is as yet very incomplete; but we may provisionally distinguish on the one hand *high-temperature minerals*, such as olivines, pyroxenes, anorthite, leucite, tridymite, and various minor accessories, and, on the other hand, *low-temperature minerals*, such as amphiboles, micas, quartz, tourmaline, and others. It will be seen that the former are among the characteristic minerals of basic rocks, especially lavas, while the latter are equally characteristic of acid plutonic rocks.

The researches of Sorby, Hartley, and others on the minute *fluid-inclusions* in quartz, topaz, beryl, etc., furnish valuable evidence from another side. They demonstrate the relatively low temperatures at which granitic rocks may crystallize, and at the same time they show the nature of the fluxes which enable such low temperatures to be attained. Sorby,[2] in his classical memoir published fifty years ago, endeavoured to determine the temperature of crystallization by means of the contraction which has occurred in the cooling of the liquid contents of a cavity

[1] Interstitial quartz in lavas may result from the transformation of original tridymite; Lacroix, *La Montagne Pelée après ses Éruptions* (1908), pp. 52-58.

[2] *Quart. Journ. Geol. Soc.*, vol. xiv. (1858), pp. 453-500.

in a crystal. The crystallization is supposed to have taken place under a pressure p in addition to that of the vapour; and it is necessary to make some assumption concerning this pressure in order to calculate the temperature of crystallization from the observed relative volume (v) of the bubble in the cavity. It is easy to see, however, from Fig. 52 that,

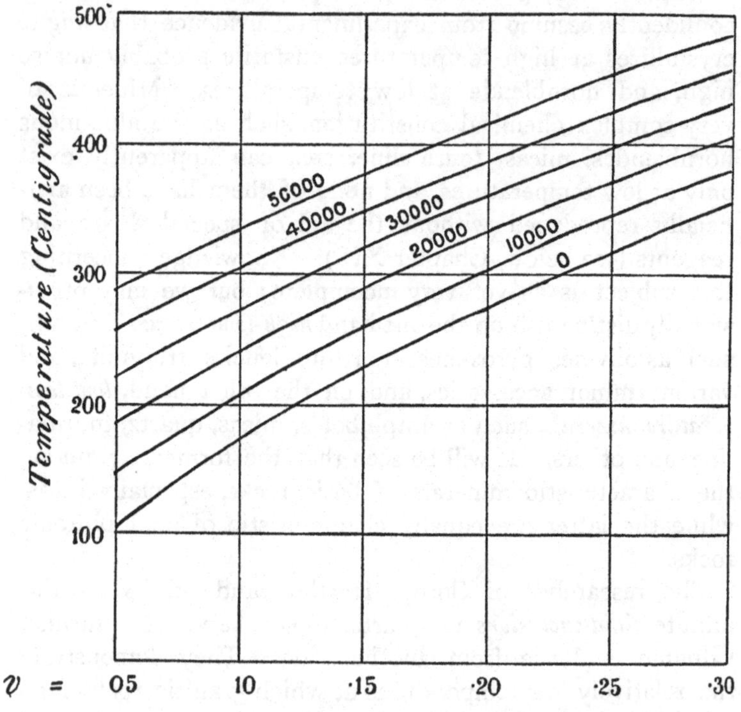

FIG. 52.—DIAGRAM SHOWING THE TEMPERATURE OF CRYSTALLIZATION OF QUARTZ, AS CALCULATED FROM v, THE RELATIVE VOLUME OF THE BUBBLE IN A LIQUID INCLUSION.

The curves, drawn from Sorby's equation, correspond with different values of the pressure p, reckoned in feet of rock of specific gravity 2·5.

whatever assumption be made relative to the pressure, the temperatures found by Sorby's method for granitic rocks cannot be high. The smallest value of v found for the Cornish granites was 0·09, and the largest only 0·20, indicating temperatures of crystallization probably between

200° C. and 350° C. The 'elvans' (quartz-porphyry dykes) of the same district gave values ranging from 0·125 to 0·25, indicating temperatures some 50° higher; unless we suppose, with Sorby, that the pressure was here much lower, which is improbable.

Again, the extent and degree of *thermal metamorphism* produced by different kinds of intrusions accords with a higher temperature for basic than for acid magmas. The plutonic rocks of the Inner Hebrides, for instance, include peridotites, gabbros, and granites. Comparing their metamorphic effects on a given rock, such as the Torridon Sandstone or the basaltic lavas, we see that the gabbros must have been hotter than the granites, and the peridotites the hottest of all. Of the granites, only that of Arran has produced any high grade of metamorphism, and this is attributable to a larger content of 'mineralising' substances, proved by the contents of the druses in the rock. The intense metamorphism found round granites and pegmatites in some other regions is to be attributed, not to high temperature, but, as the French geologists have long maintained, to the influence of mineralisers emanating from the intruded magma. In the same way, the action of a magma upon enclosed fragments of foreign rocks, in so far as it depends on temperature, is most intense in basic magmas. In acid magmas the reaction, not confined to the contact but pervading the whole, is ascribed by Lacroix[1] to mineralising agents.

The parallelism already noticed (p. 131) between the normal order of decreasing basicity in a sequence of plutonic intrusions and the order of crystallization of the constituent minerals becomes more significant when the order of intrusion of the different magmas is seen to coincide with a progressive decline in the temperature of intrusion.

[1] *Les Enclaves des Roches Volcaniques* (1893).

CHAPTER VIII

CRYSTALLIZATION OF ROCK-MAGMAS

Crystallization in a binary magma.—Pressure as affecting crystallization.—Limited miscibility in rock-magmas.—Crystallization in a ternary magma.

Crystallization in a Binary Magma.—Accepting the solution theory of rock-magmas as having at least established its claim to the rank of a working hypothesis, we proceed to examine its application to the crystallization of minerals from a magma, with special reference to the order of crystallization of the several minerals. In the present chapter we shall consider the ideal case in which crystallization begins as soon as saturation is attained in the cooling magma. The results obtained will be subject to rather important qualification, to be discussed later, for crystallization does not in general begin until a certain degree of supersaturation has been reached.

We take first the simplest case of a magma which is a mutual solution of two minerals, *A* and *B*, which are freely miscible in the liquid state. They are supposed to be definite compounds, so that each has its proper melting-point, which is also its freezing-point. It is further supposed that they do not form together either isomorphous mixed crystals or double salts. In Fig. 53 abscissæ represent the composition of the magma, in percentage by weight of *B*, and ordinates represent temperature, *a* and *b* being the melting-points of the two minerals. A cooling magma, composed wholly of *A*, will begin to crystallize when the temperature has fallen to *a*, and will continue to crystallize at that temperature until the magma is exhausted. If, however, a small proportion of *B* be dissolved in the magma, so that its composition is represented by *M*, crystallization of *A* will begin at a lower temperature *m*, and the freezing-point of *A* may therefore be said to be lowered by the presence of *B*. The amount of depression is given by Van 't Hoff's formula, which may be conveniently expressed as follows: For each part by weight of *B* dissolved in 100 of *A*, the freezing-point of the latter is lowered by an amount $T_1^2 \div 50 L_1 M_2$; where T_1 is

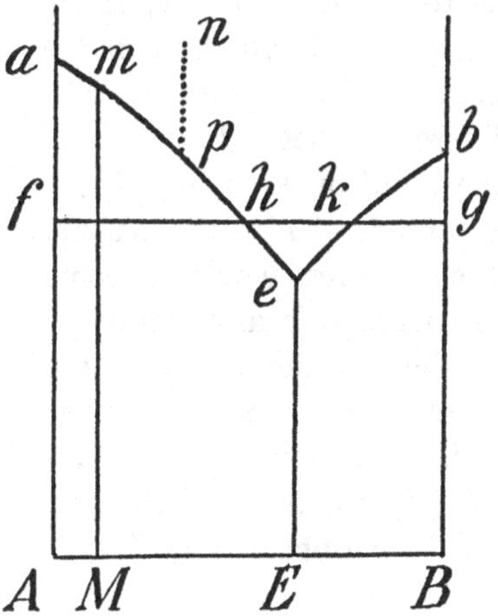

FIG. 53.—FREEZING-POINT CURVES FOR A BINARY SOLUTION.

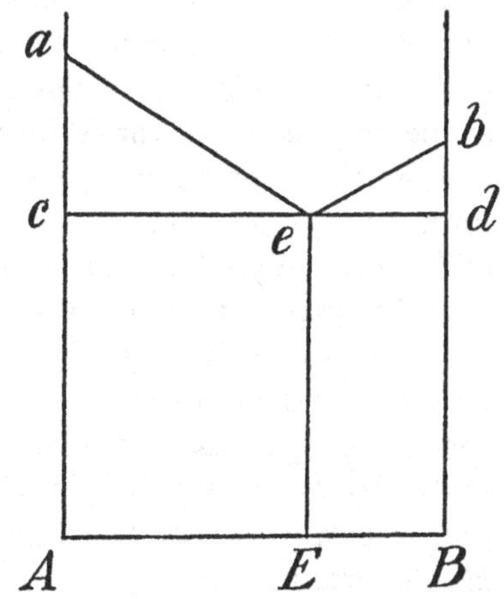

FIG. 54.—IDEAL CASE FOR THE PURPOSES OF CALCULATION.

the melting-temperature of *A* reckoned from absolute zero, L_1 its latent heat of fusion, and M_2 the molecular weight of *B*.[1] A further addition of *B* will produce a further lowering of melting-point, and in this way a curve *ame* can be traced, which may be called the *freezing-point curve* of *A* in such a binary magma, or the curve of saturation of the magma with the constituent *A*, or, again, the curve of solubility of *A* in the magma. If a similar curve be drawn for *B*, the two curves meet in a point *e*, which is the *eutectic point*. A horizontal straight line *fg*, at a certain height above the base-

[1] Ionization of *B* is here disregarded. The numerical factor 0·0198 is for brevity taken as $\frac{1}{50}$.

line, corresponds with a certain temperature. If it cuts the freezing-point curves in *h* and *k*, then *hk* represents the possible range of composition of a fluid magma composed of these two minerals at that temperature. Thus *gh* represents the greatest proportion of *A* which can be contained in the magma, and, therefore, in a certain sense, the solubility of *A* in a magma of *A* and *B* at the given temperature; and *fk* represents in like manner the solubility of *B*.

The depression of freezing-point of *A* is proportional to the amount of *B*, only so long as this amount is very small, and Van 't Hoff's law therefore gives the inclination of the curve *ame* only in the neighbourhood of the point *a*. The shape of the curve varies in different cases. If we could assume the depression of freezing-point of *A* to be, without limitation, proportional to the percentage of *B* in the magma, and conversely, the freezing-point curves would become straight lines (Fig. 54). We will make this assumption in order to obtain a rough idea of the manner in which the eutectic proportions of the two minerals depend on their specific properties.[1] The eutectic proportions are represented by *de* and *ce* in Fig. 54. Using suffixes to distinguish the properties of the two minerals, we obtain from Van 't Hoff's formula:

$$\frac{ac}{ce}=\frac{T_1^2}{50L_1M_2I_2}; \quad \frac{bd}{de}=\frac{T_2^2}{50L_2M_1I_1};$$

the factor *I* being introduced to allow for partial ionization. Its value is $1+(k-1)\alpha$, where α is the proportion of dissociated molecules and *k* the number of ions into which a molecule is dissociated. Observing now that:

$$ac=T_1-T_0, \quad bd=T_2-T_0,$$

where T_0 is the eutectic temperature reckoned on the absolute scale, we obtain the relation:

$$\frac{de}{ce}=\frac{T_2-T_0}{T_1-T_0}\cdot\frac{T_1^2}{T_2^2}\cdot\frac{L_2}{L_1}\cdot\frac{M_1}{M_2}\cdot\frac{I_1}{I_2}.$$

The five factors on the right hand of the equation indicate

[1] Vogt has given a more elaborate investigation on similar lines; *Silikatschmelzlösungen*, II. (1904), pp. 128-135.

the manner in which the eutectic proportions are affected by the relative melting-points, latent heats, molecular weights, and ionizations of the two constituents of the magma. To see the influence of melting-point, suppose *A* and *B* to differ only in respect of this, so that only the first and second of the five factors remain. These operate in opposite directions; but it is easy to see that the first factor is the more important one. Therefore, other things being equal, *de* is greater or less than *ce*, according as T_2 is greater or less than T_1—that is, the eutectic point lies towards the mineral with the lower melting-point. Considering the other three factors separately, it is evident that, *ceteris paribus* in each case, the eutectic point lies nearer to the mineral with the lower latent heat of fusion, or the higher molecular weight, or the greater degree of ionization.

We return to the actual freezing-point curves (Fig. 53). A point on the diagram, representing a certain state of the magma as regards both composition and temperature, may be called the indicating-point. Such a point as *p*, on the freezing-point curve for *A*, represents a state of equilibrium between the magma and crystals of *A*. If the magma cools, without change of composition, from a state represented by *n* to a state of equilibrium, the indicating-point descends vertically from *n* to *p*, and the change is represented by the line *np*. According to our supposition, crystals of *A* now begin to form. The equilibrium between magma and crystals, once adjusted, is maintained so long as the two coëxist in contact, and consequently the further change in the magma will be represented by a movement of the indicating-point along the curve *ae*. The vertical part of this movement corresponds with fall of temperature, and the horizontal part with the relative enrichment of the magma in *B*, owing to the continued abstraction of *A* by crystallization. When the indicating-point reaches *e*, there is equilibrium for *B* as well as for *A*, and *B* accordingly begins to crystallize. The two minerals crystallize in eutectic proportions, so that the composition of the magma undergoes no further change. The temperature is also stationary, so long as any magma remains, the loss of heat by conduction being balanced by the latent heat liberated in crystallization. The whole process consists, then, of four parts: cooling of the fluid magma until saturated (*np*); crystallization of *A*, with impoverishment of the magma in that constituent and fall of temperature (*pe*); eutectic crystallization of *A* and *B* with stationary conditions (*e*) until the magma is exhausted; and, finally, cooling of the crystalline rock (*eE*).

Pressure as affecting Crystallization.—It is proper to inquire to what extent the considerations set forth in the preceding section are affected by such differences of pressure as may be assumed between deep-seated and superficial conditions of crystallization. Bunsen[1] long ago suggested that pressure, by altering the relative fusibility of different minerals, may alter the order of their crystallization from a magma, and this suggestion has been revived by more than one writer.[2] We have already seen (p. 163) that the effect of pressure in raising the melting-point of a mineral is not great; and its differential effect as between two minerals must be much less. The effect on the eutectic ratio for the two minerals, into which the melting-points enter together with other factors, can be estimated only very roughly, but must certainly be quite inconsiderable. Thus from Thomson's equation (p. 163) we can calculate that a pressure of 450 atmospheres, equivalent to a column of rock a mile deep, will raise the melting-point of anorthite 1·7° and that of diopside 8·9°; while the eutectic point for these two minerals, from such rough data as are obtainable, would probably be displaced laterally to the extent of about 0·4 in 100 divisions. In other words, the eutectic proportions for these minerals, being, according to Vogt, about 65 : 35 at atmospheric pressure, would become 65·4 : 34·6 under a pressure of 450 atmospheres. Vogt has deduced from analyses of presumably eutectic aggregates in plutonic and volcanic rocks that pressure has little influence on the proportions, and

[1] *Pogg. Ann.* (3), vol. xxi. (1850), pp. 562-567.

[2] Stromeyer, *Mem. Manch. Lit. Phil. Soc.*, vol. xliv. (1900), No. 7; Cunningham, *Proc. Roy. Dubl. Soc.* (N.S.), vol. xvi. (1901), pp. 383-414.

Van 't Hoff had already come to the same conclusion with reference to aqueous solutions of salts.

A further question relates to the effects of a *sudden relief of pressure during the progress of crystallization.* It is evident that, by lowering the melting-points of the minerals, a diminution of pressure may be expected to cause a partial resorption of crystals already formed. The case of a binary magma is illustrated in Fig. 55. Suppose that the mineral *A* has begun to crystallize, the point indicating the com-

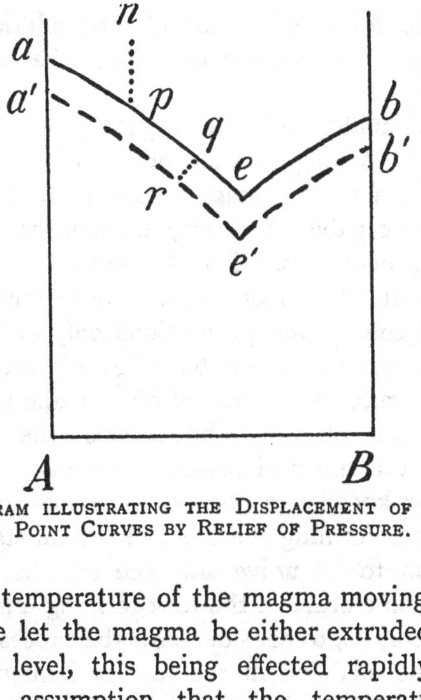

FIG. 55.—DIAGRAM ILLUSTRATING THE DISPLACEMENT OF THE FREEZING-POINT CURVES BY RELIEF OF PRESSURE.

position and temperature of the magma moving from *p* to *q*. At this stage let the magma be either extruded or intruded at a higher level, this being effected rapidly enough to warrant the assumption that the temperature remains unchanged. The curves *ae* and *be* are now displaced to the positions a^1e^1 and b^1e^1; and the magma, represented by *q*, is no longer saturated. A part of the crystals will accordingly be resorbed, the magma being enriched in *A* and falling in temperature (by absorption of latent heat as well as by conduction) until equilibrium is re-established at *r*. In reality, the displacement of the curves must be supposed to be very small, and it appears that resorption from this cause cannot be very considerable. Moreover, it will be followed immediately by renewed crystallization of the same mineral *A* (along re^1), and it is possible that the corroded crystals may be repaired by the new growth.

It is to be observed that the resorption here considered is that due simply to the solution of crystals by an unsaturated magma. There may be resorption, to which no narrow limit can be set, due to the instability under a low pressure of molecular arrangements built up under high pressure. The partial or total destruction of intratelluric hornblende and biotite in many volcanic rocks[1] may be in part due to this cause.

Limited Miscibility in Rock-Magmas.—We have hitherto assumed that, so long as they are in the fluid state, the several constituents of a magma have perfect mutual solubility—*i.e.*, are capable of making homogeneous mixtures in any relative proportions. This, however, is not a property of all liquids at all temperatures. For instance, water and phenol mix freely in any proportions only at temperatures above 68° C. At any lower temperatures there can exist water with a limited admixture of phenol, and phenol with a limited admixture of water; but between these limits there is a hiatus in the series of possible homogeneous mixtures.[2] So far as our knowledge goes, there is perfect miscibility among the rock-forming silicates in the fluid state; but this does not seem to be universally the case as between the silicates and some other of the rock-forming minerals. It is necessary, therefore, to consider how the process of crystallization will be modified in the case of limited miscibility. As before, we take, for the sake of simplicity, an ideal binary magma.

Suppose, then, a magma consisting of two minerals, *A* and *B*, which mix freely in all proportions only above a

[1] Washington, *Journ. Geol.*, vol. iv. (1896), pp. 257-282.

[2] Rothmund, *Zeits. phys. Chemie*, vol. xxvi. (1898), pp. 433-492. For other examples see Schreinemaker, *ibid.*, vol. xxiii. (1897), pp. 416-422; Klobbie, *ibid.*, vol. xxiv. (1897), pp. 614-621.

certain temperature—that represented by *m* in Fig. 56. At a temperature (*hk*) below this the only possible homogeneous mixtures are those represented by points between *h* and *p* on the one hand, and between *q* and *k* on the other. Such a mixture as *r* cannot exist, but would at once separate into two immiscible portions, represented by *p* and *q*—viz., a saturated solution of *B* in *A*, and a saturated solution of *A* in *B*. At a higher temperature the hiatus becomes

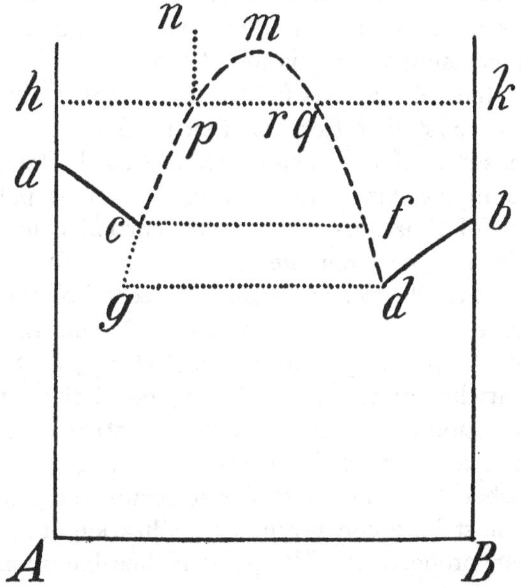

FIG. 56.—SOLUBILITY-CURVES FOR A BINARY MAGMA WITH LIMITED MUTUAL MISCIBILITY.

smaller, and disappears at *m*. Such points as *p* lie on a certain curve (*cpm*), which represents the solubility of (liquid) *B* in *A*; and such points as *q* lie on a curve (*dqm*), which represents the solubility of (liquid) *A* in *B*. The two are really parts of one continuous curve, which terminates downward at the freezing-point curves (the solubility-curves of the solid constituents). In the figure we suppose the two branches to be cut off by the two freezing-point curves respectively. If both are cut off by the same freezing-point curve (say that for *A*), there will be a eutectic point (to the right).

Taking the case illustrated in the figure, suppose the initial magma to be at a high enough temperature (*n*) to be homogeneous, and to have some composition between *c* and *d*. Then, with falling temperature, the indicating-point will meet the solubility-curve at some point *p*, and the magma will begin to separate into two parts or layers, represented at first by *p* and *q*. The two indicating-points then move down the two branches of the curve, their divergence corresponding with an increasing differentiation of the two partial magmas. When the temperature has fallen to the level *cf*, the mineral *A* begins to crystallize from the first partial magma. Since the composition of this magma as well as the temperature must remain constant (viz., at *c*, an invariant point), the abstraction of *A* is balanced by a concurrent conversion of part of the magma from the *c* to the *f* composition. We suppose that conduction maintains the second partial magma at the same temperature as the first. When the first is exhausted, the second proceeds to cool along *fd*. Since this implies enrichment in *B*, there must be at the same time some conversion to the complementary (first) kind of magma, with concurrent crystallization of *A* along *cg*. In like manner, when the mineral *B* at last begins to crystallize (at *d*), there must be a concurrent crystallization of *A* (at *g*) in the proper proportion. The point *d* thus has some of the properties of a eutectic point.

In the case supposed (with no true eutectic point) the two minerals do not at any stage crystallize simultaneously from the same magma, but give rise ideally to two distinct rocks—one composed wholly of *A*, and the other wholly of *B*. It is easy to see, however, that there must be more or less mechanical entanglement of one in the other, and especially of *A* in *B*. Under the conditions supposed, the two rocks may arrange themselves in an upper and a lower layer with more or less regularity, according to their relative densities and the viscosity of their magmas. We cannot always assume, however, that conduction maintains the whole mass at a uniform temperature, and there may be a tendency for the first rock to segregate towards the margin of an intrusive body, where the temperature is lowest (Figs. 57, 58).

The only case of limited miscibility in rock-magmas admitted by Vogt[1] is that of the *sulphides* relatively to silicates, etc. That there is in this case only a very slight mutual solubility is proved by the ordinary process of smelting, in which an almost complete separation is effected by mere fusion. The mutual solubility, however, increases with rise of temperature. Vogt supposes that in basic

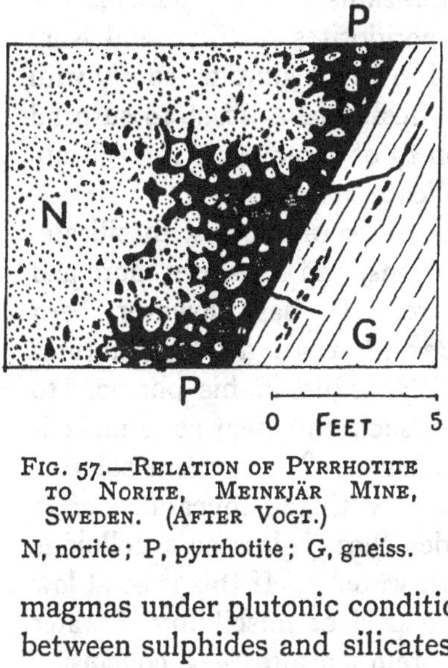

FIG. 57.—RELATION OF PYRRHOTITE TO NORITE, MEINKJÄR MINE, SWEDEN. (AFTER VOGT.)

N, norite; P, pyrrhotite; G, gneiss.

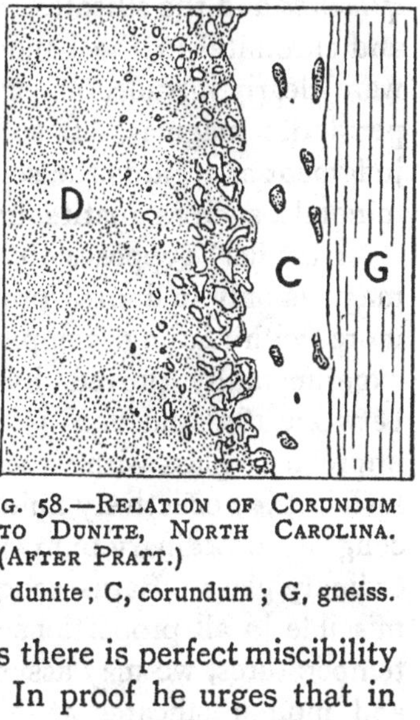

FIG. 58.—RELATION OF CORUNDUM TO DUNITE, NORTH CAROLINA. (AFTER PRATT.)

D, dunite; C, corundum; G, gneiss.

magmas under plutonic conditions there is perfect miscibility between sulphides and silicates. In proof he urges that in the pyrrhotite-bearing gabbros and norites the sulphide is found in all relative proportions from 1 to 100 *per cent.*, affording a gradual transition from a pure silicate-rock to a pure pyrrhotite-mass. The actual relations as described (Fig. 57) seem, however, to be sufficiently explained by the mechanical entanglement to which we have adverted. As already remarked (p. 186), the temperature of a norite-magma must be presumed lower, not higher, than that of a molten slag of like composition, and it follows that the

[1] *Silikatschmelzlösungen*, I. (1903), pp. 96-101, and II. (1904), p. 155.

miscibility of sulphide and silicate will be still more narrowly limited.

It appears to the present writer that the principle of limited miscibility, with its consequences, may have a somewhat wider application to rock-magmas than is allowed by Vogt. Thus, the mode of occurrence of *corundum* in association with peridotites[1] is strongly suggestive of a very limited mutual solubility between molten corundum and molten olivine (Fig. 58). It seems possible, too, that there may be only a limited miscibility, though with a wider range, between the *spinellids* and the silicates. The seams of nearly pure picotite and chromite[2] in the banded peridotites of Skye and Rum were clearly intruded as partial magmas simultaneously with peridotite magmas. It seems desirable at least to suspend judgment as regards the solubility of some other non-silicate minerals, such as apatite, in silicate-magmas.

Arrhenius[3] supposes that *water* is miscible with silicate-magmas only in limited proportions, and separates more and more with falling temperature. It has, however, been experimentally proved by Barus[4] that, at temperatures between 185° and 200° C., it is possible, in his phrase, "to impregnate glass with water to such an extent as to make it fusible [*i.e.*, effectively fluid] below 200°." Such a solution congeals at ordinary temperatures with the appearance of an ordinary glass. Barus concludes that "glass as a colloid is miscible in all proportions with water." If this is so at low temperatures, we may assume a perfect miscibility of water and molten silicates at high temperatures[5]—a conclusion already reached from the geological side by Van Hise.[6]

[1] Pratt, *Amer. Journ. Sci.* (4), vol. vi. (1898), pp. 49-65, and vol. viii. (1899), pp. 227-231.

[2] Compare Vogt, *Zeits. prakt. Geol.*, 1894, pp. 389-393, with figures; Pratt, *Amer. Journ. Sci.* (4), vol. vii. (1899), pp. 281-283.

[3] *Geol. Fören. Förh. Stockholm*, vol. xxii. (1900), pp. 395-419.

[4] *Amer. Journ. Sci.* (4), vol. vi. (1898), p. 270; see also vol. ix. (1900), pp. 161-175.

[5] The case of two liquids freely miscible only *below* a certain temperature is known, but seems to be rare (*e.g.*, water and triethylamine). Such a supposition would not serve the argument of Arrhenius.

[6] 16*th Ann. Rep. U.S. Geol. Sur.*, I. (1896), p. 687.

Crystallization in a Ternary Magma.—When a magma consists of three constituents, *A*, *B*, *C*, the diagram used to illustrate the binary magma is no longer applicable, and we must use instead a solid figure. First it is to be remarked that all possible mixtures of three constituents can be represented by points within an equilateral triangle, the angular points of which represent the three constituents respectively (Fig. 59). Divide *BC* at *L* in the ratio $\gamma : \beta$, and divide *AL* at *P* in the ratio $\beta + \gamma : \alpha$; then *P* will represent a mixture of α parts of *A*, β of *B*, and γ of *C*. Draw lines *MN*, *QR*, *ST*

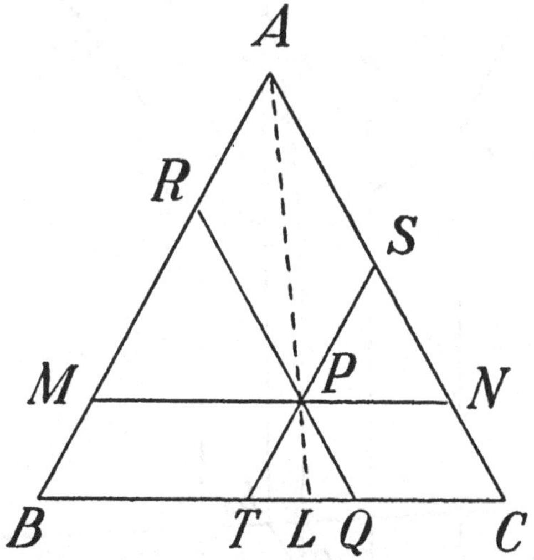

FIG. 59.—A TERNARY MIXTURE REPRESENTED BY A POINT WITHIN A TRIANGLE. (SEE TEXT.)

through *P*, parallel to the three sides of the triangle; then, if each side be taken as 100 units, and α, β, γ, be percentages of the whole, it is easy to show that each side of the triangle *PTQ* is α, of *PNS* β, and of *PRM* γ.

To represent at once composition and temperature, we take an equilateral triangle as the base of a solid figure, and erect vertical ordinates to represent temperatures (Fig. 60). Instead of freezing-point curves we have freezing-point surfaces, which meet, two by two, along binary eutectic lines, e_1e, e_2e, e_3e, and all meet in the ternary eutectic point e. If we start with the indicating-point at *n*, it will descend as the magma cools, until it meets one of the freezing-point surfaces. Suppose it to meet the surface belonging to *A* at the point *p*. Then (supersaturation being disregarded) the mineral *A* will begin to crystallize out, and the indicating-

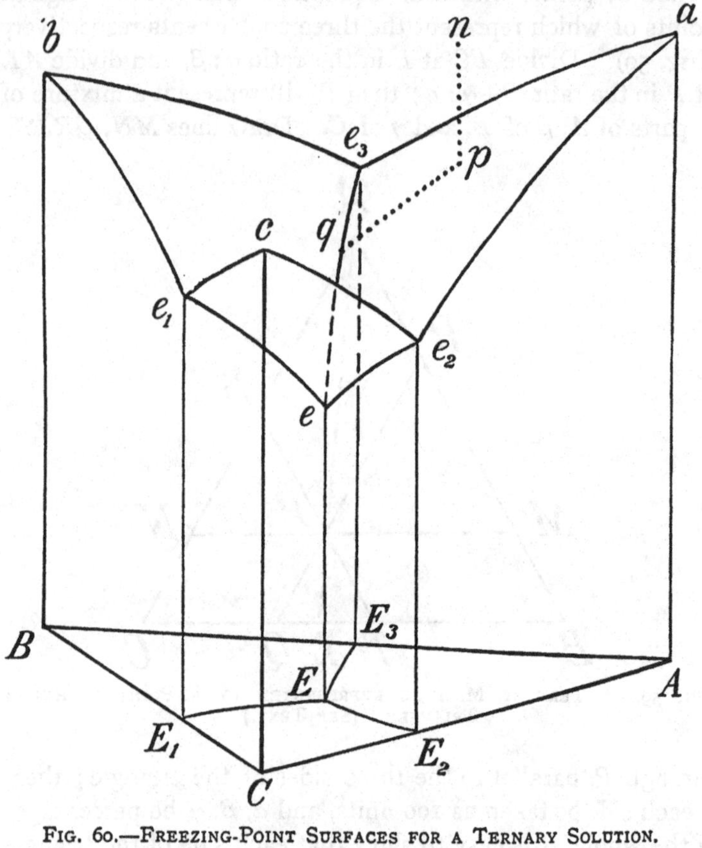

FIG. 60.—FREEZING-POINT SURFACES FOR A TERNARY SOLUTION.
Here *a*, *b*, *c* represent the melting-points (or freezing-points) of the pure minerals, *A*, *B*, *C*; e_1, e_2, e_3 the binary eutectic points for the pairs *B-C*, *C-A*, *A-B*; and *e* the ternary eutectic point for the three constituents.

point will travel along the surface on a line *pq*, directly away from *a*, until it meets one of the eutectic lines. If it meets the line e_3e (the *A-B* eutectic) at *q*, then *B* will begin to crystallize in eutectic proportions with *A*, and the indicating-point will travel down *qe*. When it reaches *e*, the three minerals will proceed to crystallize together at stationary temperature and in definite proportions, which are those of the ternary eutectic.

These relations, excepting as regards temperature, may be exhibited in a plane figure by projecting the various points and lines on the basal plane (Fig. 61). Further, let straight lines *AE*, *BE*, *CE* be drawn so that the whole triangle is divided into six smaller triangular fields. Now, if we represent the crystallization of the mineral *A* by (A), eutectic crystallization of *A* and *B* by (AB), and so on, the order of events is as follows, according as the point representing the initial composition of the magma falls into one or other of the six fields:

Field AE_3E	(A); (AB); (ABC).
Field AE_2E	(A); (AC); (ABC).
Field BE_3E	(B); (AB); (ABC).
Field BE_1E	(B); (BC); (ABC).
Field CE_2E	(C); (AC); (ABC).
Field CE_1E	(C); (BC); (ABC).

If the eutectic ratio of any pair of constituents remained the same in the ternary magma as in the binary, E_1E would be a straight line passing, when produced, through *A*, and similarly E_2E through *B* and E_3E through *C*. The geometrical condition that three straight lines, such as AE_1, BE_2, CE_3, shall meet in one point is:

$$\frac{AE_3}{E_3B} \cdot \frac{BE_1}{E_1C} \cdot \frac{CE_2}{E_2A} = 1;$$

and this condition is not fulfilled by the eutectic ratios of three minerals taken in pairs. For instance, taking the approximate values found by Vogt for olivine, diopside, and anorthite, we have:

$$\frac{32}{68} \cdot \frac{35}{65} \cdot \frac{70}{30} = 0{\cdot}59.$$

As a corollary, we may note incidentally that it is not possible to calculate the eutectic proportions of the pair *B-C* from those of the pairs *A-B* and *A-C*. The eutectic proportions between two constituents, *A* and *B*, are not only altered by the presence of a third constituent *C*, but they are altered differently by different proportions of *C*; so that the binary eutectic lines in the ternary diagram are curved (compare also Fig. 51 above).

It follows from the theory of solutions, as developed especially by Nernst, that the relations between two constituents are disturbed by the presence of a third in consequence of ionization. This action comes into play more especially in dilute solutions, and may, therefore, have an important application to the accessory minerals of igneous rocks. The chief case is that in which the third constituent

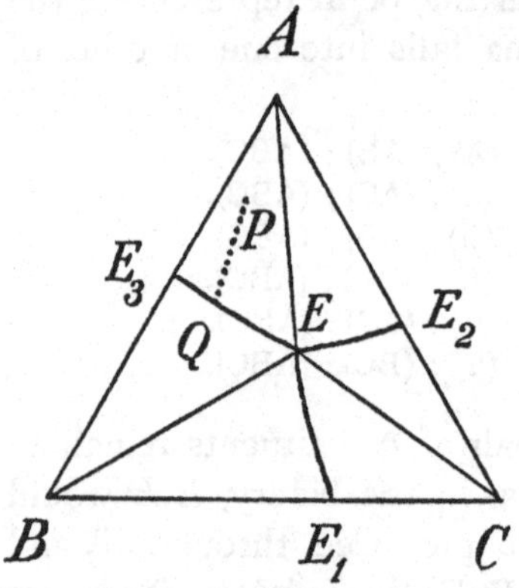

FIG. 61.—PROJECTION OF FIG. 60 ON ITS BASAL PLANE.

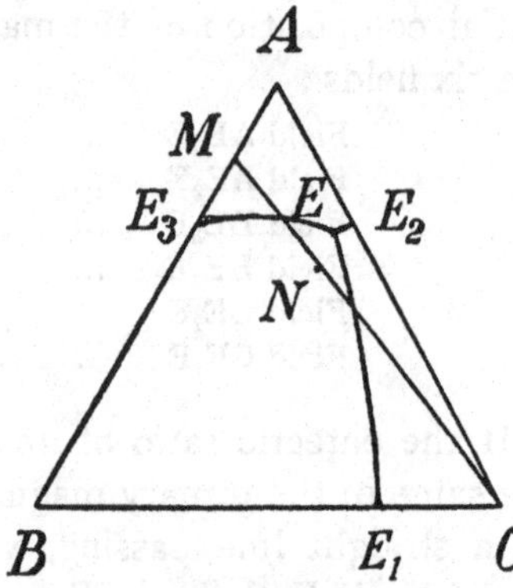

FIG. 62.—DIAGRAM TO ILLUSTRATE THE EFFECT OF A COMMON ION BETWEEN *B* AND *C*.

has *one ion in common* with one of the other constituents. A salt in aqueous solution has its solubility reduced by the addition of another salt or acid having a common ion with it—*e.g.*, sodium chloride is precipitated from a saturated solution by the addition of hydrochloric acid. Vogt[1] has applied this principle to the case of igneous rock-magmas. Suppose that the properties of two minerals, *A* and *B*, are such that the eutectic point for mixtures of these two (E_3 in Fig. 62) lies nearer to *A*. This will be the case, other things being equal, if *B* has a much higher melting-point than *A*. If *C* be a third constituent which has an ion in common with *B*, then,

[1] *Silikatschmelzlösungen*, II. (1904), pp. 151-156.

from what has just been remarked, it appears that the eutectic line E_3E must turn sharply towards the side AC of the triangle, and the ternary eutectic point E may be expected to lie very near to that side. From a mixture of A and B alone, very rich in A, so that the point M representing its composition falls between E_3 and A, the mineral A would crystallize first. But if any noteworthy amount of C be present in addition, the composition is altered to that represented by such a point as N, lying on the other side of E_3E. The entrance of C will thus cause B to crystallize first, even when its total amount in the magma is very small.

Vogt illustrates the principle by the case of spinel. This has a very high melting-point (about 1900° according to Brun), and for binary solutions of spinel with a silicate, such as felspar, the eutectic point doubtless lies near the latter mineral. Nevertheless, the eutectic mixture can scarcely contain less than 10 *per cent.* of spinel. In igneous rocks, however, spinel has always crystallized before felspar, even when present to the extent of 1 *per cent.* or less. This is ascribed to the presence of such minerals as olivine and pyroxenes, which have one ion (Mg) in common with spinel. Similar reasoning applies to apatite, perofskite, or sphene in presence of calcium-silicates, zircon in presence of another orthosilicate, and other cases. Vogt does not limit the application of the principle to the minor accessory minerals, but supposes that the crystallization of olivine may be accelerated by the presence of pyroxene in the magma, or that of biotite by the presence of orthoclase. The application of the law of the common ion to rock-magmas has been criticized, but provisionally it affords an explanation of facts which would otherwise appear anomalous. The ultrabasic plutonic rocks of the Isle of Rum illustrate this. We have remarked (p. 171) that, in those varieties which consist essentially of olivine and felspar, the order of crystallization depends on the relative proportions of the two minerals according to the eutectic law. If, however, pyroxene is present in noteworthy amount, the olivine has invariably crystallized first.

It is to be observed that the effect of the common ion is such as to throw the eutectic composition still farther away from a constituent having a high melting-point, and thus to augment the influence of melting-point in determining the eutectic ratios (p. 193). It is thus a factor which tends to make the order of crystallization conform *more nearly with the inverse order of fusibility*. Our very slender information concerning ionization in molten rock-magmas does not warrant any more dogmatic assertion. It may be assumed, however, that, in so far as this action can be counted efficient, it will be especially important in the case of accessory minerals, which, being in very dilute solution, may be expected to have a relatively high degree of dissociation. Clearly it cannot apply to a mineral (such as quartz) which is incapable of ionization. It is interesting to observe that, in so far as the action here considered is operative, the more complex the composition of a rock, the more will it tend to exemplify Rosenbusch's empirical order of crystallization.

CHAPTER IX

SUPERSATURATION AND DEFERRED CRYSTALLIZATION

Supersaturation.—Crystallization in a supersaturated magma.—Rate of growth of crystals.—Grain of igneous rocks.—Viscosity in rock-magmas.—Glassy rocks.—Devitrification.

Supersaturation.—In discussing as illustrative cases the progress of crystallization in binary and ternary magmas, and the modifications introduced by imperfect mutual solubility of the constituents in the fluid state, we have simplified the question by assuming that when the magma becomes saturated with a given mineral, that mineral forthwith begins to crystallize out. The temperature of saturation is that at which crystals and magma will be in equilibrium with one another; but we know that in the absence of the crystalline phase equilibrium is not, in fact, attained, and that, unless crystals be present to inoculate the transformation, some degree of *supersaturation* is necessary to initiate crystallization. In other words, the indicating-point in Fig. 53 must fall below the freezing-point curve before crystallization will set in. The curve *ae*, representing saturation of the magma with the mineral *A*, divides the region of under-saturation above it from that of supersaturation below it. Taking the two curves for a binary magma, and prolonging them[1] through *e*, we divide the diagram into four fields (Fig. 64), which represent different conditions of the magma in respect of saturation; thus:

[1] Meyerhoffer, *Zeits. physik. Chem.*, vol. xxxvi. (1902), pp. 593-597.

I. Under-saturated with both *A* and *B*.
II. Under-saturated with *B*, supersaturated with *A*.
III. Under-saturated with *A*, supersaturated with B.
IV. Supersaturated with both *A* and *B*.

There is some difference of opinion among physicists concerning the precise conditions which obtain in an under-cooled liquid or a supersaturated solution. The production of a new phase must always begin, as Willard Gibbs pointed out, at isolated centres. These centres may be numerous, and are not necessarily equally distributed. In a cooling liquid the centres of crystallization will be most numerous at the boundary. The number of centres of crystallization initiated in unit volume during unit time is taken by Tamman[1] as a measure of the *power of spontaneous crystallization*. He has investigated this quantity, and its dependence on the degree of undercooling, for a number of organic bodies, and Doelter[2] has experimented in the same way upon some of the rock-forming minerals. As the temperature falls below the melting-point, the power of crystallization is found to be at first extremely small. It increases with further cooling, and rises rapidly to a maximum value at a certain temperature. It then declines, at first rapidly, and then more gradually, to zero (Fig. 63, I).

The very slow generation of crystals with a small degree of undercooling and the more rapid generation with a greater degree correspond with the *metastable* and *labile* states of a solution as understood by Ostwald. Miers,[3] however, has been led by experiments on aqueous solutions of salts to the

[1] *Kristallisieren und Schmelzen* (1903), pp. 148-156. See also *Zeits. für physik. Chem.*, vol. xxv. (1898), pp. 441-479.

[2] *Physikalisch-chemische Mineralogie* (1905), pp. 111, 112.

[3] Address to Geol. Section Brit. Assoc., *Geol. Mag.*, 1905, pp. 520-523; Miers and Chevalier, *Min. Mag.*, vol. xiv. (1906), pp. 123-133; Miers and Miss Isaac, *Proc. Roy. Soc.* (A), vol. lxxix. (1907), pp. 322-351; Miers, *Science Progress*, vol. ii. (1907), pp. 121-134. Subsequent papers illustrate the application to a ternary solution and to an isomorphous series; Miss Isaac, *Trans. Chem. Soc.*, vol. xciii. (1908), pp. 384-411; Miers and Miss Isaac, *ibid.*, pp. 927-936.

conclusion that the power of spontaneous crystallization in the metastable state is *nil* (Fig. 63, II). He found that, inoculation being prevented, crystals appeared only when the solution reached the labile state, and were then suddenly generated in large numbers. On this view crystallization will begin in a binary solution only when the indicating-point reaches one of the 'supersolubility curves,' the super-solubility curve for either constituent lying at a certain

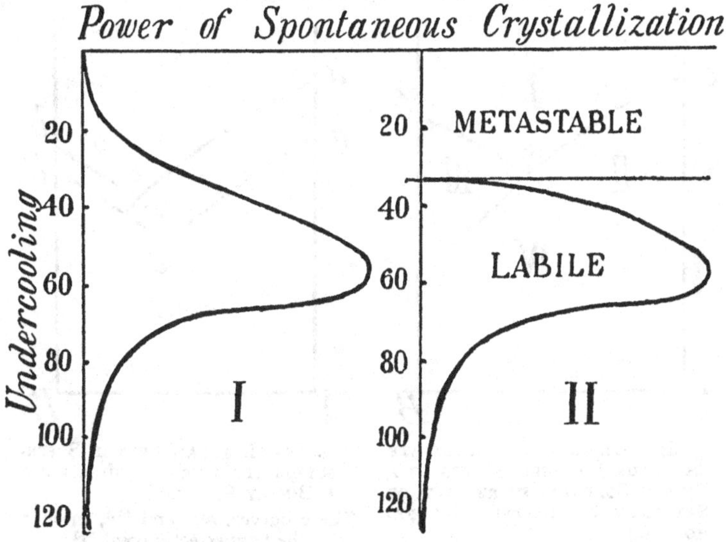

FIG. 63.—CURVE SHOWING THE VARIATION OF THE POWER OF SPONTANEOUS CRYSTALLIZATION OF AUGITE WITH DIFFERENT DEGREES OF UNDER-COOLING.

I, As obtained by Doelter; II, as modified to accord with Miers' theory. Temperature is measured downward from the melting-point, and power of spontaneous crystallization is measured to the right.

distance below the solubility (or freezing-point) curve, and sharply dividing the metastable from the labile region (Fig. 65).

Physicists who reject the rigid distinction proposed by Miers maintain that, while the power of spontaneous crystallization in the metastable region may be very small, yet the initiation of crystals in a solution, however slightly supersaturated, is only a matter of time. In most cases with which the petrologist is concerned ample time may be assumed, and the actual characters of igneous rocks certainly accord with Tamman's view better than with that of Miers. The evidence from the structures of rocks seems, indeed, to indicate that only under rather exceptional conditions have the magmas attained a sufficient degree of undercooling to approach the maximum of spontaneous generation of crystals, and so to exhibit something analogous to the suddenly

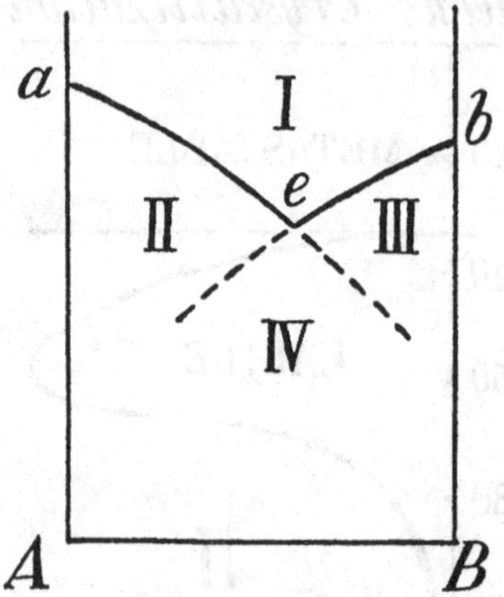

FIG. 64.—DIAGRAM TO ILLUSTRATE THE FOUR POSSIBLE STATES OF A BINARY SOLUTION IN RESPECT OF SATURATION. (AFTER MEYERHOFFER.)

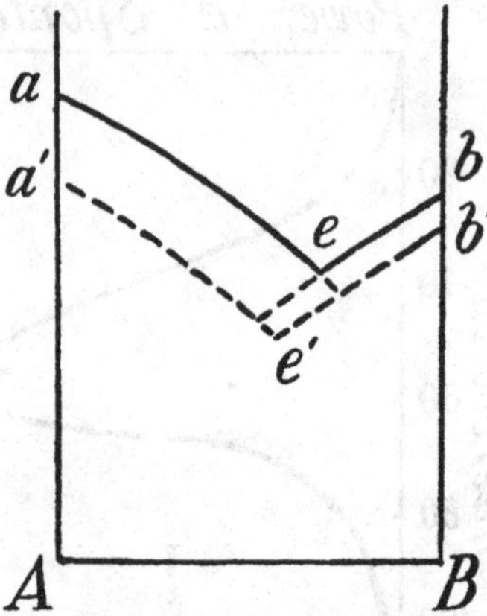

FIG. 65.—DIAGRAM OF THE SUPERSOLUBILITY CURVES (MIERS) FOR A BINARY SOLUTION.

These curves, a^1e^1 and b^1e^1, meet in the 'hypertectic point,' e^1.

formed cloud of minute crystals which, in Miers' experiments, signalises entry on the labile state.

It is easy to see that inoculation cannot, in the general case, have any important effect on the course of crystallization in igneous rock-magmas. A body of rock-magma is of a very different order of dimensions from the contents of a test-tube, and—saving such possibilities as those arising from 'stoping' action and the inclusion of 'xenoliths'—only the immediate margin of the mass of liquid can come in contact with extraneous solid crystals. The viscosity of the magma must in general prevent any distribution of such foreign elements through the general mass; and, indeed, the experiments of Day and Allen on the felspars show that viscosity may be an effectual check upon inoculation even within the limits of a crucible.

Crystallization in a Supersaturated Magma.—Doelter's experiments show that for a given mineral, crystallizing from simple fusion, the power of spontaneous crystallization (*i.e.*, the rate at which new crystals are generated) depends upon the degree of undercooling, attaining a maximum at a certain distance below the melting-point—*e.g.*, for augite at about 55° of undercooling. Comparative experiments show that this maximum value is very different for different minerals. The order of relative facility of crystallization for some common minerals was found to be—(*a*) hæmatite and spinel; (*b*) magnetite, olivine, hypersthene, and augite; (*c*) nepheline and anorthite; (*d*) labradorite and leucite; (*e*) diopside; to which may be added (*f*) albite and orthoclase, which cannot be made to crystallize from 'melts' of their own composition.

These results apply to the minerals taken separately, and their power of spontaneous crystallization in mixed rock-magmas is doubtless different. Tamman's[1] researches on organic bodies prove that even a small amount of another substance may considerably increase the power of spontaneous crystallization, and lower the temperature at which it is reached. Even an insoluble powder may have this effect.

We have seen sufficient grounds for believing that, unless the cooling of a magma be extremely slow, a noteworthy degree of supersaturation may be reached before crystals form in any considerable number, and that the requisite degree of supersaturation may differ widely for different minerals. This modifies to a greater or less extent the progress of crystallization as discussed in the preceding chapter, and may even affect the order of sequence of the several minerals. The magma, especially if of composition not very far from the eutectic, may become supersaturated with more than one constituent simultaneously, passing (in the case of a binary magma) into region IV. of Meyerhoffer's diagram (Fig. 64). The crystallization of one or other con-

[1] *Kristallisieren und Schmelzen* (1903), Figs. 53, 54.

stituent first may thus depend upon other considerations than their relative amounts—a case which seems to be realised in some of Doelter's experiments on mixtures of labradorite and augite.[1] In Miers' view (Fig. 65), the order of the *beginning* of crystallization would depend on the composition with respect to the 'hypertectic,' not the eutectic. Briefly, a mineral may be promoted in the order of crystallization if it crystallizes comparatively readily. The order of facility of crystallization, as given by Doelter's experiments, agrees generally with Rosenbusch's empirical order of crystallization, and helps to explain why this so often holds good.

Consider now the case of a binary magma, which cools not too slowly to permit of noteworthy supersaturation. Let there be an excess of the constituent *A* as compared with the eutectic proportions. There will be no appreciable crystallization of this mineral until the indicating-point has fallen, say, to *q*, at a distance below *ae* (Fig. 66) which depends both on the specific properties of the mineral and on the rate of cooling. Crystallization is accompanied by a liberation of heat, which eventually more than counterbalances the loss by conduction, so that the temperature rises. When the indicating-point meets the curve *ae* (at *r*), equilibrium is established between crystals of *A* and the magma, and the crystallization of *A* proceeds, with falling temperature, along *re*. Further, *A* continues to crystallize alone for some distance beyond *e*, until the magma is sufficiently supersaturated with *B* to start an appreciable crystallization of that constituent, say at *s*. The consequent rise of temperature causes the magma to become *under-saturated* with *A*, and there ensues a *resorption* of some part of the crystals of that mineral already formed. This is at first limited by the fact that the heat thus rendered latent must be supplied by the crystallization of *B*, which must also raise the temperature. When equilibrium for *B* is established, at *t*, the crystallization

[1] *Physikalisch-chemische Mineralogie* (1905), p. 131. Lenarčič found that from a mixture of 1 part of labradorite with 2 of augite the former mineral crystallized first if the magma was undisturbed, but the latter with stirring. See *Centralbl. für Min.* (1903), pp. 705-722, 743-751.

of this and the resorption of *A* proceed, with falling temperature, along *te*. At *e* there is now equilibrium for both constituents, since the magma contains crystals of both, and the crystallization proceeds thenceforth in eutectic proportions with stationary temperature.

We have remarked that, if the composition of the magma is not far from the eutectic, and if the constituents differ considerably in respect of specific power of spontaneous crystallization, there may even result a reversal of order. This is

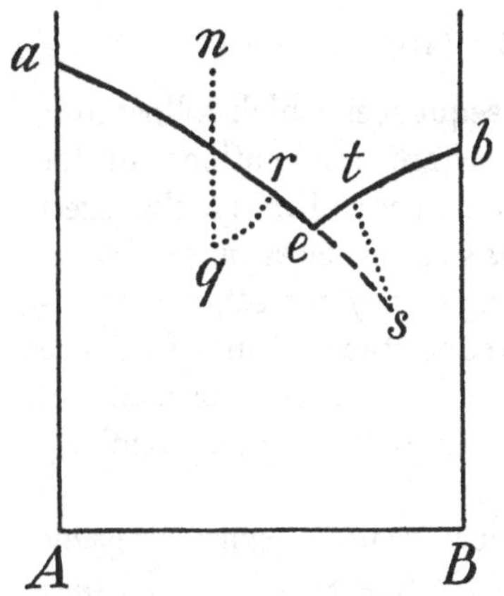

FIG. 66.—PROGRESS OF CRYSTALLIZATION IN A BINARY MAGMA WITH SUPERSATURATION.

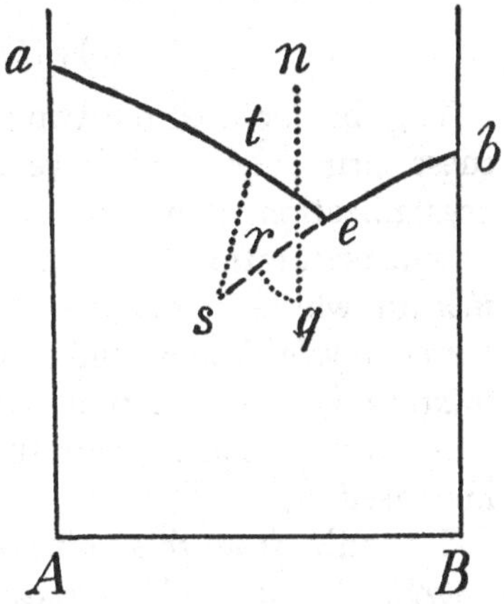

FIG. 67.—ORDER OF CRYSTALLIZATION REVERSED IN CONSEQUENCE OF SUPERSATURATION.

illustrated for a binary magma in Fig. 67. Here it is supposed that there is initially an excess of *A* over eutectic proportions, but that a notably greater degree of supersaturation is needed to cause appreciable crystallization of *A* than of *B*. The successive stages are then: crystallization of *B* from a supersaturated solution (*qr*), and then from a saturated solution (*rs*); crystallization of *A* (with partial resorption of *B*) from a supersaturated solution (*st*), and then from a saturated solution (*te*); finally, eutectic crystallization of *A* and *B* (at *e*).

The foregoing results may be represented summarily as follows: Let (A) stand for crystallization of the mineral *A*,

(−A) for resorption of the same, and (AB) for eutectic crystallization. Then, for a binary magma having an excess of *A* over eutectic proportions, the progress of crystallization, if supersaturation were negligible, would be:

(A); (AB);

but, owing to the intervention of supersaturation, the actual order of events is:

(A); (B)(−A); (AB);

or in certain circumstances it may even be:

(B); (A)(−B); (AB).

Vogt has pointed out two consequences which follow from these principles, and have important applications to the crystallization of natural rock-magmas. Firstly, the effect of supersaturation is to *extend the stage of successive crystallization* (in whatever order) *at the expense of the stage of simultaneous crystallization*, and so to reduce the amount of eutectic mixture in the resulting rock. In a magma consisting of three or four constituents the effect will be correspondingly increased.

Secondly, from this point of view we must regard *magmatic resorption as a regular incident of the progress of crystallization* in a magma; so that it does not necessarily argue any discontinuous change of external conditions, such as a sudden relief of pressure.

Rate of Growth of Crystals.—Distinct from the 'power of spontaneous crystallization' of a mineral—*i.e.*, the rate at which new centres are started—is what Tamman terms the 'speed of crystallization'—*i.e.*, the rate at which crystals, when once initiated, continue to grow. This may be considered with reference either to volume or to linear dimensions; but we shall have regard more particularly to the latter, the rate of voluminal increase being more difficult to evaluate in practice. The rate of linear growth of a crystal is obviously a vector quantity, differing for different directions. It is a function of the temperature, or, more precisely, the degree of undercooling, and the rates of growth in different directions are different functions of the undercooling. If this were not so, the form and habit of a crystal (of a given mineral) would be independent of the temperature of its formation, which is clearly not the case. Experiment shows that crystals of salts slowly built up in a solution only slightly supersaturated differ in habit from those which grow rapidly in a solution considerably supersaturated. The former are commonly much richer in faces, while the latter are of simpler habit and characteristically acicular. It is remarkable that even salts crystallizing in the regular system, such as alum,[1] assume the acicular habit when formed rapidly in a strongly supersaturated solution. The explanation is probably that given by Miers.[2] The solution in immediate contact with a growing crystal is reduced in concentration, and within this envelope of weaker supersaturation growth can proceed only slowly. But, if at any point the growing crystal pierces this envelope, it may proceed to grow out in that direction in the more strongly supersaturated solution, so rapidly that the envelope cannot close round it. The crystallites so common in many glassy rocks doubtless arise in this way, and are indicative of a high degree of supersaturation in the magma. They appear to have been initiated suddenly, in immense number, at a stage of undercooling when the power of spontaneous crystallization increased very rapidly, and perhaps correspond with the cloud of little crystals produced in Miers' experiments on the passage from the 'metastable' to the 'labile' condition.

The principal vector rate of growth, or linear 'speed of crystallization,' depends on the degree of undercooling in somewhat the same way as the 'power of spontaneous crystallization' does. It is at first very small, then rises to a maximum, and finally falls off to zero. It has been studied for various organic bodies by Tamman,[3] and for some of the rock-forming minerals by Doelter.[4] Minerals grow

[1] Chevalier, *Min. Mag.*, vol. xiv. (1906), pp. 134-142.
[2] *Science Progress*, vol. ii. (1907), pp. 128, 129.
[3] *Kristallisieren und Schmelzen* (1903), pp. 131-148.
[4] *Physikalisch-chemische Mineralogie* (1905), pp. 105-110.

very slowly, the maximum speed for augite being about 0·001 to 0·002 mm. per minute. This is one of the most rapidly crystallizing minerals, the relative order for some of the others being: (*a*) augite and olivine rich in iron; (*b*) labradorite and olivine poor in iron; (*c*) nepheline; (*d*) leucite; (*e*) magnetite. These results apply to minerals crystallizing from pure 'melts': in a mixed rock-magma the rate of growth of crystals may be considerably modified. Thus, in a mixture of augite, labradorite, and olivine, Doelter found that the augite crystals grew more slowly than those of labradorite.

While the comparative results obtained by Doelter are instructive, the actual rates of growth observed must depend largely on the conditions of the experiments, and it seems very doubtful whether the absolute values thus found can safely be used to calculate the rate of growth of minerals and rocks in nature. Doelter calculates that the large leucite-crystals in the lavas of Vesuvius might form in from 800 to 1,600 hours, supposing that the temperature did not fall more than about 10° during the time. A fine-grained basalt, with crystals 0·10 to 0·25 mm. in diameter, might consolidate in twenty-four hours, and a medium-grained gabbro in about ten days, assuming the rate of cooling to be suitably adjusted. The assumption in the latter case is certainly inadmissible.

Grain of Igneous Rocks.—The size of the individual crystals in igneous rocks varies between wide limits. In a given rock the crystals of one and the same mineral may belong to two different orders of magnitude; but, setting aside for the present this case (porphyritic structure), the crystals of any particular mineral in a given rock do not usually vary much from an average size. For different minerals in one rock the average size may be very different; and this must be ascribed to the specific properties of the minerals and the different proportions in which they enter. Greater differences are found when we compare one rock with another, even of like mineral composition; and such differences, affecting more or less all the principal constituents, must be referred to the conditions under which crystallization was effected. The average size of the crystals of the principal minerals of a rock determines what is understood as its *grain*—coarse or medium or fine in different rocks.

In the simplest ideal case, that of a rock composed wholly of one mineral, it is clear that, if n be the number of crystals in unit volume of the rock, and v the average volume of a crystal, then $nv = 1$. The grain of the rock, therefore, depends solely, in inverse proportion, upon the total number of centres of crystallization initiated during the process of consolidation. This number depends in rather a complex manner upon the specific properties of the mineral and the rate of cooling. Suppose a body of magma to cool gradually by conduction, the heat thus lost by unit mass in unit time being K. Let us call the amount of magma which passes into the crystalline state (also for unit mass and unit time) the 'rate of crystallization,' and denote it by r. This depends on the rate at which new crystals are started, and also on the rate at which they continue to grow; and both of these quantities at first increase with undercooling, so that r likewise increases. The temperature will continue to fall until the loss of heat by conduction is balanced by the heat liberated in crystallization—that is, until $rL = K$, L being the latent heat of fusion. Then, as the heat liberated becomes greater than that lost by conduction, the temperature will rise. When it reaches the melting-point, equilibrium is established between crystals and magma; and thenceforth the existing crystals will continue to grow, but no new ones will be initiated.

The connection of the grain of the resulting rock with the rate of cooling of the magma thus becomes plain. The less the degree of supersaturation reached, and the sooner equilibrium is established, the fewer will be the centres of crystallization started, and the larger the completed crystals. The deeper the temperature falls into the region of relatively rapid spontaneous formation of crystals (Fig. 63, I), and the longer it remains there, the more numerous will be the distinct centres of crystallization, and the smaller the resulting crystals. It is evident from the foregoing considerations that *slow cooling will cause larger crystals and more rapid cooling smaller crystals.*

A rock composed of several minerals presents a less simple case. As regards the chief *essential minerals,* it is to be remarked that each has its specific properties modified by the presence of the other minerals. The power of spontaneous crystallization is presumably increased, as in Tamman's experiments; but it seems probable that the rate of growth of crystals may often be lessened by viscosity, as a result of the lowering of melting-points. This implies a finer grain in more complex than in simpler rocks, otherwise comparable, and the facts seem to be in accord with this generalisation. At the same time, the differential influence of specific properties is often very apparent on a comparison of different minerals in the same rock. The conspicuous size of the olivine crystals in many basalts is an example. These are not porphyritic crystals, for there is in general no second generation. Olivine stands high in the list of minerals as regards both power of spontaneous crystallization and rate of growth, and probably it does not reach any high degree of supersaturation.

The *accessory minerals* stand on a somewhat different footing. They are in dilute solution in the magma, if, indeed, they are always freely soluble in it, and no noteworthy degree of supersaturation can be supposed. There are apparently no experimental data relative to the effect of dilution on the power of spontaneous crystallization or the rate of growth of crystals. Such minerals as apatite, zircon, chromite, sphene, etc., commonly occur in very minute crystals even in coarse-grained rocks; but, where such a mineral is exceptionally abundant, it builds at the same time exceptionally large crystals. Vogt[1] makes a like observation in the case of slags. Spinel to the extent of 0·5 *per cent.* in a slag gave crystals averaging 0·01 mm. in diameter, but when the mineral amounted to 6·5 *per cent.*, the crystals

[1] *Silikatschmelzlösungen*, II. (1904), p. 164.

averaged 0·20 mm. We have for such a case, not $nv=1$ (p. 217), but $nv=p$, where p is the proportion by volume which the mineral makes in the whole rock. If v increases more rapidly than p, as in Vogt's slags and apparently in natural rocks, it follows that n diminishes as p increases—*i.e.*, the total number of crystals formed is increased by dilution, which is probably true.

From what has been said, it is evident that the grain of an igneous rock depends in part on its composition, but more on the conditions of its consolidation. The clearest proof of this is that one and the same rock-body may vary greatly in grain in different parts. In particular, a mass which is relatively coarse in the interior often has a *fine-grained margin*, and sometimes apophyses of still finer texture. Here the composition and initial temperature of the magma were the same throughout, and no significant difference of pressure can be supposed; so the *rate of cooling* has clearly been the ruling condition. The relatively rapid cooling at the margin of an intrusion depends on a great initial difference of temperature between the magma and the contiguous solid rocks. If the country-rocks were already heated prior to the intrusion, there is no such effect; and in this case the intrusive rock may exhibit, for other reasons, a coarser instead of a finer texture at the margin (Chapter XII.). As an illustration of a relatively fine-grained margin, we take the great intrusive sill of the Palisades, near the Hudson River. Queneau examined specimens taken at different distances from the contact-surface, and measured the areas of the crystals in thin slices. The results are presented graphically in Fig. 68. They extend to about 100 feet in a total thickness of 700 or 800 feet.

The subject has been discussed especially by Lane,[1] who

[1] *Bull. Geol. Soc. Amer.*, vol. viii. (1896), pp. 403-407; *Geol. Rep. on Isle Royale* (*Geol. Sur. Mich.*, vol. vi., 1898), pp. 106-151; *Amer. Journ. Sci.* (4), vol. xiv. (1902), pp. 393-395; *Bull. Geol. Soc. Amer.*, vol. xiv. (1903), pp. 369-406; *Ann. Rep. Geol. Sur. Mich.* for 1903 (1904), pp. 205-237; *Do.* for 1904 (1905), pp. 147-153, 163; *Amer. Geol.*, vol. xxxv. (1905), pp. 65-72; *Journ. Canad. Mining Inst.*, vol. ix.; 'Die Korngrösse

treats the problem simply as one of conduction. He assumes that an intruded sill or dyke cools in the same way as a slab of metal heated to a uniform temperature and then placed in contact with colder bodies. Such an ideal case might represent closely enough the cooling of a thoroughly viscous magma or of a solid rock; but it presents no just parallel to a highly superheated magma (which Lane supposes possible),

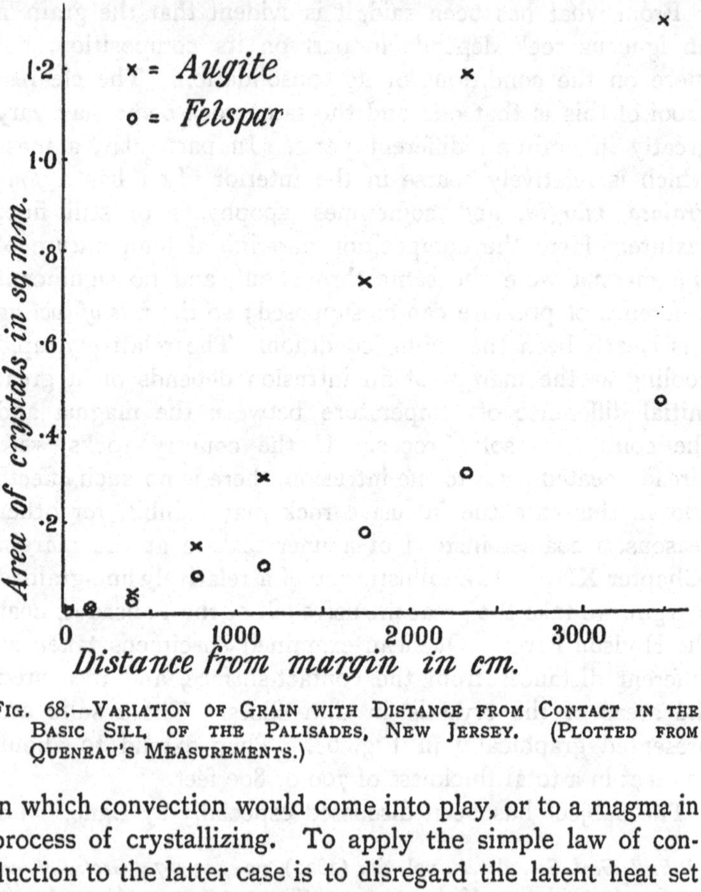

FIG. 68.—VARIATION OF GRAIN WITH DISTANCE FROM CONTACT IN THE BASIC SILL OF THE PALISADES, NEW JERSEY. (PLOTTED FROM QUENEAU'S MEASUREMENTS.)

in which convection would come into play, or to a magma in process of crystallizing. To apply the simple law of conduction to the latter case is to disregard the latent heat set free, which plays, as we have seen, a very essential part. If,

der Auvergnosen,' *Suppl. to Rosenbusch Festschrift* (1906). Also Queneau, *Sch. of Mines Quarterly*, vol. xxiii. (1902), pp. 181-195 : this paper requires numerous corrections; see Lane, *loc. cit. sup.* (1902).

for instance, we may suppose a given magma to crystallize within a temperature-range of, say, 300°, and if we take the average specific heat within that range as $\frac{1}{3}$, and the average latent heat of fusion as 100 calories, then only one half of the heat lost by conduction during that range represents cooling, while the other half is the heat liberated in crystallizing.

Lane discusses the different cases which may arise, according as the temperature-range of crystallization of a given mineral is related to the initial and final temperatures, and compares the results of calculation and observation; but it does not appear that such measure of agreement as is found can be regarded as significant.

Viscosity in Rock-Magmas.—We have seen that the power of spontaneous crystallization, after attaining a maximum value at some degree of undercooling, diminishes with further fall of temperature, and at last becomes insensible. Since the anisotropic state must become increasingly unstable, the checking of crystallization must be ascribed to growing viscosity. This checks likewise the growth of crystals already formed, and is to be regarded as a very important property in relation to the crystallization of rock-magmas or their consolidation as glasses.

The viscosity of a fused mineral under given conditions—say at its melting-point—is a specific property which varies greatly, the extreme being found in such minerals as the alkali-felspars. Mixed magmas have not such extreme viscosity under ordinary conditions, though they are doubtless less fluid than aqueous solutions of salts. From the rate of flow of basaltic lavas at Kilauea, Becker[1] deduced a viscosity fifty times as great as that of water. The viscosity, of course, depends on the composition of the magma, and in this connection Vogt's[2] observations on silicate-slags are of interest. He found that in general viscosity increases

[1] *Amer. Journ. Sci.* (4), vol. iii. (1897), p. 29.

[2] *Zeits. prakt. Geol.*, 1893, p. 275. For a fuller investigation see Greiner, *Ueber die Abhängigkeit der Viskosität in Silikatschmelzen von ihrer chemischen Zusammensetzung*, Inaug. Diss., Jena, 1907.

with acidity, and rather rapidly when the silica-percentage exceeds 58 or 60. On the other hand, ferrous oxide promotes fluidity strongly, and magnesia in a less degree. Lime, and probably soda also, tend to promote fluidity; but even a small content of potash, in an acid or moderately acid magma, imports a noteworthy degree of viscosity. A small content of alumina has little influence, but with a high percentage of alumina in any of the more acid magmas the viscosity is very marked. Translating these results into mineralogical terms, we may expect that in a rock-magma free silica, potash-felspar, and felspathoid minerals will tend to viscosity, while augite, and especially olivine and magnetite, will tend to fluidity. These indications, further enforced by some of Doelter's[1] experiments, seem to be borne out by natural lavas; for basalts in general flow more freely than rhyolites, while some trachytic and phonolitic lavas are extremely viscous.

It must be remembered, however, that the silicate-magmas of Vogt and Doelter were without water or other 'mineralising agents,' and that natural lavas must lose much of their content of those constituents under the conditions of extravasation. Magmas contained in intercrustal reservoirs and intruded among solid rocks are, in this respect, in different circumstances, and we cannot safely extend to them inferences deduced from the behaviour of slags or even of natural surface outpourings. It cannot be doubted that the water in rock-magmas is a highly important factor in reducing viscosity, and that in some magmas other volatile substances contribute to the same result. On such action depends, at least in part, the efficacy of these substances in the artificial reproduction of minerals otherwise refractory. If, as there is reason to suppose, these 'mineralisers' or fluxes are, under intratelluric conditions, more richly present in acid than in basic magmas, it is probable that their presence affects the relative as well as the absolute viscosity of the magmas.

Phenomena familiar to all field-geologists make it clear

[1] *Tscherm. Min. Petr. Mitth.* (2), vol. xxi. (1902), p. 217.

that *intruded rock-magmas* are not necessarily very viscous. It may often be seen that a magma has penetrated narrow fissures in the rocks invaded, or even traversed them in a system of almost microscopic threads, thus evincing a high degree of fluidity; and this behaviour seems to be more common in acid than in basic magmas. An equally strong argument, which need not be elaborated here, may be based on the phenomena of magmatic differentiation. Becker has urged that the viscosity of rock-magmas must always be so great as to be prohibitive of differentiation by diffusion, but this conclusion is at variance with plain facts. Brögger[1] considers that Becker's argument is justified as regards superficial lavas, which do not, in fact, show any noteworthy effects attributable to differentiation subsequent to extrusion, and he contrasts these with abyssal and hypabyssal rocks, in which such effects are widespread. The fact that differentiation is practically restricted to intruded magmas, as distinguished from extruded, he ascribes partly to the slower cooling of the former, but chiefly to their greater fluidity, the result of their retaining their natural content of water and other special fluxes.

Glassy Rocks.—It has been pointed out (p. 217) that the fall of temperature in a supersaturated magma is arrested only when the liberation of heat by crystallization becomes sufficient to counterbalance the loss of heat by conduction. If the cooling be too rapid, this condition is never realised, and the temperature continues to fall, passing from the range most favourable to the generation and growth of crystals into a lower range where these processes cease to be effective. In this case a larger or smaller proportion of the magma does not crystallize, but passes without any discontinuous change of state into a glass, viscosity becoming so high that it is indistinguishable from rigidity. Vitreous or partially vitreous rocks, then, are the result of *rapid cooling*.

Evidently it is to be expected that the glassy residuum in a partially vitreous rock will *approximate to eutectic composition*. But it also appears that an approximately eutectic composition

[1] *Eruptivgesteine des Kristianiagebietes*, III. (1898), pp. 336-339.

of the initial magma, other things being equal, favours bodily consolidation in the glassy form. For instance, varieties essentially glassy are much commoner among the rhyolites, which consist principally of alkali-felspar and quartz, than among the typical trachytes, which consist essentially of alkali-felspar alone. A glance at the analyses of those obsidians which contain only very small amounts of the ferro-magnesian minerals shows that the great majority have silica-percentages in the neighbourhood of 73–75, agreeing with a eutectic mixture of orthoclase and quartz. Vogt[1] remarks that, since in mixed magmas there is a mutual lowering of melting-points, deferring crystallization to lower temperatures, and since viscosity increases with fall of temperature, the formation of glass must be favoured by complexity of composition, and especially by an approximation to eutectic proportions in the magma. It is true that in a mixed magma there is not only a lowering of melting-points, but also a reduction of viscosity: the latter apparently is not sufficient to counterbalance the former. The special tendency of eutectic mixtures to yield glasses with rapid cooling is well illustrated by Vogt[2] in the case of slags consisting essentially of melilite and anorthite. Under like conditions mixtures in the neighbourhood of the eutectic consolidate as glass, while those richer in either constituent yield products more or less crystalline according to their composition. Vogt's graphical representation of the results is reproduced in Fig. 69. The melilites, being regarded as isomorphous mixtures of åkermanite, $(Ca,Mg)_4Si_3O_{10}$, and gehlenite, $(Ca,Mg)_3Al_2Si_2O_{10}$, give a line on the diagram; and the curves representing different degrees of crystallinity thus tend to be parallel to the melilite-line and concentric with reference to the anorthite point.

Most glassy rocks contain at least scattered crystals, and many contain innumerable minute crystallites. The former doubtless crystallized in the earlier stages of supersaturation, and the latter at a much later stage. This is sometimes

[1] *Silikatschmelzlösungen*, II. (1904), p. 166.
[2] *Ibid.*, I. (1903), pp. 69-77, Pl. II.

indicated by their mineralogical nature. In the well-known pitchstones of Arran, the ferro-magnesian phenocrysts are of the high-temperature mineral augite (sometimes also enstatite), but the little green crystallites are usually of the low-temperature mineral hornblende. The clear ring sur-

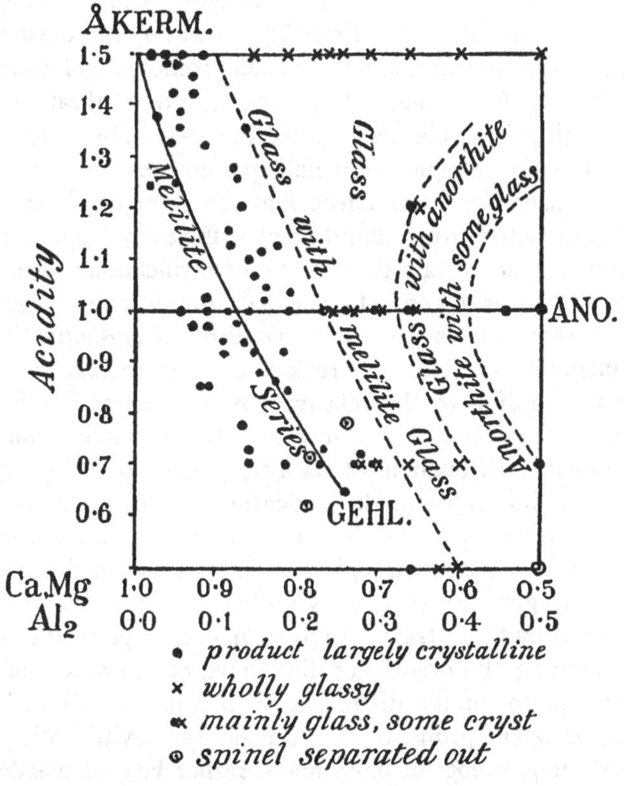

FIG. 69.—DIAGRAM FOR MELILITE-ANORTHITE-SLAGS. (AFTER VOGT.)

Explanation in the text. The ordinates represent acidity—*i.e.*, oxygen-ratio of silica to bases—and the abscissæ represent the atomic proportions of Al_2 and (Ca, Mg).

rounding each arborescent group bespeaks a considerable degree of viscosity at the crystallitic stage.

Devitrification.—Although a glassy rock is not theoretically in a stable condition, it may remain in the vitreous state even throughout geological ages. We know, indeed, very

15

few glassy rocks of Palæozoic age, though there are many which, by perlitic and other characteristic structures, are recognisable as devitrified glasses. This may perhaps indicate that the power of spontaneous crystallization at low temperatures, though infinitesimally small, is not absolutely nil, but may have a sensible effect during an immense lapse of time. Probably, however, the devitrification of these ancient glasses has been promoted by favourable conditions. Experiment shows that the devitrification of a silicate-glass is accelerated by heating it—*i.e.*, by restoring it to a temperature-region in which spontaneous crystallization is more sensible. The effect may be brought about by a moderate temperature maintained sufficiently long. Again, pressure must certainly promote devitrification, since the change involves in general a decided diminution of volume.

Devitrification is merely a *long-deferred crystallization* of the magma. In a complex rock it appears to take place in stages, according to the relative power of crystallization of the several minerals. For instance, the magnetite which is disseminated through most basaltic glasses probably represents the first stage of devitrification. The recent lavas of Kilauea, with 15 or 20 *per cent.* of iron-oxides, are quite clear, as is the artificial glass from the olivine-dolerite of Rowley Regis. Further, it is probable that crystallization, hampered by the extreme viscosity, may take place at various stages during the original cooling down of the rock; and we find no place for the distinction, which has sometimes been made, between 'primary' and 'secondary' devitrification.

Certain petrologists have made rather large demands on devitrification to account for the existing structures of crystalline rocks. Judd[1] has expressed the belief that crystals in igneous rocks have often grown, after the effective solidification of the mass, at the expense of a glassy groundmass, which may eventually have been wholly exhausted by this process. The fringing growths of micropegmatite, which

[1] 'On the Growth of Crystals in Igneous Rocks after their Consolidation,' *Quart. Journ. Geol. Soc.*, vol. xlv. (1889), pp. 175-186, Pl. VII. (with numerous references to literature).

surround felspar crystals in many acid rocks, he assigns to devitrification of this kind, and he would apparently regard micrographic and spherulitic structures in general as 'secondary' in this sense. The cavities in drusy granites he attributes to "the contraction which a more or less glassy ground-mass has undergone during devitrification." The growth of relatively large crystals, however, implies a freedom of molecular movement which is difficult to reconcile with the ordinary conception of a glass, and the phenomena on which Judd grounds his theory seem to be capable of quite other interpretation. In particular, he describes crystals of labradorite in an andesitic rock from Mull which have been corroded and fissured, and have acquired secondary glass-cavities, after which a clear border of new felspar-substance has been added. This he supposes to indicate a later outgrowth at the expense of a glassy ground-mass; but the peculiarities described are exactly those belonging to the 'xenocrysts' which are so common in these British Tertiary intrusions, the new border being formed from the new magma (Chapter XIV.). It is true that crystals of quartz, hornblende, felspar, etc., sometimes develop a new growth at ordinary atmospheric temperatures; but these truly secondary outgrowths are formed at the expense of other minerals, not of glass. They occur in sandstones, conglomerates, and tuffs, as well as in igneous rock-masses, and are not connected with devitrification, but with solution and chemical reactions.

CHAPTER X

ISOMORPHISM AND MIXED CRYSTALS

Isomorphism and mixed crystals. — Crystallization of isomorphous minerals in a rock-magma.—Roozeboom's five types of mixed crystals.—Double salts.—Isomorphism of the plagioclase felspars. —Imperfect isomorphism of orthoclase and plagioclase.—Relation of felspar crystals to magma. — Crystallization in a magma of felspars and quartz.—Spontaneous changes in crystals on cooling.

Isomorphism and Mixed Crystals.—In discussing crystallization from a binary magma under the simplest conditions, we supposed that the two minerals, *A* and *B*, could form only two kinds of crystals—pure *A* and pure *B*. We will now suppose that *A* and *B* are capable of forming mutual solutions (*i.e.*, homogeneous mixtures) in the crystalline state. There are two cases, according as the miscibility is unlimited or limited:

(i.) In *perfect isomorphism*, *A* and *B* form mixed crystals in any proportions, so that there is a complete series of possible varieties between the two end-members. The properties of these are continuous functions of the composition—say, of the percentage of *B* in the mixture.

(ii.) In *imperfect isomorphism* only certain mixtures are possible—viz., of *A* with a limited admixture of *B*, and of *B* with a limited admixture of *A*—so that there is an interrupted series, the hiatus in the middle part being greater or less in different cases.

Perfect isomorphism is found only between two minerals which agree very nearly both in molecular volume and in crystallographic elements. Usually, also, the two are of the same general chemical type (*e.g.*, $MgSiO_3$ and $FeSiO_3$, forming the enstatite series), but there may be perfect isomorphism without this (*e.g.*, between albite, $NaAlSi_3O_8$, and anorthite, $CaAl_2Si_2O_8$). Imperfect isomorphism is found between two minerals which differ somewhat more as regards either molecular volume or crystallographic elements. There may, indeed, be a certain miscibility, very narrowly limited, between two minerals which can scarcely be brought, even approximately, into the geometrical comparison implied in the term 'isomorphism.' In so far as mixed crystals arise in this way, we may regard them as an extreme case of imperfect isomorphism.

In some isomorphous series, one of the end-members (say *B*) is a 'labile' form; and the only stable mixtures are those with a preponderance of *A*, the other end of the series being wanting. A special case is that of 'isodimorphism.' Here we have to do with four minerals, related thus:

A	isomorphous with	(*B*)
dimorphous with		dimorphous with
(*A'*)	isomorphous with	*B'*.

The two forms indicated by parentheses are labile, so that the only mixed crystals which can exist are, on the one hand, *A* with an admixture of *B*, and, on the other, *B'* with an admixture of *A'*. They belong to the opposite ends of two different series, not, as in imperfect isomorphism, to opposite ends of the same series; but this distinction is of theoretical rather than practical value.

Isomorphism and imperfect isomorphism (whatever its theoretical significance) play a very important part among the rock-forming minerals. For instance, corresponding magnesian and ferrous compounds (metasilicates, orthosilicates, aluminates, etc.) appear to be perfectly isomorphous with one another, while between these, on the one hand, and the corresponding calcic compounds there is only imperfect isomorphism. It is primarily in virtue of isomorphous relations that so many of the rock-forming minerals fall into *groups*—such as the pyroxene group, the spinel group, etc.

Isomorphism serves in some degree to account for the relatively small number of distinct minerals found in igneous rocks. For instance, $MgSiO_3$ and $FeSiO_3$, in whatever proportions they bear in a molten magma, give rise by crystallization, not to two, but to one mineral. In particular, isomorphism may be regarded as providing a way of disposing of small quantities of the less common elements without necessitating the formation of special minerals—*e.g.*, manganese goes into pyroxenes, barium into felspars, lithia into micas, etc. In such cases a very imperfect degree of isomorphism is sufficient.

Crystallization of Isomorphous Minerals in a Rock-Magma.—We proceed to consider the progress of crystallization in the ideal case of a binary magma, consisting of two constituents, *A* and *B*, between which there is either perfect or imperfect isomorphism. It has been discussed from the theoretical standpoint by Bakhuis Roozeboom,[1] and the application of his results to the rock-forming minerals has been made, with such detail as the data permit, by Vogt.

In the case of perfect isomorphism, there are not two distinct freezing-point curves meeting in a eutectic point, but one continuous freezing-point curve (in Roozeboom's terminology the *liquidus*), as shown diagrammatically in Fig. 70. Further, the melting-point curve of the mixed crystals does not coincide with the freezing-point curve of the solution or magma, but lies below it as a distinct curve (the *solidus*), meeting the former one at *a* and *b*. It follows that a mixed crystal does not melt at a stationary temperature, but has a certain temperature-interval of melting instead of a fixed melting-point. It follows also that, for equilibrium between the magma and a mixed crystal, there must be a difference of composition between them. If the points *p* and *q*—the one on the liquidus and the other on the solidus—represent the composition of magma and crystals respectively, then for equilibrium these points must be on the same horizontal line, at a height representing the common temperature of magma and crystals. A crystallizing magma will therefore

[1] *Zeits. physik. Chemie*, vol. xxx. (1899), pp. 385-429.

produce *crystals of different composition from the magma* itself, the relation between the two being indicated by corresponding points on the liquidus and solidus.

Suppose, now, the magma, of composition represented by *n* (Fig. 70, I), cools until the indicating-point meets the liquidus at *p*, so that crystallization is imminent. The actual progress of crystallization, as the cooling proceeds, will depend upon the conditions. If crystals form at once, they will be of the composition represented by *q*; but as the indicating-point moves from *p* towards *r*, the first-formed crystals will not be in equilibrium with the changed magma. If the cooling be supposed sufficiently gradual, the composition of the crystals may conceivably be changed concurrently, so as to adapt themselves to the changing magma. This will be effected by '*fractional resorption*'—that is, a gradual resorption of part of the one constituent (*A*) in the crystals, and its simultaneous replacement by the other constituent (*B*). Thus, the indicating-point for the crystals will move along *qs*, as that for the magma moves along *pr*. The point *s* will be reached only when the whole magma has crystallized; for it is clear that, when the crystals are wholly of the composition of the initial magma, there can be no residual magma of a different composition. The final result in this case will be homogeneous mixed crystals of the same composition as the initial magma.

The adjustment of equilibrium by fractional resorption, which we have supposed, cannot in general be continuously maintained. There will therefore be a deposition upon the first-formed crystals of new layers, differing in composition according to the changing magma from which they are formed. Since the average composition of the crystals when consolidation is complete must be the same as that of the original magma, and the kernels of the crystals are richer in one constituent, the outer layers must be richer in the other. Their composition will be represented by some such point as *v* (Fig. 70, II), and that of the final residual magma by *t*, the magma therefore undergoing a greater change during the progress of crystallization than in the former case.

In the process just described we see the origin of the *zonary structure*, so common in rock-forming minerals which belong to isomorphous groups. The deposition on the original nucleus of a crystal of material continuously varying in composition causes in the completed crystal a gradual change in optical properties from centre to margin. More often resorption has from time to time during the growth of the crystal intervened to restore temporary equilibrium, and in that case successive zones of the crystal are sharply

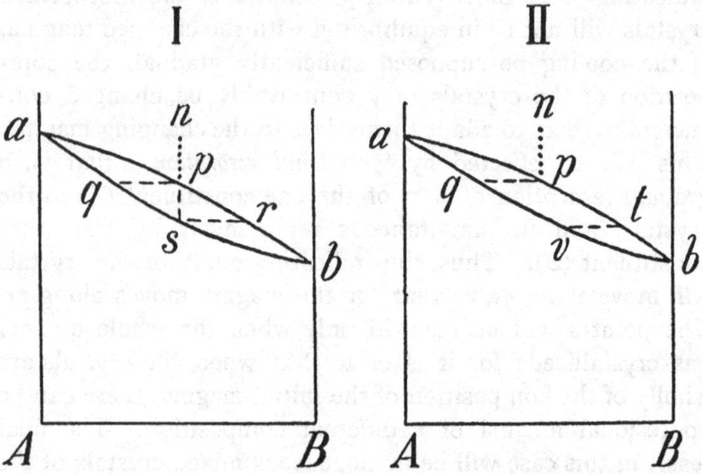

FIG. 70.—DIAGRAM TO ILLUSTRATE THE CRYSTALLIZATION OF ISOMORPHOUS COMPOUNDS.

Explanation in text. The upper curve in each case is the liquidus or freezing-point curve, and the lower is the solidus or melting-point curve.

divided. If this periodic resorption has attacked the angles and edges of the crystals more than the faces, the zonary banding will have more or less rounded outlines; and if it has attacked especially the equatorial region of a columnar crystal, an 'hourglass' structure is the result, the division being in this case always sharply marked. The abstraction of a given amount of material by crystallization will (other things being equal) produce more change in the composition of the residual magma in proportion as this magma diminishes in quantity. Hence the composition of the magma (and so of the crystals) varies most rapidly towards the end of the process, and the zonary structure is always most strongly marked in the border of a crystal.

Absence of zonary structure may be due to continuous adjustment of equilibrium by fractional resorption in the manner first described; but it may be, on the other hand, a consequence of supersaturation.[1] If a considerable degree of supersaturation be requisite to initiate crystallization, this may not come about until the temperature has fallen below

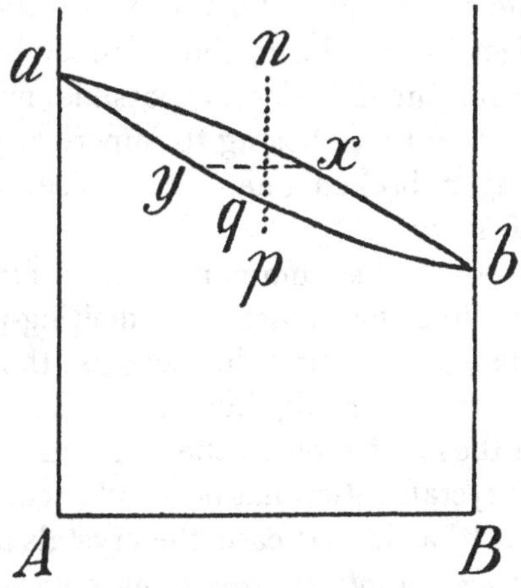

FIG. 71.—DIAGRAM TO ILLUSTRATE THE CRYSTALLIZATION OF ISOMORPHOUS COMPOUNDS WITH SUPERSATURATION.

the region in which change of composition can occur, and there may be formed only homogeneous crystals of the same composition as the magma. This was the result obtained in the experiments of Day and Allen on the plagioclase felspars.

If the undercooled magma begins to crystallize at a temperature not greatly below the region specified (p in Fig. 71), the heat liberated by rapid crystallization may raise the temperature again into that region. Equilibrium will then not be attained until the magma and the crystals (or their outer layers) have reached conditions represented by such points as x and y on one horizontal line. Mean-

[1] Compare Day and Allen, *The Isomorphism and Thermal Properties of the Felspars* (1905), pp. 69, 70.

while, the change in composition of the crystals in successive layers has followed the order *q* to *y*, which is a *reversal* of the normal order. Further, cooling may now give rise to layers following the normal order. Some irregularities observed in the zonary banding of crystals may be explained in this way.

Roozeboom's Five Types of Mixed Crystals.—As regards their actual behaviour in the process of crystallization, the various examples of isomorphism fall under five possible types, a distinction first made by Roozeboom on general thermodynamic principles.[1] Of these, the first three types belong to perfect isomorphism, the end-members being connected by a continuous series of possible mixed crystals; while the other two types belong to imperfect isomorphism, the series being a broken one. These several cases are illustrated in Fig. 72.

Type I.—In this, the commonest case for perfect isomorphism, the freezing-points and melting-points of the various intermediate mixtures lie between those of the end-members, and the curves, liquidus and solidus, decline continuously from the less fusible to the more fusible constituent. The course of crystallization has been sufficiently considered, and it is evident that in this case the crystals are constantly *enriched in the less fusible constituent* as compared with the magma from which they form. The difference is greatest in the middle of the series, and may be very considerable. It is greater (for a given vertical distance between the two curves) when the curves are less steeply inclined—that is, when the difference of melting-point between the two pure end-members is not very great. To this type belong numerous isomorphous series of rock-forming minerals, such as enstatite-hypersthene (the ferrous end-member labile), diopside-hedenbergite, jadeite-acmite, diopside-jadeite, forsterite-fayalite, probably akermanite-gehlenite, and others, including, perhaps, albite-anorthite.

Type II.—There is a maximum freezing- (and melting-) point for some intermediate composition (*m*), the two curves

[1] Compare Whetham, *A Treatise on the Theory oj Solutions* (1902), pp. 62-68.

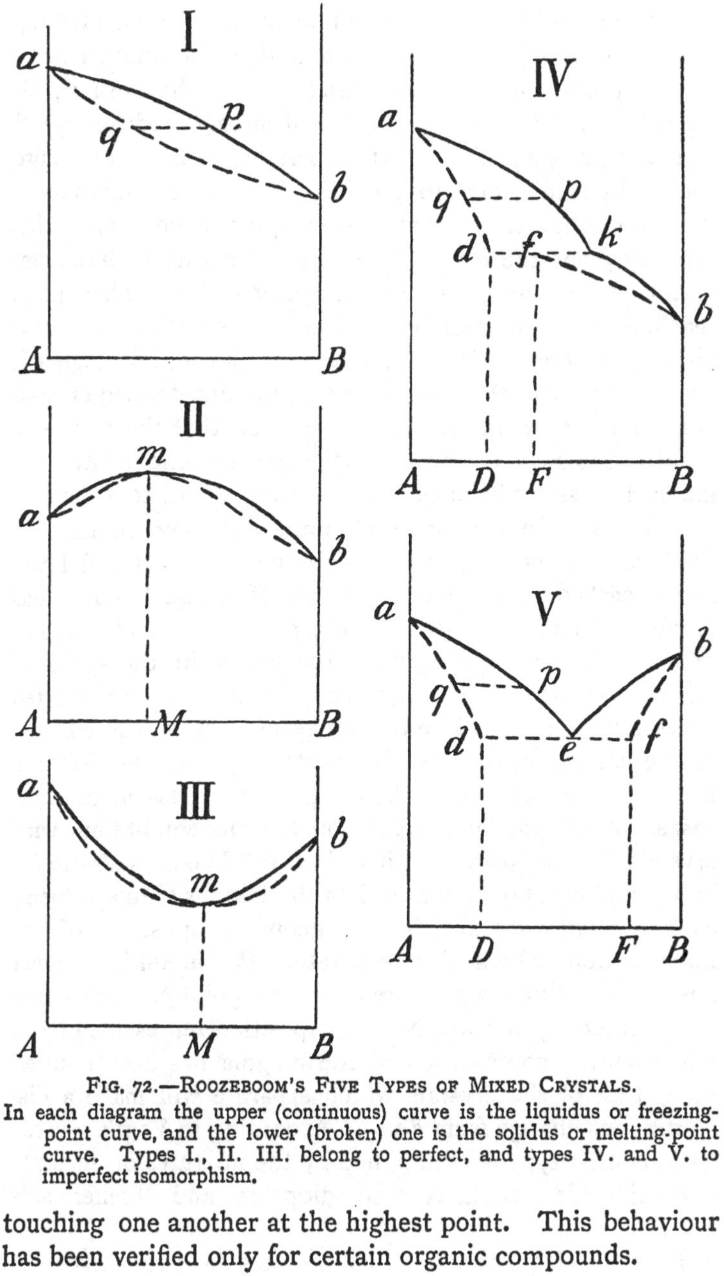

FIG. 72.—ROOZEBOOM'S FIVE TYPES OF MIXED CRYSTALS.

In each diagram the upper (continuous) curve is the liquidus or freezing-point curve, and the lower (broken) one is the solidus or melting-point curve. Types I., II. III. belong to perfect, and types IV. and V. to imperfect isomorphism.

touching one another at the highest point. This behaviour has been verified only for certain organic compounds.

Type III.—There is a minimum freezing- (and melting-) point, the liquidus and solidus touching one another at the lowest point (*m*). If the initial magma has exactly the composition represented by *m*, it will merely produce crystals of the same composition. If it is richer in *A*, and so represented by some point to the left of *m*, the crystals formed will be richer in *A* than the magma from which they separate; but the difference will diminish as both magma and crystals approach the composition *m*. That point reached, crystallization will be completed without further change of composition. Similarly, if the initial magma is richer than *m* in the constituent *B*, the first-formed crystals will be enriched in *B*. It will be seen that the minimum point *m* has some of the properties of a eutectic point. The plagioclase series belongs either to this type or to Type I.

Type IV.—In this type of imperfect isomorphism the liquidus lies (as in I.) wholly between the upper and lower levels marked by the melting-points of *A* and *B*, but it has an abrupt change of direction at a point *k*. This kink-point (*Knickpunkt*) corresponds with the break in the series of mixed crystals. The solidus consists of two disconnected curves, *ad* and *fb*. Here the range of possible composition in the mixed crystals is represented by *AD* and *FB*, the hiatus in the series (*DF*) being greater or less in different cases. A magma of composition *k* is in equilibrium with crystals of composition either *d* or *f*. The crystals are (as in Type I.) constantly enriched in the less fusible constituent, as compared with the instantaneous composition of the magma from which they separate. If the initial magma has a composition represented by some point *p*, above *k*, the first-formed crystals will be of composition represented by *q*; but when the composition of the magma has been reduced to *k*, that of the crystals which separate will make a discontinuous change from *d* to *f*. According to Vogt's investigations, this type is exemplified by the mixtures of enstatite and diopside; perhaps also diopside and Tschermak's silicate.

Type V.—Here the liquidus consists of two curves declining from *a* and *b* and meeting at a eutectic point *e*. The corresponding curves for the solidus are *ad* and *bf*. As before, the only mixed crystals which can exist are those of composition between *A* and *D* and between *F* and *B*. The first-formed crystals will be enriched in *A* or *B*, according as the initial magma is richer in the one constituent or the other as compared with the eutectic—*i.e.*, according as the indicating-point is to the left or right of *e*. A magma of composition *p* will give rise to crystals of composition *q*; and, as the magma changes from *p* to *e*, the crystals which separate from it will change from *q* to *d*. When the magma is reduced to eutectic composition (*e*), crystals of two kinds will proceed to form simultaneously, having the compositions represented by *d* and *f*, and crystallizing together in the eutectic proportions. The most important case falling under this type is that of orthoclase and albite. Here also belong the imperfect series hedenbergite-rhodonite and hypersthene-rhodonite.

Double Salts—In discussing the progress of crystallization in an ideal binary magma (p. 190), we supposed that the two constituents *A* and *B* were such as do not enter into combination with one another. Suppose, now, that *A* and *B* may unite to form a compound such as *AB* (or A_2B or some other formula), so that there are three kinds of crystals which may possibly be formed—viz., *A*, *B*, and *AB*. Here the compound *AB* is not intermediate in its physical properties between *A* and *B*, as an isomorphous mixture would be; and, if mixed crystals can be formed, they will be mixtures, not of *A* and *B*, but of *A* and *AB* on the one hand, or *B* and *AB* on the other (Fig. 73).

In this case there may be two eutectics, one between *A* and *AB* and the other between *AB* and *B*, and the freezing-point curves will then have some such form as that shown in Fig. 74, where the place of the definite compound is marked by a summit. If *A* and *B* form two different compounds, such as A_2B and *AB*, there may be two summits, corresponding with these two compounds, and three eutectics. An example of this is the 'lime-silica series,' as worked out by Day and Shepherd,[1] in which two intermediate bodies (not in this case double salts) occur between lime and silica —viz., the orthosilicate and the metasilicate.

With sufficient data, the freezing-point curve could be constructed for any given case, and from this the progress of crystallization from the magma would be easily seen. Thus, for the case represented in Fig. 74, if the initial composition of the magma lie between e_1 and *ab*, then the first crystals to separate will be of the *AB* type, but with some dissolved *A* (or isomorphous admixture of *A*). When the

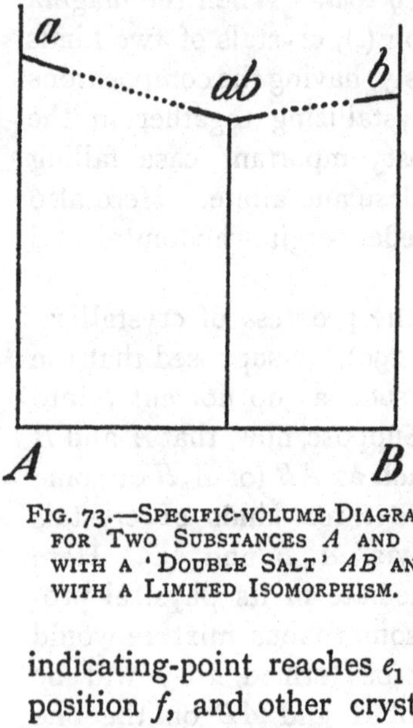

FIG. 73.—SPECIFIC-VOLUME DIAGRAM FOR TWO SUBSTANCES *A* AND *B* WITH A 'DOUBLE SALT' *AB* AND WITH A LIMITED ISOMORPHISM.

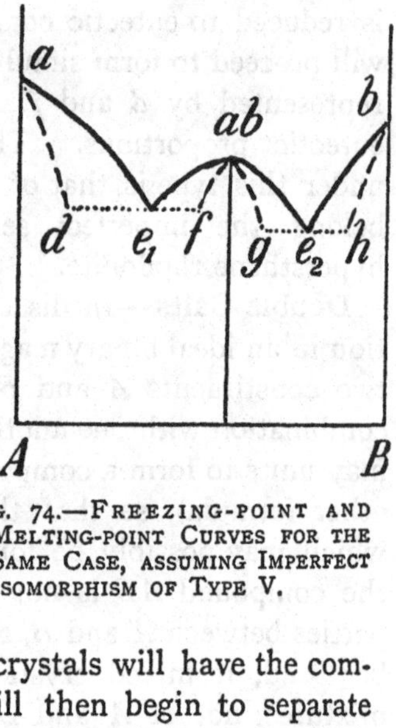

FIG. 74.—FREEZING-POINT AND MELTING-POINT CURVES FOR THE SAME CASE, ASSUMING IMPERFECT ISOMORPHISM OF TYPE V.

indicating-point reaches e_1 these crystals will have the composition *f*, and other crystals will then begin to separate having the composition *d*—*i.e.*, crystals of the *A* type with the maximum content of *AB*, the two kinds of crystals proceeding to separate in eutectic proportions. Where there is imperfect isomorphism of the Type IV of Roozeboom, a corresponding modification will follow in the form of the freezing-point curve and in the process of crystallization, and further complication may be introduced by dimorphism.

[1] *Amer. Journ. Sci.* (4), vol. xxii. (1906), pp. 265-302; *Tscherm. Min. Petr. Mitth.* (2), vol. xxvi. (1907), pp. 169-232.

This is illustrated by the case of diopside, one of the most important double salts among the rock-forming minerals (Fig. 75). It may be regarded as a compound of enstatite (or clino-enstatite) and wollastonite (or pseudo-wollastonite). According to Vogt, the pair enstatite and diopside belong to Roozeboom's Type IV., diopside and pseudo-wollastonite to Type V. Also it appears that the eutectic temperature for the latter pair is lower than the inversion-point between

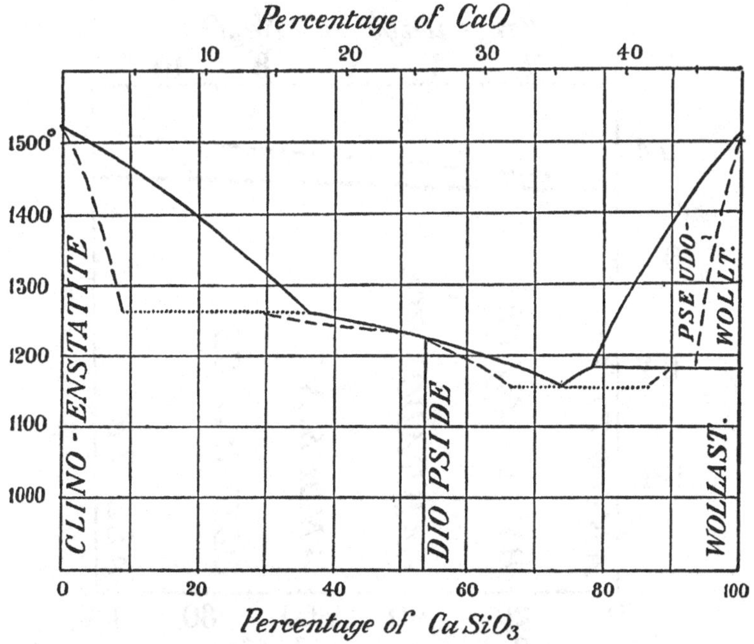

FIG. 75.—FREEZING-POINT AND MELTING-POINT CURVES FOR CLINO-ENSTATITE AND PSEUDO-WOLLASTONITE, WITH DIOPSIDE AS A DOUBLE SALT.

The melting-points of the two simple metasilicates and the inversion-point of wollastonite are taken from the American physicists, the other data from Vogt.

pseudo-wollastonite and wollastonite; from which we may infer that the right-hand branch of the liquidus will show a discontinuous change at the latter temperature (about 1180°). Since clino-enstatite is the theoretically stable form of $MgSiO_3$ for all temperatures, no such dimorphism is indicated in the left-hand part of the diagram. The corresponding diagram with the more easily fusible $FeSiO_3$ in place of $MgSiO_3$, and hedenbergite ($CaFeSi_2O_6$) in place of the magnesium diopside, would, according to Vogt's data, show a different character. He assigns the pair hypersthene and hedenbergite to Type V., and the pair hedenbergite and pseudo-wollastonite to Type IV.

Isomorphism of the Plagioclase Felspars.—That the plagioclase felspars constitute a continuous series of 'mixed

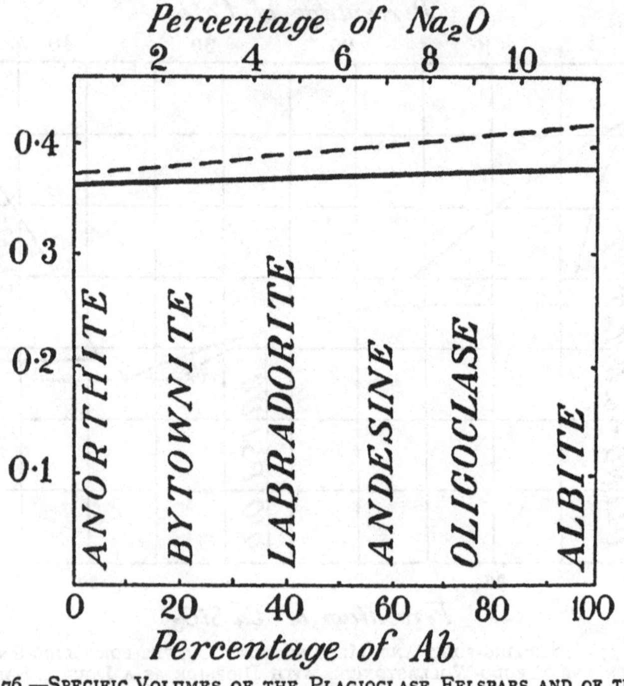

FIG. 76.—SPECIFIC VOLUMES OF THE PLAGIOCLASE FELSPARS AND OF THEIR GLASSES. (AFTER DAY AND ALLEN.)

The full line corresponds with the crystalline and the broken line with the glassy state.

crystals,' between anorthite and albite as end-members, was rendered practically certain by Schuster's optical studies, and may be considered finally established by the synthetic experiments of Day and Allen.[1] From chemically pure

[1] *The Isomorphism and Thermal Properties of the Felspars*, 1905.

material these physicists reproduced by fusion anorthite and five varieties of mixed crystals, ranging from Ab_1An_5 to Ab_3An_1. The specific volumes of these, and of pure albite, plotted as a function of the composition, give a sensibly straight line (Fig. 76). The melting-point of anorthite was determined as 1532°, and the different varieties of mixed crystals gave successively lower temperatures, down to 1340° for Ab_3An_1, the results being expressed graphically by the broken curve in Fig. 77. The more sodic plagioclases, which form extremely viscous 'melts,' could not be investigated in this way; but, if the curve be prolonged on the assumption that it maintains the same general character, the extrapolated value of the melting-point of albite will be about 1230° It is thus clear that the anorthite-albite series belongs either to Type I. of Roozeboom or to Type III., with a minimum-point very near to the albite end. If the curve is to be prolonged in the manner supposed, the series falls under Type I., and the crystals which separate from the magma must always be enriched in anorthite, as compared with the magma from which they form. Further, crystals formed from a given magma at successive stages will become increasingly richer in albite and poorer in anorthite.

Various well-known facts relative to the plagioclase felspars of igneous rocks become significant in this light. It was noticed by Fouqué so long ago as 1879 that, when a rock contains plagioclase felspars of two distinct generations, the later crystals are of more acid composition (*i.e.*, richer in albite) than the earlier; and this has been recognised as a general law by Michel-Lévy, Rosenbusch, and others. Again, where crystals of plagioclase are built up by successive zones of varying composition, it is the general rule that the outer zones are progressively more acid than the inner. Becke,[1] in 1897, pointed out how this arrangement appeared to connect itself with Küster's investigations on the melting-points of isomorphous mixed crystals.

Day and Allen assign the anorthite-albite series to Roozeboom's first type, and Vogt has come to the same con-

[1] *Sitz. deuts. naturw.-med. Ver. Böhmen*, '*Lotos*,' 1897, No. 3.

clusion, although he had originally[1] placed it doubtfully under the third type, with a minimum melting-point near the albite end. On the latter supposition, a magma very rich in albite should yield crystals still richer in that constituent. Some facts suggest that this possibility cannot be wholly excluded. Certain remarkable experiments recorded by Day and Allen seem to be consistent with a considerably

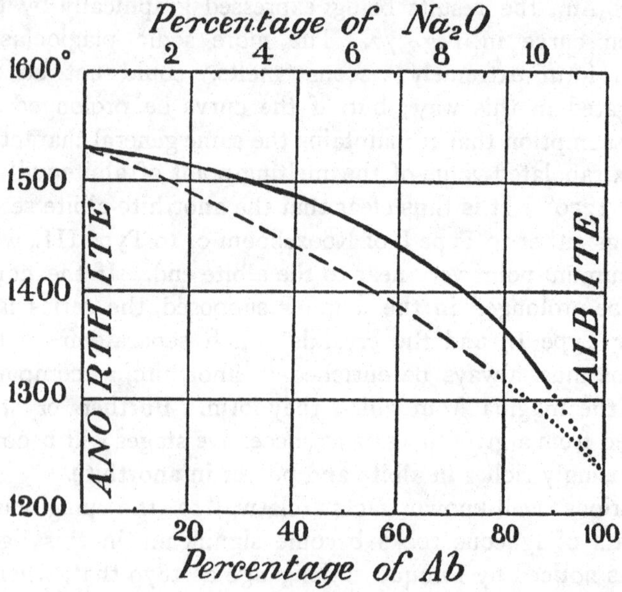

FIG. 77.—MELTING-POINT AND FREEZING-POINT CURVES FOR THE PLAGIOCLASE FELSPARS.

The lower curve (broken line), taken from Day and Allen, is the solidus or melting-point curve. The upper (full line), from Vogt, is the liquidus or freezing-point curve. Both curves are conjectural towards the albite end.

higher melting-point for albite than that found by extrapolation as in Fig. 77. The petrological evidence bearing on this point is scanty, since few rocks very rich in albite are known. The rarity of pure or nearly pure albite in igneous rocks is accounted for by this constituent entering usually into an orthoclase, as was pointed out by Becke. For albite to crystallize by itself, it is necessary that the

[1] *Silikatschmelzlösungen*, I. (1903), p. 154.

magma shall be very poor in potash as well as in lime, the granite of Croghan Kinshelagh, in Wexford, being an example.[1] Heddle[2] has described a rock from Beinn Bràghaid, in Sutherland, which consists essentially of porphyritic crystals of albite in a ground-mass also of albite. He gives analyses of both, from which we may calculate that the ratio (by weight) of Ab to An is about 95·2 : 4·8 in the ground-mass, but 95·9 : 4·1 in the porphyritic crystals. The difference is not great, but, if it can be regarded as significant, it indicates a certain enrichment in the albite constituent of the first crystals as compared with the magma. Again, it appears from Becke's observations that, when there is a marked zonal structure in crystals of the more acid varieties of plagioclase, the arrangement of the zones, at least in some cases, is the reverse of that seen in the basic felspars, the inner being richer in albite than the outer.

The temperature determined by Day and Allen in each experiment indicates the point at or near which crystallization from fusion actually occurred. Owing to the viscosity of these purely felspathic magmas, crystallization did not begin at the freezing-point curve, but at a lower point in the neighbourhood of the melting-point curve. This is proved by the circumstances of the experiments, and in particular by the fact that the mixed crystals obtained were homogeneous, without zonary structure. The curve plotted by Day and Allen may therefore be taken as giving, at least approximately, the solidus. It is clear that from this we can lay down the liquidus too, provided we know for a sufficient number of points on the former curve the corresponding points on the latter. Vogt[3] has used for this purpose the comparative analyses of Lagorio and others, calculating the relative proportions of Ab and An—(i.) in the first-formed felspar crystals, and (ii.) in the total rock. These correspond roughly with points at the same horizontal

1 Haughton, *Quart. Journ. Geol. Soc.*, vol. xii. (1856), p. 183.

2 *Min. Mag.*, vol. v. (1884), p. 141.

3 *Tscherm. Min. Petr. Mitth.*, (2), vol. xxiv. (1906), pp. 511-514.

level on the solidus and liquidus respectively, and in this way is constructed the curve shown by a full line in Fig. 77. It is only an approximation, for, besides experimental errors, it is to be remembered that the mutual relations of albite and anorthite are altered to some extent by the presence of other constituents in the mixed magma. It is clear, however, that there must be a considerable vertical distance between the two curves in the middle part of the series, the temperature-range between the melting- and freezing-points of andesine being about 65°.

Imperfect Isomorphism of Orthoclase and Plagioclase.—The relation of orthoclase (or microcline) to albite or to anorthite is clearly of a different kind from the relation of these last two to one another. Analyses of orthoclase in general show a certain content of soda and to a less extent of lime, and analyses of plagioclase show a certain content of potash; but there is not a continuous series of mixtures connecting pure orthoclase with pure albite or anorthite. We have to do here with an interrupted series (or, rather, two such series—orthoclase-albite and orthoclase-anorthite) with a wide hiatus in the middle part.[1] Since in nature the An molecule enters only in very minor amount into mixed crystals of the orthoclase type, or Or into lime-felspars, we shall have in mind more especially mixtures of Or and Ab, in which a part of the latter may be replaced by An.

The mixed crystals of orthoclase and albite must belong, then, either to Type IV. or to Type V. of Roozeboom's classification. There can be no doubt, as Vogt[2] has shown, that they fall under Type V. This might be inferred from the fact that the melting-points of orthoclase and albite appear to differ but little. It is clearly proved by the existence of a eutectic point, and the fact that the mixed crystals, compared with the magma from which they form, are enriched in one or the other constituent, according as they fall on one or the other side of the eutectic (Fig. 78).

[1] The case of 'anorthoclase' will be considered below.

[2] *Silikatschmelzlösungen*, I. (1903), p. 157, and II. (1904), pp. 180-185; *Tscherm. Min. Petr. Mitth.* (2), vol. xxiv. (1906), pp. 519-522, etc.

Vogt has elaborately discussed numerous analyses by Lagorio and others of igneous rocks and of the first-formed felspars in them. He shows that from a felspathic magma the first crystals separated are of the orthoclase kind or the plagioclase kind, according as the ratio of Or to Ab (with An) is greater or less than a certain (eutectic) ratio; that the crystals are not pure orthoclase or pure plagioclase, but

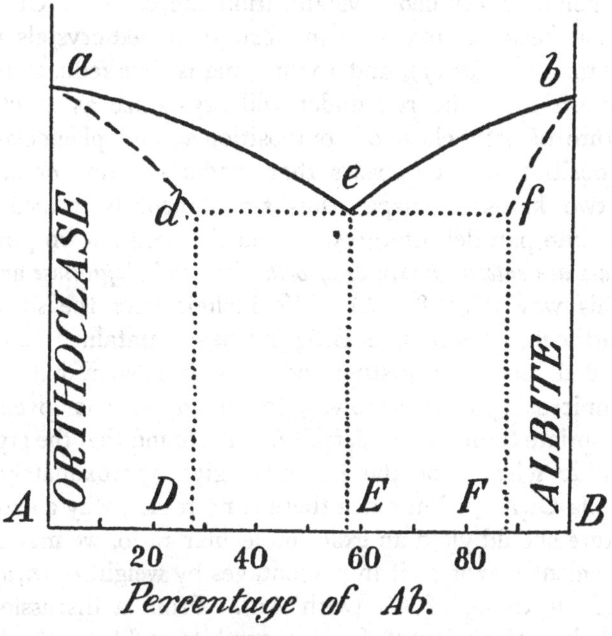

FIG. 78. — IMPERFECT ISOMORPHISM OF THE ORTHOCLASE-ALBITE SERIES. (AFTER VOGT.)

The full lines are the two branches of the liquidus or freezing-point curve, and the broken lines the two branches of the solidus or melting-point curve. As regards vertical distances (temperatures), the figure is only diagrammatic.

contain an admixture of Ab (and An) in the former case and of Or in the latter; and that, as compared with the magma from which they form, the orthoclase cystals are enriched in Or and the plagioclase crystals in Ab (with An). The eutectic ratio is approximately 40–44 parts of Or to 60–56 of Ab (with An). Orthoclase crystals may contain an admixture of Ab (and An) up to a limit of about 28 *per cent.*, and plagioclase crystals may contain Or up to about 12 *per cent.* The points *E*, *D*, and *F* in Fig. 78 are thus approximately fixed. The relations between orthoclase and anorthite are doubtless of the same general kind, though the higher melting-point of the latter mineral will presumably throw the eutectic point somewhat nearer to orthoclase.

When by continued crystallization the excess of Or (or of Ab) has been eliminated in the form of mixed crystals along the branch *ad* (or *bf*), and the magma is thus reduced to the composition *e*, the remainder will crystallize as a eutectic mixture of orthoclase of composition *d*, and plagioclase of composition *f*. It appears that, under certain conditions, the two kinds of felspar thus simultaneously formed may enter into parallel intergrowth, and the well-known *perthitic felspars are eutectic mixtures of orthoclase and plagioclase* arising in this way (Fig. 85, A). We include here the so-called 'anorthoclase,' which, as Brögger[1] has maintained, is to be regarded, not as a distinct potash-soda-felspar, but as an ultramicroscopic intergrowth, to which he has given the appropriate term 'cryptoperthite.' He found that the crypto-perthitic felspars of the larvikites give approximately the formula Or_2Ab_3; but since there is no reason why a eutectic mixture should yield an exact molecular ratio, we may more conveniently express it in percentages by weight—viz., about 45 of Or to 55 of Ab (with An). From a discussion of a number of analyses of cryptoperthite and 'anorthoclase,' Vogt arrives at the ratio 42 : 58. This agrees well with the approximate value 40-44 : 60-56, obtained by comparing the compositions of magmas from which orthoclase and plagioclase respectively first crystallize. The analyses of perthites give probably the closer approximation, and this value has been adopted in Fig. 78.

It is to be remarked that the eutectic intergrowth may be formed round the earlier separated crystals, whether of orthoclase or of plagioclase, with the usual crystallographic relation. An analysis made on a crystal of perthite with

[1] *Zeits. Kryst.*, vol. xvi. (1890), pp. 524-551.

a kernel of orthoclase or plagioclase (composition *d* or *f*) will, of course, not give the true eutectic ratio.

We have supposed that, when the magma has been reduced to the eutectic composition by the separation of mixed crystals of one kind (say orthoclase, final composition *d*), crystals of the other kind (plagioclase of composition *f*) thereupon begin to separate together with the orthoclase. But, in accordance with what has been said in an earlier chapter (p. 207), this will not immediately happen in the absence of crystals of *f* or isomorphous with *f*; and the orthoclase *d* is not in this sense isomorphous with the plagioclase *f*, since no mixed crystals of intermediate composition are possible. This case, then, is the same as that illustrated in Fig. 66, and, as a consequence of supersaturation, crystals of both kinds may separate before eutectic crystallization is reached, and the first-formed crystals may show resorption effects (p. 214).

Relation of Felspar Crystals to Magma. — We shall obtain a clearer conception of the relation which felspar crystals bear, in respect of composition, to the magma from which they form by considering a ternary magma with orthoclase, albite, and anorthite as its three constituents. Natural rock-magmas are, of course, not of so simple a composition, though some approximate to it, and by disregarding the non-felspathic constituents we shall necessarily introduce some errors. This remark applies to the two preceding sections, and is to be remembered where discrepancies are found between theory and observation. For instance, since the eutectic ratio between orthoclase and albite is altered to some extent by the presence of another independent constituent, eutectic cryptoperthite or 'anorthoclase' from different rocks will vary somewhat in composition.

Suppose, now, that we have a mineralogical analysis of an igneous rock, which can often be calculated with sufficient approximation from the chemical analysis. By rejecting the non-felspathic constituents and recalculating to a total of 100, we can express the felspathic part of the rock (and so of its initial magma) as a percentage mixture of Or, Ab, and An. Its composition can then be represented by a point on a triangular diagram (Fig. 79). Thus, the point *m* represents in this way the initial magma of an andesite from Costa Rica, analysed by Lagorio, the proportions being:

19·5 Or, 43·5 Ab, 37 An.

The porphyritic crystals of plagioclase, which were the first felspar to form from this magma, have the composition:

7 Or, 35 Ab, 58 An,

which is represented by the point *p*. The relation is marked by joining the points *mp* and drawing an arrow.[1] It is evident at a glance that the crystals are, relatively to the magma, enriched in Ab + An and impoverished in Or, and also that they are enriched in An as compared with Ab. In the same way the point *r* represents a granite-porphyry from Rödö, in Sweden, analysed by Holmquist, the composition of its felspathic part being:

52·5 Or, 41 Ab., 6·5 An.

The point *s* represents the porphyritic crystals of orthoclase in this rock, with composition:

70 Or., 26 Ab, 4 An.

Here there is an enrichment in Or relatively to Ab + An, and a slight enrichment in Ab relatively to An, and the arrow is therefore directed nearly towards *A*, but slightly to the left.

After laying down in this way the data derived from a number of comparative analyses, Vogt shows that a line *EE′* can be drawn, which sheds the arrows this way and that; so that a magma will first produce orthoclase or plagioclase crystals, according as the point representing its composition lies on one side or the other of this eutectic line. On or near this line cluster the points representing those magmas in which crystallization begins with orthoclase and plagioclase simultaneously, either in separate crystals or

[1] This mode of representation is taken from Vogt, *Silikatschmelzlösungen*, II. (1904), pp. 180-188, and *Tscherm. Min. Petr. Mitth.* (2), vol. xxiv. (1906), pp. 536, 537.

in perthitic intergrowth. All homogeneous crystals of the orthoclase type are represented by points lying between *A* and a certain curve *DD′*, which marks the maximum possible admixture of Ab+An in orthoclase crystals. Similarly, all homogeneous crystals of the plagioclase type are represented by points lying between *BC* and a certain curve *FF′*. The points *D*, *E*, *F* here are the same as those so lettered in Fig. 78, and correspond approximately with 28, 58, and 88 *per cent.* of Ab respectively; but the positions of *D′*, *E′*, *F′* cannot be precisely fixed for want of data.

The diagram in Fig. 79 may be used to show Becke's explanation[1] of the rarity in igneous rocks of felspars approaching pure albite in composition. The abstraction of felspar crystals of composition *p* from a magma of composition *m* will alter the latter to some near point *n* on the prolongation of *pm*. If crystallization proceeds continuously, the changing composition of the residual magma will be represented by such a curve as *mnt*, and *mp* is evidently the tangent at *m* to this curve. The changing composition of successive zones of a crystal (or of the whole crystal if we suppose equilibrium to be continuously adjusted) is represented by such a curve as *pq*. Unless the magma be previously exhausted, the two curves will be cut off simultaneously at *t* and *q* respectively, *tq* being the tangent to the former curve at the point *t* where it meets the eutectic line. Consequently, no felspar richer in Ab than *q* can be formed from a magma of initial composition *m*. For the final formation of a felspar approaching albite it is necessary that the initial magma should be very rich in Ab relatively to An, and also relatively to Or.

Crystallization in a Magma of Felspars and Quartz.—A very important case is that which can be represented ideally by a ternary magma having orthoclase, albite, and quartz as its three constituents. Many natural magmas approximate more or less closely to this standard, the presence in minor amount of the anorthite molecule and

[1] *Sitz. deuts. naturw.-med. Ver. Böhm.*, '*Lotos*' (1897), No. 3, and Vogt, *Silikatschmelzlösungen*, II. (1904), p. 187.

various independent constituents probably importing no very serious modification of the conditions.

In the triangular diagrams of Fig. 80, *A*, *B*, *D*, *F* are the points so lettered in Fig. 78, and *C* represents quartz. From a number of analyses of graphic intergrowths, Vogt[1] concludes that the eutectic proportions are about 72·5 felspar to 27·5 quartz, and approximately the same for different kinds

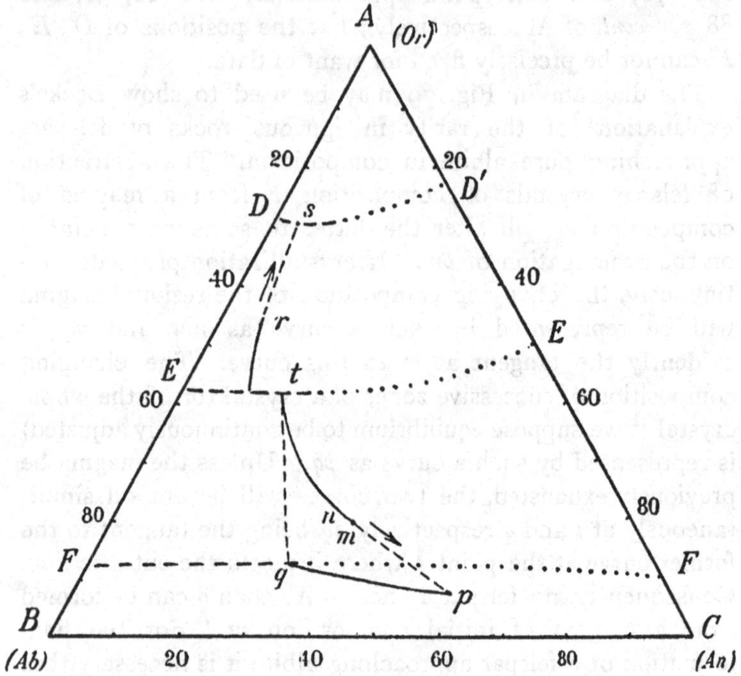

Fig. 79.—Diagram illustrating the Relation, in Respect of Composition of Felspar Crystals to the Magma from which they Form.

of felspar. If this can be assumed, the points E_2 and E_1 are fixed, and it appears that the curves E_2E and E_1E make nearly one straight line. If we further assume as an approximation that the relations between orthoclase and albite are not materially disturbed by the presence of quartz,

[1] *Silikatschmelzlösungen*, II. (1904), pp. 120, 121; *Tscherm. Min. Petr. Mitth.* (2), vol. xxv. (1906), pp. 361, 362, 383-385.

the curve E_3E becomes nearly a straight line directed towards C. The eutectic lines E_1E, E_2E, E_3E divide the triangle into three fields; and orthoclase (mixed crystals), plagioclase (mixed crystals), or quartz will first form from the magma according as the initial composition of the latter falls into one or other of these three fields.

By tracing the further progress of crystallization, we are enabled to divide the triangle into six fields, which will indicate the second as well as the first mineral which will separate out. This answers to Fig. 61, but with the

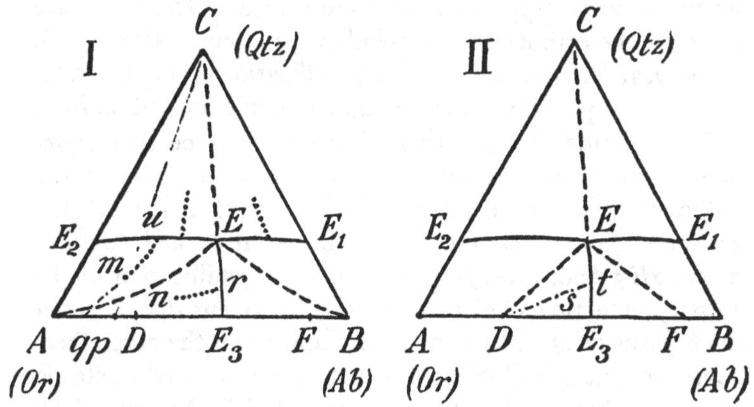

FIG. 80.—DIAGRAM TO EXHIBIT THE PROGRESS OF CRYSTALLIZATION IN A TERNARY MAGMA OF ORTHOCLASE, ALBITE, AND QUARTZ.

I, for zoned felspar crystals; II, for continual adjustment of equilibrium.

modification introduced by partial isomorphism between two of the constituents. There are two cases to be distinguished (see p. 231), according as (I.) the mixed crystals of felspar are built up by successive zones, or (II.) they are continually altered by fractional resorption, so as to be always homogeneous and in equilibrium with the magma.

Taking the former case, suppose, for instance, that the initial composition of the magma is represented by such a point as m, within the area AE_2EE_3, and not far from AC. If a straight line through C and m meets the base in p, then p represents the composition of the felspathic part of the magma. The orthoclase mixed crystals which first form are relatively enriched in Or, and have the composition *q*, where *q* is related to *p* in the same way as the points so lettered in Fig. 72, V. The indicating-point must move directly away from *q*, so that *qm* is the tangent at *m* to the path of that point. Since successive zones of the crystals contain an increasing proportion of Ab, the path must gradually curve away from *B*. It meets the eutectic line E_2E at some point *u*, the felspar substance deposited having then some composition between *q* and *D*. Now quartz begins to separate in eutectic proportions with the orthoclase, as the indicating-point moves from *u* to *E*. The successive zones of the orthoclase meanwhile vary in composition to *D*. Finally, at *E*, there is eutectic crystallization of quartz, orthoclase of composition *D*, and plagioclase of composition *F*.

If the initial composition of the magma be that represented by such a point as *n*, then the pat *nr* of the indicating-point will meet E_3E instead of E_2E; eutectic crystallization of *D* and *F* will begin at *r*, and continue along *rE*; and, finally, there will be crystallization of the ternary eutectic as before. Now, it is evident that, between such paths as *mu* on the one hand and *nr* on the other, there must be one possible path which meets the two eutectic lines together at *E*. If the area AE_2EE_3 be divided by a curve *AE* taking this course, then the second mineral to form will be quartz or plagioclase, according as the indicating-point falls initially within one or other of the fields AE_2E and AE_3E. A corresponding curve *BE* may be drawn to divide the plagioclase field into two, and a straight line *CE* to divide the quartz field.

In Case II., where the felspar mixed crystals are continually adjusted to equilibrium with the magma, it is easy to see that, in order to reach the eutectic line E_3E, the indicating-point must start from some point within the triangle *DEF*. For, if it reaches the line from the orthoclase side at the point *t*, the crystals are then wholly of the composition *D*, and, consequently, the initial magma must have had the composition indicated by some point *s*, between *D* and *t*. In this case, therefore, we have merely to draw straight lines, *DE*, *FE*, *CE*, to have the required division into six fields.

Spontaneous Changes in Crystals on Cooling.—Hitherto it has been tacitly assumed that the characters of igneous rocks as directly observed are those which they acquired in the process of consolidation from their parent magmas. Only in the case of the devitrification of glasses—a deferred crystallization in highly supersaturated magmas—have we adverted to the possibility of important changes in solid (or quasi-solid) rocks. It is necessary at this point to make a few remarks concerning changes *in the solid* in crystalline igneous rocks—meaning, not the class of transformations (metamorphism) dependent on external conditions, but *spontaneous rearrangements*, which are in the nature of readjustment of equilibrium consequent upon falling temperature.

It is well known that metallic alloys, when slowly cooled, often undergo radical changes in constitution and micro-structure at temperatures hundreds of degrees below those of consolidation. Valuable investigations on this subject have been published during recent years,[1] and they are highly instructive to the student of igneous rocks. It is true that there are important differences between alloys and minerals. The definite compounds of metals with one another are of relatively simple composition, most of them being binary compounds of the type $X_m Y_n$; though, on the other hand, there may be a number of such compounds between two given metals X and Y. The metals have much lower latent heats than the rock-forming minerals, and in most cases notably lower melting-points and specific heats, while their thermal conductivity is much higher. The metals, again, crystallize from fusion far more readily than the ordinary constituents of igneous rocks, so that metallic glasses are scarcely known. Especially, there is among metals and metallic alloys a very general mutual solubility

[1] See especially Heycock and Neville, 'On the Constitution of the Copper-Tin Series of Alloys,' *Phil. Trans.* (A.), vol. ccii. (1903), pp. 1-69, Pl. 1-11.

in the crystalline state, though this is often limited in a manner controlled by temperature. On this, and on polymorphism, depend the changes in the solid which are so commonly found in metallic alloys when cooled slowly enough to allow the establishment of equilibrium. Changes comparable with these, and arising from the same two causes, are found to occur in the rock-forming minerals, but, so far as our present knowledge goes, much less commonly. This may be due in part to the greater difficulty of establishing true equilibrium, a consequence of the specific properties of the silicate-minerals. We have already had occasion to remark that the existence of a given form, even throughout geological ages, is not a conclusive proof of its theoretical stability.

A well-known case of mineralogical change in the solid arising from *dimorphism* is that of leucite. This crystallizes at high temperatures in the regular system (α-leucite), and on cooling changes into the rhombic form (β-leucite) at about 450° C. The strain set up by unequal contraction in cooling below this temperature is relieved by a complex system of twin lamellæ, or, more properly, gliding lamellæ. Perofskite presents an analogous case, though the inversion at a definite temperature has not been experimentally proved. Day and Shepherd have shown that tridymite and quartz stand to one another in the same enantiotropic relation, with an inversion-point at about 800°; but the change from one form to the other is brought about with great difficulty. It follows that tridymite is theoretically unstable or metastable at ordinary temperatures.

There are many indications that dimorphous and other rearrangements in crystalline bodies may sometimes be long deferred or 'suspended,' so that metastability may be scarcely distinguishable in effect from true stability. External conditions are not without their influence on the adjustment of equilibrium. It seems probable that the deformation and condition of stress set up in the crystals of solid rocks in consequence of crust-movements have an effect in some measure analogous to the stirring of a liquid; and possibly some of the transformations incident to dynamic metamorphism are not strictly caused, but only precipitated by that agency. It is certain at least that pressure will promote the change from a metastable to a stable form or system whenever this change involves a condensation of volume. Again, the transformation may be promoted by the presence of an independent substance, and in particular accelerated by water or some other solvent, the action of which is analogous to that of a catalyser. Suppose, for example, that the solubility-curves for two dimorphous forms (α and β)

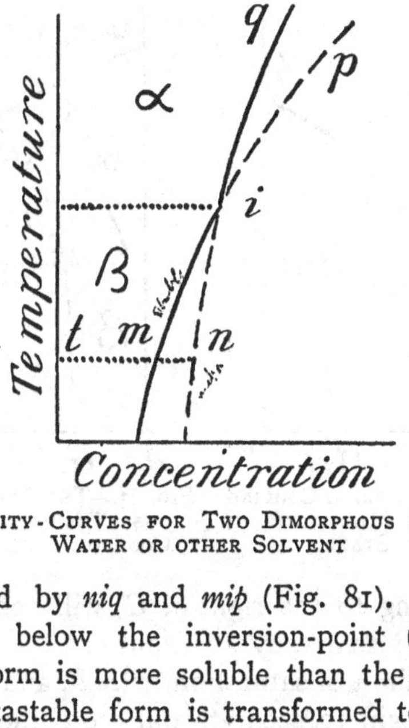

FIG. 81.—SOLUBILITY-CURVES FOR TWO DIMORPHOUS SUBSTANCES WITH WATER OR OTHER SOLVENT

are represented by *niq* and *mip* (Fig. 81). Then at any temperature *t* below the inversion-point (*i*) the higher (metastable) form is more soluble than the lower (stable) one. The metastable form is transformed to the stable in solution, and a solution saturated for the former is supersaturated for the latter, which therefore crystallizes out. Even with a low degree of solubility and a small quantity of the solvent this continuous process may be important.

Changes in the solid may arise also in connection with *imperfect isomorphism*. Isomorphous mixed crystals are often spoken of as 'solid solutions,' and they, in fact, resemble liquid solutions in the characteristic property of presenting homogeneous mixtures of varying composition. Mixed crystals constituting a continuous series (Types I., II., III. of Roozeboom) correspond with liquid solutions in which the constituents are miscible in all proportions, while interrupted series (Types IV. and V.) correspond with the case of limited miscibility discussed on pp. 196-198. Taking Type V. for illustration, suppose that *dG* (Fig. 82) is the curve of solubility of *B* in *A* in the crystalline state, and that the solubility diminishes with temperature, so that the curve slopes to the left downward. Mixed crystals of composition *p* (any

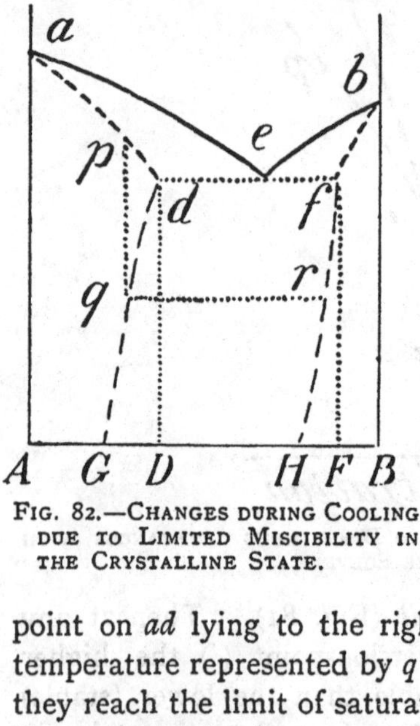

FIG. 82.—CHANGES DURING COOLING DUE TO LIMITED MISCIBILITY IN THE CRYSTALLINE STATE.

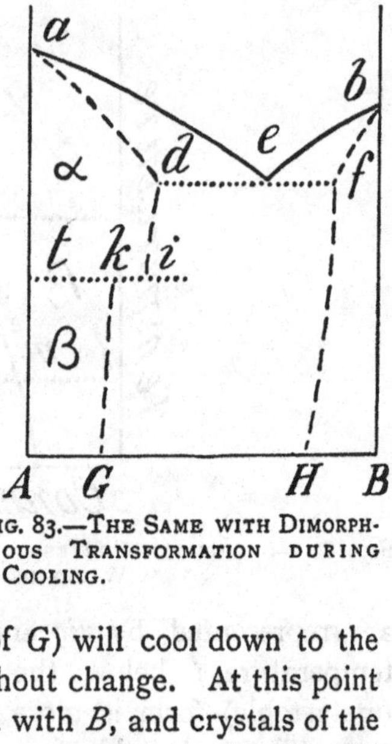

FIG. 83.—THE SAME WITH DIMORPHOUS TRANSFORMATION DURING COOLING.

point on *ad* lying to the right of *G*) will cool down to the temperature represented by *q* without change. At this point they reach the limit of saturation with *B*, and crystals of the *B* type must begin to separate out from them, having the composition represented by *r*, the corresponding point on the opposite branch of the solubility-curve. With falling temperature the composition of the original mixed crystals changes along the curve *qG*, more of the type rich in *B* being thrown out continuously. The latter, being set free within the original crystals, will be found as inclusions in them, and may enter into regular intergrowth. If the original mixed crystals had the composition *d*, the change would begin at once with falling temperature. The final result would then be crystals of composition G with inclusions of crystalline material of the opposite type, of composition H, the proportions of the two being in the ratio $HD : DG$. If the solubility of B in A (in the crystalline state) increased with fall of temperature, there would be no change in the mixed crystals. This case is shown, for the sake of illustration, on the opposite side of the figure, the solubility-curve of A in B being supposed to slope away from the point B.

A modification of the preceding case arises where the constituent A is dimorphous, with a point of inversion at some temperature (t) below the temperature-range of crystallization (Fig. 83). Suppose that the solubility of B in the higher form (α) of A is greater than in the lower form (β). Then the curve dG is replaced by two detached portions, di and kG, and there is an *abrupt change* of solubility, from ti to tk, at the temperature of inversion. If the original composition of the mixed crystals lie between d and i (that is, between verticals through those points), the elimination of material rich in B with fall of temperature will take place in three stages: (1) continuously along part of the curve di to i; (2) abruptly at the inversion-point, from i to k; (3) continuously along the curve kG. If the original composition be between i and k, the first stage will be wanting; if between k and G, the second stage also will be wanting; if to the left of G, there will be no change. We have assumed the temperature of inversion to be constant. If it is lowered by the admixture of B, the horizontal straight line tki should be replaced by a curve starting from t and declining to the right.

These theoretical considerations are not without application to the subject in hand. The minute '*schiller*' *inclusions*, common in many high-temperature minerals of plutonic rocks, probably afford an example. As is well known, Judd[1] regards these as of secondary origin, and attributes them generally to the agency of solution under pressure. Without entering into this question, it may be suggested that certain

[1] *Quart. Journ. Geol. Soc.*, vol. xli. (1885), pp. 374-389.

types of inclusions embraced under this comprehensive name may be connected with limited solubility as a function of temperature. If, for instance, olivine, at the temperature of its crystallization, is capable of containing magnetite in solid solution up to a certain limit, and this limit is lowered with falling temperature,[1] there may be in consequence an elimination of magnetite, now represented by the well-known dendritic inclusions (Fig. 84, A).

Another interesting case is that of the *perthitic intergrowths* (distinct from the eutectic perthites already considered),

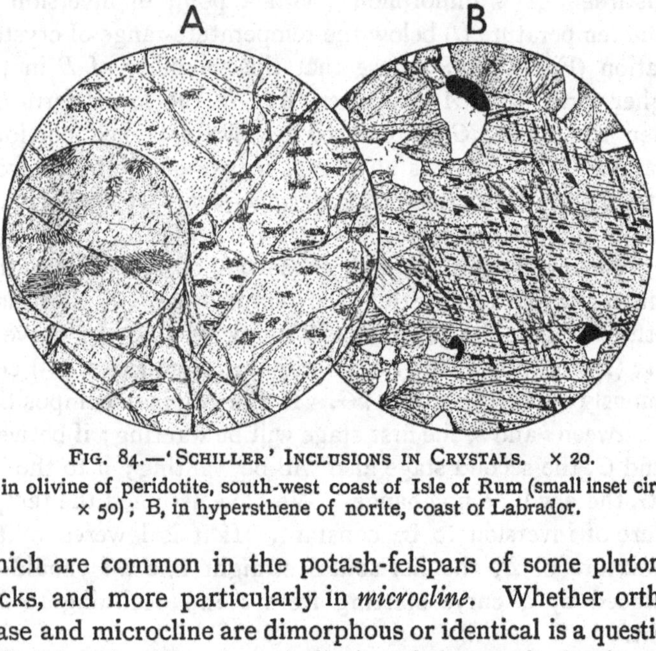

FIG. 84.—'SCHILLER' INCLUSIONS IN CRYSTALS. × 20.
A, in olivine of peridotite, south-west coast of Isle of Rum (small inset circle × 50); B, in hypersthene of norite, coast of Labrador.

which are common in the potash-felspars of some plutonic rocks, and more particularly in *microcline*. Whether orthoclase and microcline are dimorphous or identical is a question still debated. The close similarity of the two in density and in crystallographic angles, which causes Groth to place this case under the head of 'polysymmetry' as distinguished from polymorphism, does not appear to be in any wise conclusive: it is paralleled by the cases of enstatite and

[1] Vogt has recently made a like suggestion for the minute inclusions of titanic oxide in felspars and pyroxenes, *Quart. Journ. Geol. Soc.*, vol. lxv. (1909).

clino-enstatite,[1] zoisite and clino-zoisite, and probably others. If orthoclase and microcline are dimorphous, the latter must clearly be the lower form. Where it occurs with the apparent characters of a primary mineral, it is the latest product of crystallization, and it is characteristic of the most acid granites, and especially of pegmatites, doubtless consolidated at relatively low temperatures. There are often features, however, which suggest that this mineral has been formed from primary orthoclase. The characteristic fine lamellation

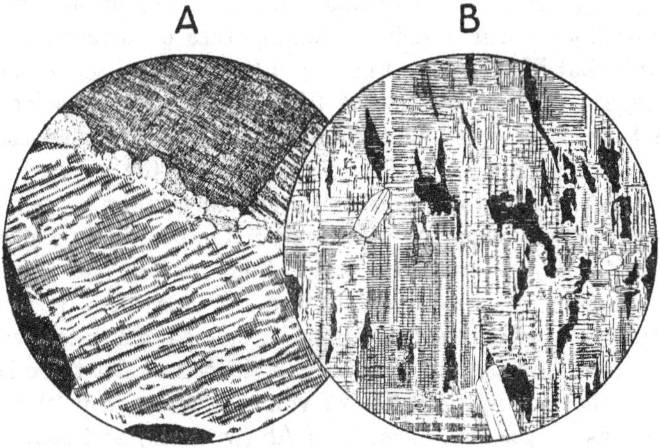

FIG. 85.—MICROPERTHITIC STRUCTURES IN FELSPARS. ×20, WITH CROSSED NICOLS.

A, primary (eutectic) microperthite of albite and orthoclase in alkali-granite, Rockport, Massachusetts; B, secondary microperthitic intergrowth of albite (dark patches) in microcline of nepheline-syenite, Miask, Urals. The small patches are in parallel intergrowth with the microcline, while enclosed primary crystals of plagioclase have no such arrangement.

is not uniformly distributed, but is especially marked in the neighbourhood of perthitic intergrowths of plagioclase. Several petrologists have supposed that the two peculiarities are related to one another, and have maintained the secondary origin of both. The conversion of orthoclase to microcline, or the setting up of the microcline structure in orthoclase, has been attributed to dynamic causes, and the

[1] On the analogies between the pyroxene and felspar groups see Wahl, *Oefversigt af Finska Vet.-Soc. Förh.*, 1906-07, No. 2 (1908).

vein-like intergrowths of plagioclase have been regarded as occupying mechanically-formed fissures in the crystals, but the phenomena seem to be sufficiently accounted for by spontaneous changes consequent upon fall of temperature.[1] In this view the lamellation in microcline may be strictly analogous with that in leucite, which it closely resembles, and the intergrowths of plagioclase are to be regarded as eliminated from the cooling crystals in consequence of diminishing solubility of plagioclase in the potash-felspar (Fig. 85, B). If orthoclase and microcline are truly dimorphous, with a definite temperature of inversion, the case is that illustrated in Fig. 83; if there is no discontinuous change, we have the case of Fig. 82. To fix the point *G*, we have Vogt's estimate that the intergrown plagioclase may constitute up to 15 or 20 *per cent.* of the whole, from which it is easy to calculate that the maximum content of Ab in potash-felspar at atmospheric temperatures is 10 or 15 *per cent.*, as compared with about 28 *per cent.* at the temperature of crystallization from the magma. It is in accordance with this that adularia, formed at low temperatures from aqueous solution, is poor in soda. On the other hand, the sanidine crystals of volcanic rocks may carry as much as 28 *per cent.* or more of Ab (with An) and show no sign of heterogeneity. Here we may perhaps suppose that the cooling was too rapid to permit the breaking-up of the mixed crystals, which are, in respect of the dissolved albite, in a metastable condition. In plutonic rock-masses the rearrangement may doubtless be aided by mechanical causes as an incidental result of dynamic metamorphism.

Intergrowths comparable with those of plagioclase in microcline are found in other imperfectly isomorphous series of minerals. Thus, the rhombic pyroxenes may contain diopside up to a certain limit (12 or 15 *per cent.*); but the crystals often show under the microscope that a part of their contained diopside, at atmospheric temperatures, is in the form of very fine intergrown lamellæ. This 'pyroxene-perthite,' as Wahl terms it, is attributed by Vogt

[1] Vogt, *Tscherm. Min. Petr. Mitth.* (2), vol. xxiv., pp. 537-541.

to the elimination of a part of the dissolved diopside in cooling. In this case, since the series falls under Type IV. of Roozeboom, no eutectic perthite is possible.

A like secondary origin may, perhaps, be assigned in some cases to *graphic intergrowths*. In the perfectly fresh charnockites of Southern India, for example, the felspar often contains numerous little intergrown inclusions of quartz, sometimes with a vermicular habit ('quartz de corrosion' of Lacroix). This mineral is in much too small amount for eutectic proportions; it may, perhaps, have been dissolved in the felspar at a relatively high temperature and eliminated on cooling.

CHAPTER XI

STRUCTURES OF IGNEOUS ROCKS

Porphyritic structure.—Characteristic structures of plutonic rocks.—Graphic intergrowths.—Spherulitic structures.—Structures related to movement.

Porphyritic Structure.—In this chapter we shall discuss briefly some of the chief structural characters of igneous rocks, with special reference to their origin and the light which they throw on the manner of consolidation of rock-magmas. From this point of view the most significant structure is the porphyritic. By this we understand the occurrence of some constituent of a rock (or of more than one constituent) in *two distinct generations*, referable to different stages in the consolidation of the magma. Where a mineral showing such recurrence is a member of an isomorphous series, there is a difference of composition between the two generations, in accordance with principles already discussed (p. 231). Apart from this, there are differences as regards crystal-habit and relations to other constituents of the rock. The porphyritic elements or phenocrysts—that is, the crystals of the earlier generation—are, as a rule, of notably larger dimensions. The occurrence of conspicuously large crystals of a particular mineral, such as the olivine in many basalts, does not, however, constitute in itself a porphyritic structure, the essential feature of this being the recurrence in a second generation.

In the *volcanic rocks* the phenocrysts are doubtless, at least in the main, of intratelluric origin—that is, they were formed prior to the extrusion of the lava—while the ground-mass, including the later generation of the same minerals, crystallized after extrusion. The two generations were, therefore, separated by a discontinuous change of physical conditions—viz., an abrupt acceleration of the rate of cooling, a sudden diminution of pressure, and probably some loss of the volatile constituents of the magma. Partial resorption of the phenocrysts may result, but we have seen that this may be brought about equally without any discontinuous change of external conditions (p. 212).

In *hypabyssal rocks* it is probable that the phenocrysts have in some cases been formed, or at least begun to be formed, when the magma was contained at a lower level in the earth's crust; but in general their formation and growth seems to have occurred 'in place' under the same general conditions as the crystallization of the enclosing ground-mass.[1] It is not uncommon to find abundant large porphyritic crystals in the central part of an intrusive sheet or laccolite, while the marginal part carries only a few small phenocrysts, or is wholly devoid of them. The phenocrysts in an intrusive rock do not always conform with the direction of flow-structure, and they sometimes take part in radiate and other groupings which give evidence of formation in place. In some cases the porphyritic crystals enclose, not only the earliest elements, such as apatite, but also small crystals or grains like those of the ground-mass, these occurring sometimes only in the marginal part of a phenocryst, but sometimes also in the interior. We may conclude that the continued growth of the porphyritic crystals, if not their initiation, has been subsequent to the intrusion of the magma, and in part simultaneous with the crystallization of the ground-mass. The facts, therefore, demand some other explanation than that offered by Rosenbusch, which covers most of the phenomena in the case of extruded lavas.

The view of rock-magmas as solutions leads directly to

[1] Zirkel, *Lehrbuch der Petrographie*, 3rd ed. (1893), vol. i., pp. 737-747; Cross, *14th Ann. Rep. U.S. Geol. Sur.* (1895), p. 231; Pirsson, *Amer. Journ. Sci.* (4), vol. vii. (1899), pp. 271-280; Crosby, *Amer. Geologist*, vol. xxv. (1900), p. 299.

an explanation of the porphyritic structure, which doubtless represents its general significance in this application. We have seen that, in the ideal case, the crystallization of a mixed magma, which is not of eutectic composition, takes place in two stages—(*a*) separation of the excess of one or more constituents over eutectic proportions, and (*b*) simultaneous crystallization of the residual minerals as a eutectic mixture. The actual conditions may import some modifications of this simple process. In particular, we have seen that, in consequence of supersaturation, the first stage of crystallization may be prolonged at the expense of the second. Making allowance for these disturbing factors, we may expect that, in a hypabyssal rock, the phenocrysts often represent, with some degree of approximation, the excess over eutectic proportions, and the ground-mass the quasi-eutectic residuum. In volcanic rocks this distinction will be obscured by the effects of the discontinuous change of physical conditions at the time of extrusion; for the interruption of crystallization which this implies will not in general coincide with the division between the stages (*a*) and (*b*). In this case, therefore, the phenocrysts must bear a different interpretation. It may be remarked that in hypabyssal rocks the porphyritic elements usually make up only a small part of the whole, while in volcanic rocks their relative amount varies greatly, and sometimes, as in the trachytic rocks termed sanidinites, the ground-mass is reduced to very small proportions.

The suggestion that the ground-mass of a porphyritic rock tends to approximately eutectic composition seems to be due in the first instance to Becker.[1] Vogt[2] has discussed the subject in some detail with special reference to the more acid rocks, in which felspars and quartz are the chief constituents. The approximate constancy of composition of the so-called 'microfelsite,' 'petrosilex,' etc., which makes the ground-mass of so many acid rocks, has long been

[1] *Twenty-first Ann. Rep. U.S. Geol. Sur.*, III. (1901), p. 519.

[2] *Silikatschmelzlösungen*, II. (1904), pp. 169-180; *Tscherm. Min. Petr. Mitth.* (2), vol. xxv. (1906), pp. 371-378.

recognised, and has, indeed, been responsible for the creation of more than one mineral species, now abandoned. Thus the 'krablite' and 'baulite' of Forchammer, regarded by Sartorius von Waltershausen as highly silicated felspars with oxygen-ratios 1 : 3 : 24, were shown by Zirkel to be fine-textured aggregates of felspar and quartz, and are doubtless approximately eutectic mixtures. Very many of the more acid porphyritic rocks consist mainly of felspars and quartz, with only a small amount of ferro-magnesian silicates, magnetite, etc. The rocks may differ very considerably in bulk-composition; the phenocrysts differ much more in different examples; but the ground-mass shows much less range of variation, and may be conceived as tending to a fixed standard. Vogt, amplifying one of Lagorio's generalisations, notes that in typical acid rocks, with silica-percentage 71 to 75, the ground-mass does not differ very notably in composition from the whole rock; in sub-acid rocks the ground-mass is decidedly more acid than the whole; and in ultra-acid rocks it is decidedly less acid. In short, the ground-mass of all these rocks tends to a silica-percentage of about 71 to 75, and a composition approximating to that of a felspar-quartz eutectic, such as a graphic granite. Since the rocks do not consist exclusively of felspars and quartz, the approximation is necessarily only of a rough kind. In rocks composed essentially of two minerals we may expect a closer agreement between observation and theory, and the composition of the ground-mass may then afford some estimate of the eutectic proportions of the two minerals. Zemčuzny and Löwinson-Lessing[1] have applied this method to some diorite-porphyrites, and found that theii presumably eutectic ground-mass of hornblende and plagioclase contains 45 to 50 *per cent.* by volume of the former mineral. This would give the eutectic proportions as between 50 : 50 and 55 : 45.

Miers has propounded an entirely different theory of the porphyritic structure, starting from the idea of a fundamental difference between the metastable and labile conditions of

[1] Abstr. in *Geol. Centralbl.*, vol. viii. (1906), p. 393.

supersaturation (p. 209). In his view the phenocrysts represent sporadic crystallization as the result of inoculation in the metastable region, while the ground-mass corresponds with the crowd of small crystals suddenly started in his experiments when a solution touches the labile region. We have already pointed out that in rock-magmas inoculation must be an exceptional thing. Further, there is an undoubted relation between the nature and amount of the porphyritic elements in a rock and the relative proportions of the minerals in the parent magma; while on Miers' hypothesis the crystallization of one or another constituent in the earlier stage would seem to depend upon accidental circumstances.

Characteristic Structures of Plutonic Rocks.—From the extremely slow rate of cooling under abyssal conditions, it is probable that plutonic magmas in general crystallize without attaining any great degree of supersaturation, and the relatively *coarse texture* of the rocks is in accordance with this supposition (p. 217). Again, plutonic rocks are as a rule *non-porphyritic*. It is true that Michel-Lévy[1] recognises even in normal granitoid rocks two distinct periods of consolidation. He relies chiefly upon such evidence as the fracture of early crystals and their cementation by later-formed elements; but the phenomena which he adduces seem to be capable of a different interpretation, and it is certain that plutonic rocks do not in general carry distinct phenocrysts in a finer-textured ground-mass, as is the rule in hypabyssal and volcanic rocks. There are, of course, exceptions, but they are scarcely known in any but the more acid rocks. Granites enclosing large felspar crystals are not uncommon, but the more basic plutonic rocks associated with them are non-porphyritic (compare pp. 131, 134).

If phenocrysts of certain minerals in a rock represent the excess of those minerals over the eutectic proportions, it would appear that the porphyritic structure is to be regarded as a normal character of any rock formed from a magma not

[1] *Ann. des Mines* (7), vol. viii. (1875), pp. 395-398; *Structures et classification des roches éruptives* (1889), pp. 3, 4.

of eutectic composition. From this point of view it is not the porphyritic, but the non-porphyritic, structure that calls for special explanation. Clearly we cannot suppose the magmas of plutonic rocks in general to be so nearly of eutectic composition that there is no appreciable excess of any constituent to separate out. If, however, the whole process of crystallization is effected within a moderate range of supersaturation, where the power of spontaneous crystallization is very small, there may be no sensible difference in dimensions and habit between the crystals of the earlier and later stages. In different porphyritic rocks the size of the elements of the ground-mass varies enormously, while the size of the phenocrysts (of a given mineral) varies comparatively little. It is precisely in the plutonic rocks that we should expect to find the limiting case realised, in which the crystals of the ground-mass are indistinguishable from the phenocrysts. There is, however, another point, which is, perhaps, of no less importance. It is probable that in plutonic rocks crystallization in the later stage (corresponding with that of the ground-mass in porphyritic rocks) proceeds largely by addition to crystals already existing rather than by the initiation of new centres. Of this there is sometimes direct evidence from inclusions in the crystals, and especially in the marginal parts of the crystals.

It is significant that the porphyritic granites give indications, which are wanting in more ordinary plutonic rocks, of the effects of supersaturation. The most remarkable examples are the 'rapakiwi' granites of Finland. These contain rounded felspars up to 3 or 4 inches in diameter, the rounding being doubtless due to resorption in connection with supersaturation (p. 212). The interior and principal part of each of these large crystals is of orthoclase, and this is bordered by oligoclase in crystallographic relation with it, or there may be numerous zones alternately of oligoclase and orthoclase. Some zones in this marginal part have inclusions of quartz, biotite, or hornblende. The alternations of orthoclase and oligoclase point to supersaturation of the magma with the two minerals alternately. In discussing the ideal case of a binary magma, we remarked (p. 213) that, when crystals of both minerals have been formed, and the eutectic point reached, supersaturation is no longer possible. But this does not hold if the crystals, say, of *A* are covered by a layer of *B*, or conversely. In this case the indicating-point (Fig. 86) may continue to oscillate about *e*, following the course *qrstqrst* . . ., until a number of alternating zones have been built up. This may be more clearly illustrated by a triangular diagram (Fig. 87), supposing the magma to

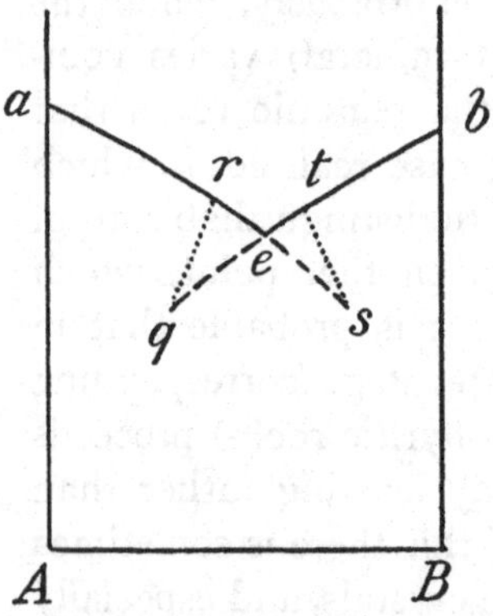

FIG. 86 —DIAGRAM TO ILLUSTRATE THE CRYSTALLIZATION OF THE LARGE FELSPARS IN THE 'RAPAKIWI' GRANITES. (COMPARE FIG. 66.)

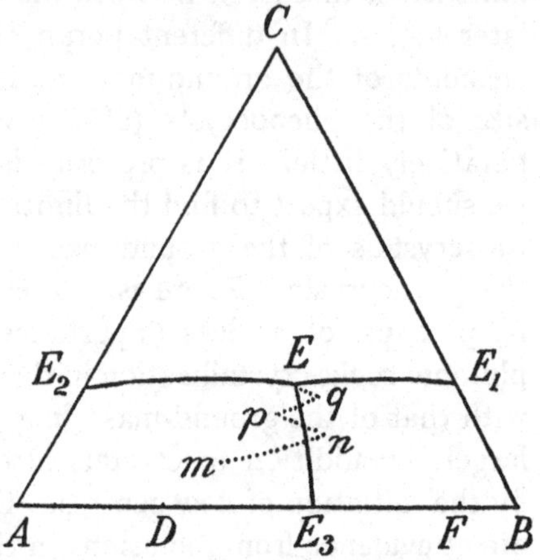

FIG. 87.—ANOTHER MANNER OF ILLUSTRATING THE SAME. (COMPARE FIG. 80.) *A* represents orthoclase; *B*, oligoclase; and *C* quartz.

consist only of felspars and quartz. The indicating-point follows the zigzag course *mnpq* . . ., oscillating about E_3E, but also approaching the ternary eutectic point *E*. Orthoclase crystallizes along *mn*, oligoclase along *np*, orthoclase again along *pq*, etc.; and it is easy to see that quartz also may be enclosed in the outermost zones of the crystal.

In very many plutonic rocks crystallization has not proceeded simply with reference to a definite eutectic composition, as in the ideal case, but its course has been modified by other factors, as already discussed. The general effect of these is to cause the several minerals to crystallize from the magma more or less distinctly in succession, the order often corresponding in a general way with Rosenbusch's law. In such a case the principles which make the porphyritic structure appear as a normal character of igneous rocks are no longer applicable. On the other hand, the crystallization of the several minerals in a definite order may give rise to other distinctive structures. Crystals originating in a magma only little supersaturated have a strong tendency to attach themselves to existing crystals of earlier-formed minerals. With a definite sequence of the several minerals,

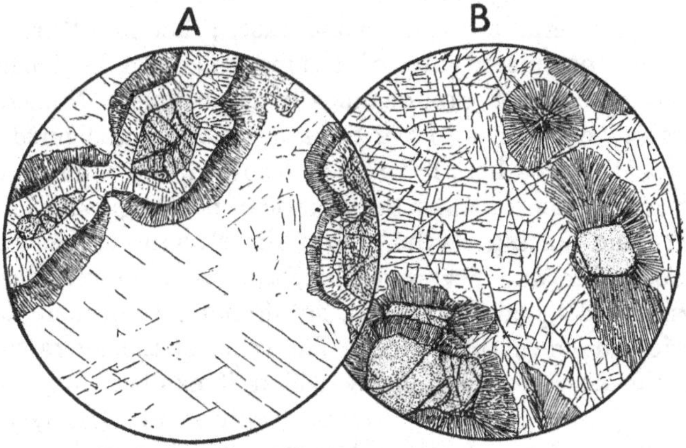

FIG. 88.—CORONA STRUCTURES. × 20.

A, norite, Risör, Norway: olivine surrounded by a double border of pyroxene, interposed between it and the felspar; B, eclogite, Chlumicek, Bohemia: garnet surrounded by a deep border of fibrous enstatite, interposed between it and the augite (omphacite).

this must give rise to a more or less pronounced arrangement of them in concentric zones about the earliest mineral as a nucleus. Such an arrangement is scarcely perceptible in the more acid plutonic rocks (apart from distinct 'spheroids'), but in basic rocks it is sometimes developed in a remarkably regular manner. This is the *corona structure*, common in gabbros and norites (Fig. 88, A). For instance, a crystal of olivine may be surrounded by a ring of enstatite grains, and this by a second ring, composed of augite, felspar, as the latest mineral, filling the interspaces between aggregates of this kind. Often there is a radiate arrangement of the crystals in these bordering growths, and sometimes two minerals have crystallized together in a radiate fibrous intergrowth. The 'celyphite' borders round the garnets of some eclogites and peridotites are structures of the same kind; they are of pyroxene, hornblende, etc., in different cases (Fig. 88, B). Corona-structures have been regarded by some petrologists as of secondary origin, a pyroxene border interposed between olivine and felspar, for example, being attributed to chemical reactions between these two minerals, and referred to dynamic metamorphism. This is doubtless the true interpretation in some cases; but in others the reactions postulated are of a kind unknown in dynamic metamorphism, or are chemically impossible. Numerous types of 'couronnes' have been described and figured by Lacroix;[1] and others have been recorded, often under the name 'reaction-rims,' by the American petrologists.[2]

Graphic Intergrowths.—Graphic structures, including micrographic and cryptographic, are found in very many abyssal and hypabyssal rocks, but do not call for more than a brief notice in this place. The scale of the intergrowth and the degree of regularity which it exhibits may vary greatly. As a rule the coarser structures are less regular than those of fine texture; but this does not apply to the coarse-grained pegmatites, which often contain intergrowths of great regularity and beauty.

The commonest case is that in which the constituents are quartz on the one hand, and orthoclase, or microcline, or oligoclase on the other; but many other associations are recorded.[3] In the Norwegian syenite-pegmatites, Brögger

[1] *Contributions à l'étude des gneiss à pyroxène et des roches à wernerite* (1889).

[2] G. H. Williams, *Bull. No.* 28, *U.S. Geol. Sur.* (1886); Bayley, *Amer. Journ. Sci.* (3), vol. xliii. (1892), pp. 515-518, and *Journ. Geol.*, vol. i. (1893), pp. 702-710; Kemp, *Bull. Geol. Soc. Amer.*, vol. v. (1894), pp. 218-221; Matthew, *Trans. N.Y. Acad. Sci.*, vol. xiii. (1894), pp. 198-201; etc.

[3] Lacroix, *op. cit. sup.;* Brögger, *Syenitpegmatitgänge* (1890), pp. 149-152; Duparc and Pearce, *Rech. géol. et pétr. sur l'Oural du Nord*, II. (1905), p. 457; etc.

mentions graphic intergrowths of alkali-felspars with diopside, ægirine, hornblende, lepidomelane, nepheline, and sodalite. In other rocks augite is intergrown with oligoclase, labradorite, anorthite, nepheline, or spinel; anorthite with olivine;[1] garnet with quartz, or with hornblende; hypersthene with labradorite; magnetite with hornblende, probably uralitized augite; quartz with tourmaline; etc.

A primary graphic intergrowth clearly proves the simultaneous crystallization of the two minerals involved, and may be interpreted as a eutectic mixture (Fig. 89, A). There

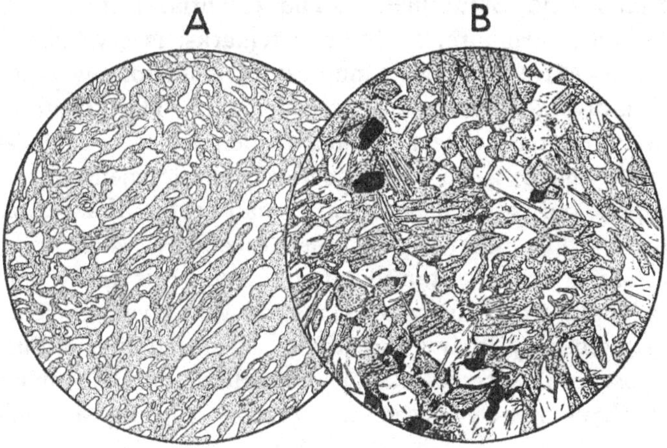

FIG. 89.—MICROGRAPHIC INTERGROWTHS. × 20.
A, felspar and quartz in pegmatite, Auchendryne, Aberdeenshire: drawn in polarised light; B, augite and nepheline in nepheline-dolerite, Löbau, Saxony.

may be, however, some uncertainty in using it to determine directly the eutectic proportions of the two minerals. For instance, in Fig. 89, B, augite and nepheline are intergrown in presumably eutectic proportions; but there are also more or less idiomorphic crystals of augite, which we must suppose to represent an excess of that mineral, and in practice it would not be easy to estimate and allow for this. Again, in a coarse-textured pegmatite it may sometimes be seen that a large crystal of felspar contains intergrown quartz in one

[1] Harker, *Geology of the Small Isles*, *Mem. Geol. Sur.* (1908), p. 88.

part and not in another. When the stage of eutectic crystallization was reached, the felspar still continued to grow in continuity with an existing crystal.

Graphic intergrowths of three minerals, presumably corresponding with ternary eutectics, have been described in some rocks. Probably some of the associations noted by Lacroix in certain pyroxene-gneisses represent ternary and even quaternary eutectics. A commoner instance of a ternary eutectic mixture is a graphic intergrowth of quartz in a perthitic felspar.

Spherulitic Structures.—The essential feature of the spherulitic structure, as Iddings remarks, is crystallization about a centre, or about a number of neighbouring centres, with a divergent or radiate arrangement. The growth may or may not terminate at a sharply-defined outer boundary. It may have a regular spherical development, or may comprise only a hemisphere or a small sector of a complete sphere. The last is the case especially where radiate crystallization has started simultaneously at many points distributed on a particular plane, so that there is freedom of growth only in two directions. The 'axiolite' of Zirkel, supposed to represent radiate crystallization about an axis, is perhaps merely a section of such a plane spherulitic growth.

It is chiefly in the acid rocks, volcanic and hypabyssal, that regular spherulitic structures are common, and they have been studied by numerous observers.[1] The spherulites of acid igneous rocks fall into two chief classes, according as the radiate growth is constituted (*a*) by graphic intergrowths of felspar and quartz, or (*b*) by felspar fibres only.

Spherulites of the first kind characterize a large group of hypabyssal acid rocks, to which Rosenbusch has applied the inappropriate name 'granophyre,' and they are also met with in rhyolitic lavas and pitchstones. A quartzo-felspathic magma with a composition approaching the eutectic, if

[1] For a historical review, and a clear presentation of the subject, see Cross, *Bull. Phil. Soc. Washington*, vol. xi. (1891), pp. 411-444, and Iddings, *ibid.*, pp. 445-464.

crystallized not too rapidly, has a marked propensity towards graphic intergrowths; and these growths tend to form about centres, with or without a porphyritic crystal as nucleus. The tendency to centric grouping, with a divergent arrangement, becomes more pronounced in proportion as the graphic intergrowth is on a more delicate scale. At the same time the sectors within which the felspar (or quartz) has a common orientation become narrower, and there is a more regularly radiate disposition

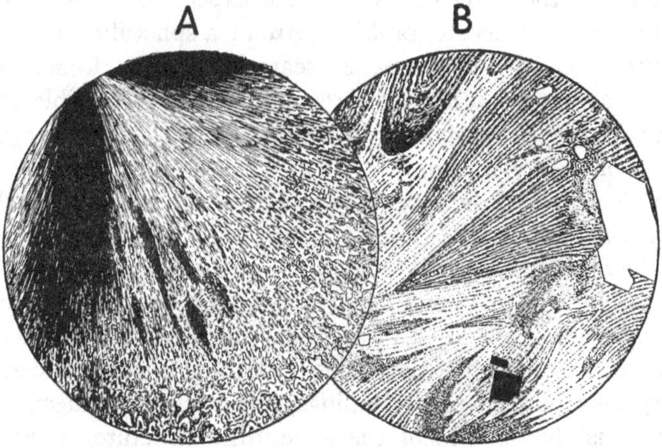

FIG. 90.—SPHERULITIC STRUCTURES IN ACID ROCKS × 20, WITH CROSSED NICOLS.

A part of a graphic spherulite in the marginal modification of a granite mass, Glas-bheinn Bheag, Skye. The centre is above and to the left, and two arms of the 'black cross' are shown. The structure graduates from cryptographic in the centre to micrographic at the margin. B, part of a large complex spherulite of felspar fibres in a dyke, apophysis of a granite, Druim an Eidhne, Skye. The centre is to the left and below the figure. The spherulite consists essentially of a number of closely-packed conical growths, originating at different centres and having a general radiate disposition. A cone in the middle of the field is cut along its axis; others in the upper part give parabolic sections.

both of these sectors and of their component elements. Finally, the structure merges, in appearance, into a simple radiate grouping of closely-packed fibres, giving between crossed nicols a more or less regular 'black cross' effect. The gradation from a micrographic structure to a cryptographic (in which the component elements can no longer be resolved by the microscope) may be seen, not only in different spherulites, but in different parts of the same spherulite (Fig. 90, A). Teall[1] has expressed the opinion that spherulites of this graphic type are of eutectic composition. Probably this is often true, but it does not appear to be a necessary conclusion in all cases. A spherulite does not crystallize at a point of time, but grows from the centre outward. Where a rock, not as a whole of eutectic composition, consists of spherulites with an interstitial groundmass, it is the latter that we should expect to represent the eutectic. The cryptographic centre of a spherulite like that figured has optical properties scarcely different from those of an aggregate of felspar fibres, and it seems possible that there is a larger proportion of felspar there than in the outer parts of the spherulite.

Spherulites of the second kind are found especially in the acid volcanic rocks. They are represented in great variety in the obsidians and other rhyolites of the Yellowstone Park, Colorado, Hungary, and Lipari; and their counterparts may be recognised, disguised by secondary changes, in the ancient acid lavas of the British Isles.[2] Although other minerals may take part in the groupings, the radiate arrangement, which is the essence of the spherulitic structure, is one of delicate felspar fibres only. The larger spherulites of this kind, sometimes measurable by inches or even feet, are always of complex structure, the radiation starting from numerous centres (compare Fig. 90, B), and the fibres frequently bifurcate, producing a tufted or plumose arrangement. Such spherulites are skeleton structures, the aggregate of felspar fibres occupying only part of the volume which they pervade. Sometimes, in fresh examples, they are embedded in glass; sometimes the interspaces are vacant

[1] *British Petrography* (1888), p. 402, and *Quart. Journ. Geol. Soc.*, vol. lvii. (1901), pp. lxxv, lxxvi.

[2] Cole, *Quart. Journ. Geol. Soc.*, vol. xlii. (1886), pp. 183-190; Miss Raisin, *ibid.*, vol. xlv. (1889), pp. 247-269; Cole, *ibid.*, vol. xlviii. (1892), pp. 443 445; Parkinson, *ibid.*, vol. lvii. (1901), pp. 211-225; Boulton, *ibid.*, vol. lx. (1904), pp. 457-463.

or partly occupied by pellets of tridymite scales and other minerals. Sometimes, again, there is a matrix of opal, and Cross assigns to this amorphous hydrous silica an important part in the formation of spherulites of this kind. In the altered rocks tridymite and opal are converted to quartz, and any empty spaces which may have existed are filled by this and other secondary products. Cross and Iddings proved that the fibres of felspar (orthoclase or soda-orthoclase) are not always elongated along the clino-axis, but often along the vertical axis of crystallography, and that the optical orientation may be either the normal or the abnormal. In the latter case ('orthose déformé' elongated parallel to the *c*-axis) the fibres are of positive character, and for this reason they have probably been often confused with quartz. The bifurcation of fibres (elongated parallel to the *a*-axis) corresponds with the Manebach twin-law.

The fibrous habit suggests that the spherulites crystallized rapidly in a highly supersaturated solution, and this was doubtless their mode of origin. When they are embedded in a glassy or devitrified matrix, the flow-lines are seen to pass uninterruptedly through the spherulites, and, indeed, the latter may sometimes be seen to have formed subsequently to brecciation of the rock. In such a case the matrix was a glass rather than a liquid when the spherulites crystallized. The crystallization at isolated centres doubtless depended upon local richness in dissolved water-vapour,[1] reducing the viscosity in those places. The porous nature of the spherulites is connected with the liberation of the steam in the process of crystallization, and in the complex chambered 'lithophyses' the steam-cavities may occupy the greater part of the volume.

Concerning spherulitic structures in basic rocks our information is much more scanty. Spherulites are here of less frequent occurrence than in the acid rocks; they are usually of very small dimensions; and their true nature may be greatly obscured by secondary changes, as in the so-called 'variolites.' Nevertheless, an examination of some of the minor intrusions, and especially the tachylytic margins of

[1] Iddings, *Amer. Journ. Sci.* (3), vol. xxxiii. (1887), pp. 36-45.

dykes and sheets, in such a region as that of the western islands of Scotland furnishes some interesting material.[1] It is probable that here, also, two principal classes of spherulites may be recognised—viz., fine intergrowths (possibly eutectic) of two minerals, and radiate groupings of fibres of a single mineral, forming skeleton growths.

The most usual spherulites in tachylytic rocks are very minute, and of too fine structure to be clearly resolved. In thin slices a radiate structure may or may not be apparent

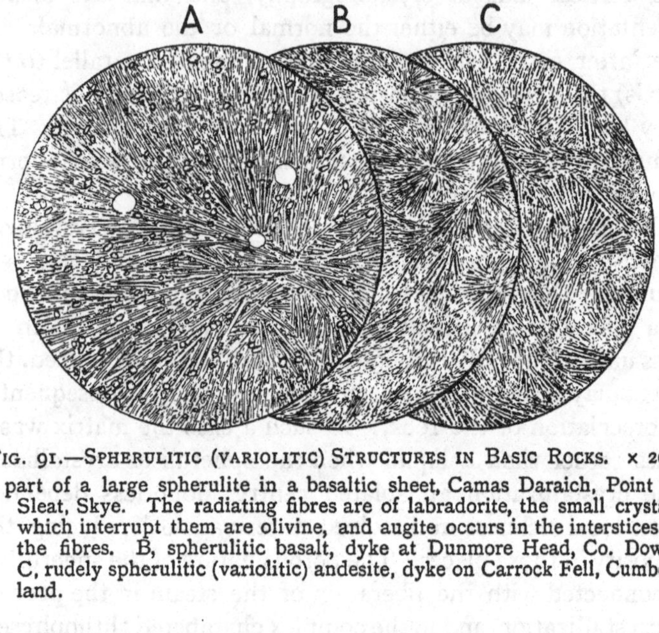

FIG. 91.—SPHERULITIC (VARIOLITIC) STRUCTURES IN BASIC ROCKS. × 20.
A, part of a large spherulite in a basaltic sheet, Camas Daraich, Point of Sleat, Skye. The radiating fibres are of labradorite, the small crystals which interrupt them are olivine, and augite occurs in the interstices of the fibres. B, spherulitic basalt, dyke at Dunmore Head, Co. Down C, rudely spherulitic (variolitic) andesite, dyke on Carrock Fell, Cumberland.

in natural light, and the depolarising effect between crossed nicols also varies in different cases. Instead of a distinct 'black cross,' more irregular dark brushes are usually seen, a natural consequence of a wide extinction-angle in the component fibres. The brown and yellow colours of many spherulites of this kind may probably be attributed to an admixture of interstitial glass between minute fibres of

[1] Harker, *Tertiary Igneous Rocks of Skye*, *Mem. Geol. Sur.* (1904), chap. xix., with references.

felspar (labradorite); but in some cases there are features which require some other explanation. For instance, the spherulites in some rocks are strongly pleochroic, which apparently points to some pleochroic mineral in fibrous or cryptographic intergrowth with the felspar.

The larger spherulites in basic rocks are of the skeleton kind. In one type the chief element is labradorite, in delicate rods or fibres with a radiate disposition. The other minerals, which do not share in this arrangement, are small crystals of olivine, little subophitic patches and granules of

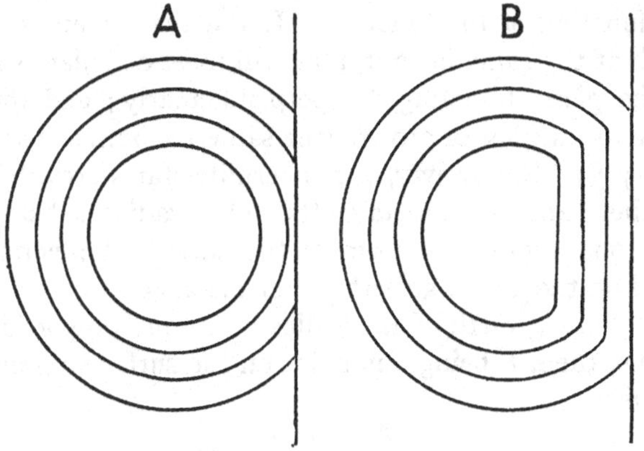

FIG. 92.—DIAGRAM OF A SPHERULITE MODIFIED IN FORM BY THE INTERFERENCE OF AN OBSTACLE. (AFTER POPOFF.) (SEE TEXT.)

augite, and some magnetite (Fig. 91, A). In other rocks the mineral which occurs in fine rods or fibres, and gives the spherulitic structure, may be augite or even magnetite. The radiate arrangement of these is often of a rude kind. This, indeed, is true of many of the spherulitic structures in basic rocks, and in some cases the radiation degenerates into a sheaf-like or fan-shaped grouping (Fig. 91, B, C).

Although spherulites in general doubtless crystallize rapidly, their growth often takes place in distinct stages, and this may be marked by concentric zones, differing perhaps in structure, or merely divided by pauses in the growth. Where a spherulite has encountered some obstacle, the concentric zones are truncated as in A, and do not adapt themselves by deformation as in B (Fig. 92). This proves that radiation from a centre is the essential thing, and a concentric structure only incidental. Popoff[1] has pointed out that, when spherulites in a rock are closely set, the manner of their interference with one another may throw light on the law of their growth. Suppose that spherulites start from two neighbouring centres A and B, at a distance a apart, and grow outward until they meet. Suppose, further, that they follow a definite law of growth, so that the radius of a spherulite after a certain time of growth is some function of the time. (i.) If the two spherulites start growth at the same instant, they will meet at a plane surface, viz., the plane bisecting AB perpendicularly; and this will be true for any law of growth (the same for both spherulites), (Fig. 93, I). If, however, one spherulite starts growth before the other, and has already attained a radius c before the second one begins, the form of the surface of junction will depend on the law of growth. For instance:

(ii.) If the growth of the radius be simply proportional to the time, then, P being any point on the surface of junction, we have

$$AP-BP=c.$$

This surface is concave towards B—*i.e.*, towards the smaller spherulite. Its trace on the plane of the figure (Fig. 93, II) is one branch of a hyperbola, with foci A and B and eccentricity $a \div c$.

(iii.) If the square of the radius (and, therefore, the surface-area of the spherulite, supposed solid) be proportional to the time, then

$$AP^2-BP^2=c^2.$$

The surface of junction is, therefore, a plane perpendicular to AB, and dividing it in the ratio $a^2+c^2 : a^2-c^2$ (Fig. 93, III).

[1] 'Beitrag zum Studium der Sphärolithbildung,' *Förh. Nord. Naturf. Helsingfors* (1902), sect. iv.; *Tscherm. Min. Petr. Mitth.* (2), vol. xxiii. 1904), pp. 153-179.

(iv.) If the cube of the radius (and, therefore, the solid content of the spherulite) be proportional to the time, then

$$AP^3 - BP^3 = c^3,$$

and it can be shown that in this case the surface is concave towards *A*—*i.e.*, towards the large spherulite (Fig. 93, IV).

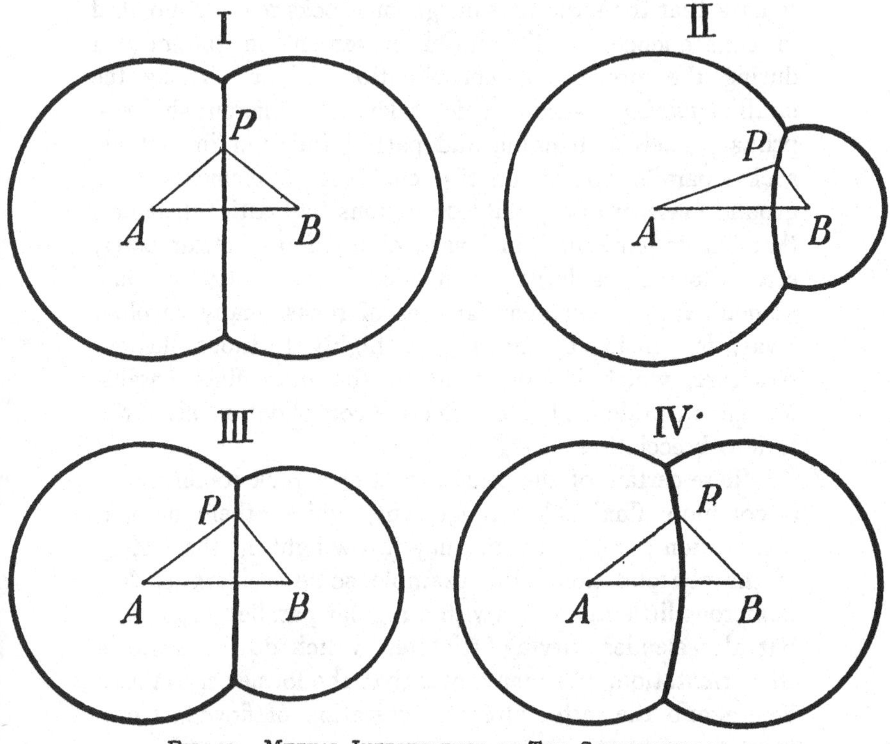

FIG. 93.—MUTUAL INTERFERENCE OF TWO SPHERULITES: I, starting growth at the same instance and following any law; II, when the radius is proportional to the time of growth; III, when the square of the radius is proportional to the time; IV, when the cube of the radius is proportional to the time.

Popoff finds the hyperbolic form (with the plane as a particular case) in several spherulitic rocks which he has examined, and in these the radial growth was presumably at a uniform rate. This, however, is by no means a universal rule. Plane surfaces of junction, between spherulites of different sizes, may be seen in various rhyolites from Lipari, Hungary, Sauerland, and Antrim,[1] and in tachylytes and variolites from Mull and Skye.[2] These presumably fall under case (iii.), the radial growth being proportional to the square root of the time.

Structures related to Movement.—Finally, we have to remark that the structures of igneous rocks may be modified in consequence of differential movement in the magma during the process of consolidation. Here belong the familiar *fluxion-structures*, which take many different shapes—primary gneissic banding and partial foliation in plutonic rocks; parallel orientation of phenocrysts, of elements of the ground-mass, or of crystallites; various 'eutaxitic' structures (banded or lenticular) in lavas; elongation of steam-pores, etc. Flow being largely controlled by viscosity, the phenomena vary in different families of rocks, many rhyolitic lavas, for instance, showing a highly tortuous fluxion-structure, which is not found in the more fluid basalts. Movement prolonged after effective consolidation gives rise to flow-brecciation.

Where certain of the constituents of a rock conform with a common fluxional arrangement, while others do not, observation of these points may throw light on the history of the consolidation. For example, some obsidians enclose numerous little microlites with a regular parallel disposition, but also slender curving 'trichites' which do not share in this orientation. We may infer that the former crystallized before, and the latter after, the cessation of flowing movement.

In a rock wholly or largely crystalline, differential movement during the period of crystallization may impart a very distinctive *granulitic* habit to some of the constituent minerals. Judd,[3] describing the Tertiary dolerites of Scot-

[1] Note also a well-known spherulitic felsite from Arran, figured in Teall's *British Petrography*, Pl. XXXIX., Fig. 1.

[2] See, *e.g.*, Clough and Harker, *Trans. Edin. Geol. Soc.*, vol. vii. (1899), p. 384.

[3] *Quart. Journ. Geol. Soc.*, vol. xlii. (1886), p. 76, and figures.

land and Ireland, distinguishes two main types of structure, determined by the ophitic or the granulitic habit of the augite, and depending on the conditions attending the consolidation. When crystallized free from disturbance, the augite has enwrapped or enclosed the felspar crystals; but with contemporaneous movement intricate shapes could not form, or were immediately broken up, and the augite appears as aggregates of granules in the interspaces between the felspars (Fig. 94). In some aplitic and other rocks all the

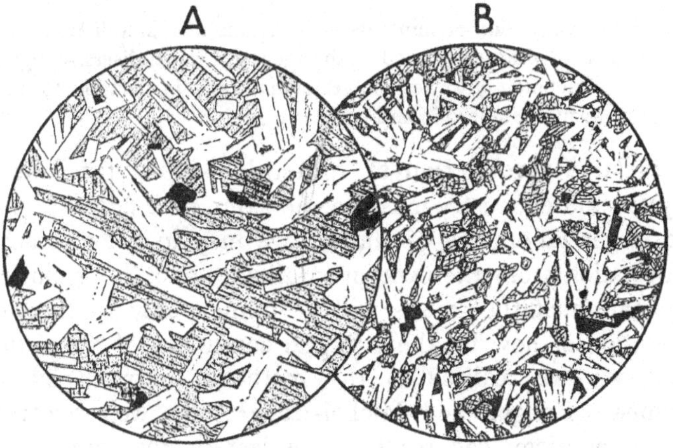

FIG. 94.—DOLERITE OF INTRUSIVE SILL, ISLE OF EIGG. ×20.
A, main body of sill, with typical ophitic structure; B, finer-textured modification, with granulitic augite.

principal constituent minerals are more or less affected in this way by the disturbing influence of movement, prolonged or frequently renewed during the period of crystallization. The resulting even-grained mosaic gives a very characteristic appearance in thin sections of such granulitic rocks. It is interesting to observe that somewhat similar structures may arise from internal movement in solid rocks subjected to powerful stresses. This secondary granulitisation is usually accompanied by effects of internal strain in the individual crystals, or by mineralogical transformations of the kind proper to dynamic metamorphism.

CHAPTER XII

MINERALISERS AND PNEUMATOLYSIS

Artificial reproduction of minerals.—Functions of mineralisers in rock-magmas.—Minerals formed by the agency of mineralisers.—Pegmatites.—Pneumatolytic action.—Pneumatolysis in relation to metamorphism.—Fumerolic and solfataric action.

Artificial Reproduction of Minerals.—Since the laws of physics and chemistry must be the same in a crucible as in the larger laboratory of Nature, we may look for valuable information from the experimental side, and the geological application of the results will be limited only by the consideration of how far the essential conditions of the natural processes are realised in the methods contrived by the chemist. Fouqué and Michel-Lévy[1] have well remarked that mineralogy has reached the last of the three stages which mark the evolution of a science—observation, analysis, and synthesis. Further information is afforded by metallurgical and other technical operations, and by various phenomena of an accidental kind.

Many mineral syntheses have been made by the French chemists Berthier, Ebelmen, Sénarmont, Ste.-Claire Deville, Daubrée, Fouqué and Michel-Lévy, Bourgeois, and Hautefeuille; and to this list we may add the names of Von Chrustschoff, Doelter, Vogt, Morozewicz, and others, especially the American experimenters whose work has already been mentioned. Some minerals have been obtained in crystals by merely fusing together their constituents, or by devitrifying a glass of the proper composition; or the

[1] *Synthèse des minéraux et des roches* (1882).

required mineral has been obtained with others by fusion of other bodies, which react in the way of 'double decomposition.' In another class of experiments water is employed as a solvent, and crystals of various minerals are obtained by prolonged digestion of appropriate substances with water in a sealed vessel and at a relatively high temperature. Other methods involve the use of various special fluxes and solvents, such as fluorides, chlorides, borates, tungstates, etc.; or, again, volatile substances of this kind have been employed in the gaseous form. Negative results may be of great value, so that the failures are sometimes as instructive as the successes.

The most important result, from our present point of view, which may be derived from this large body of data is that the rock-forming minerals divide broadly into two classes. Some crystallize more or less readily from 'dry fusion'—*i.e.*, from a 'melt' or magma not containing water or other fluxes—while others can be obtained only by the aid of some extraneous flux or solvent, or by reactions in which a body of this nature takes part.

The minerals which have been artificially reproduced *from simple igneous fusion* include the following common constituents of igneous rocks:

Olivine group (forsterite, fayalite, and intermediate varieties); various pyroxenes, including enstatite, diopside and magnesium-diopside, augites, acmite, etc.; certain garnets, in particular melanite (not grossularite or common garnet); perhaps some varieties of biotite,[1] though the brown micas which come here (occurring exceptionally in slags) may not be identical with any natural micas.

Plagioclase felspars, ranging from anorthite to oligoclase; leucite and nepheline; the melilite group (åkermanite, gehlenite, and intermediate varieties); doubtfully meionite; sillimanite and cordierite.

Tridymite; corundum and hæmatite; rutile; various spinellids, including spinel, pleonaste, hercynite, and magnetite.

[1] See especially Vogt, 'Om kunstig dannelse af glimmer,' *Christiania vidensk. forh.*, 1887, No. 6.

Apatite, though it is doubtful whether the artificial crystals (found in certain slags) have the composition of the natural mineral.[1]

These are, as a whole, what we have distinguished as high-temperature minerals. They are, also, for the most part, minerals characteristic of basic rocks rather than acid; and we may remark that the only igneous rocks which have been imitated in the laboratory by simple fusion are those of basic and ultrabasic composition.

The second category includes minerals which have never been produced from 'dry' fusion, but only *with the aid of water or other fluxes, solvents, or special reagents.* Here come, of course, all those minerals which contain water or hydroxyl, fluorine, chlorine, or boron, as part of their constitution; but the list comprises also many others, in which the special reagent employed does not form any part of the final product. Omitting for the present the rarer minerals, which are exceedingly numerous, we mention only—

Quartz, obtained by digestion with water in a sealed vessel, and also by the aid of tungstates and other special reagents.

Alkali-felspars, albite and orthoclase (sanidine, adularia), reproduced by similar methods.

Sodalite group (sodalite, haüyne, nosean).[2]

Hornblende[3] and certain biotites.[4]

Beryl; zircon; perofskite and sphene.

Some important minerals—notably muscovite, topaz, and tourmaline—have hitherto defied all attempts towards their synthesis.

The minerals here enumerated are among those which we have seen reason for regarding as low-temperature minerals.

[1] Vogt suggests that the $CaCl_2$ or CaF_2 of the natural mineral may be replaced by $CaSiO_3$. Fluor-apatite and chlor-apatite, like those in igneous rocks, have been artificially made by various special methods.

[2] Morozewicz, *Tscherm. Min. Petr. Mitth.* (2), vol. xviii. (1898), pp. 128-155.

[3] Chrustschoff, *Neu. Jahrb. Min.*, 1891, vol. ii., pp. 86-90.

[4] Chrustschoff, *Tscherm. Min. Petr. Mitth.* (2), vol. ix. (1888), pp. 55-60.

They are, as a whole, the characteristic minerals of acid rocks, and especially of acid plutonic rocks. It is impossible to disregard the significance of such a body of facts. The experimental evidence proves clearly that ordinary basic rocks may (we cannot assert that they in general do) crystallize from simple fusion of their known constituents; and, indeed, close imitations of such rocks as basalts have been artificially made in this way. No quartzo-felspathic rock has been so reproduced, and all attempts to crystallize the characteristic minerals of such rocks from simple fusion of their constituents have been fruitless. What essential factor, then, is found in the natural and not in the artificial processes? Considerations relative to pressure and rate of cooling afford no satisfactory answer to this question, but experiment offers an answer which seems to be sufficient. We are led to the conclusion that, while a basalt may represent closely enough, in total chemical composition, the magma from which it crystallized, this is not true of a granite. In the latter case constituents of some kind, which played a very essential part in the crystallization of the magma, have been lost. The special methods by which quartz, alkali-felspars, etc., have actually been reproduced in the laboratory suggest that these volatile constituents of the magma were water and certain chemically active substances of a class which figures in most experiments of this kind. The water imprisoned in the quartz-crystals of a granite proves the presence of this body in the magma, and there are also indications (such as traces of fluorine in mica) that other substances have been concerned.

There are other points on which experimental research affords information. It is clear that very important effects may be brought about by even small amounts of special reagents in a molten magma. Thus, Morozewicz found that from a simple fused silicate-magma an excess of silica separated out either as tridymite or as a new (tetragonal) form of silica; but the addition of 1 *per cent.* of tungstic acid caused the whole to crystallize as an aggregate of sanidine, quartz, and biotite.

The action of such special substances is doubtless partly physical, partly chemical. The nature of the chemical effect, where the agent does not enter as part of the crystallized product, is somewhat obscure—it is sometimes designated a *catalytic action*. As originally proposed by Berzelius, this term seems to signify a peculiar property possessed by certain bodies of inducing chemical changes in other bodies without themselves partaking in the reactions. Ostwald[1] conceives that the function of a catalyser is to accelerate the rate at which a given system reaches equilibrium; if that rate was so slow as to be imperceptible, the effect may almost be described by saying that the presence of the catalyser causes a change which would not otherwise have taken place. It must be difficult to prove in any case by demonstration that the agent in question does not itself participate in any chemical changes, and many chemists appear to hold a somewhat different view. They regard the catalytic agent as tending to form with one of the constituents of the system a compound which is unstable, and immediately breaks up, releasing the catalyser to react upon a fresh portion of the system.[2] In this way a small quantity of the catalyser may produce a large effect. To this *cyclical action* we may perhaps attribute the influence of fluorides and borates as mineralisers in cases where the resulting mineral contains no fluorine or boron.

Functions of Mineralisers in Rock-Magmas.—From the time of Elie de Beaumont[3] many petrologists, especially those of the French school, have recognised the important part which may be assigned in petrogenesis to what are often designated 'agents minéralisateurs' or *mineralisers*. These have been defined as volatile substances which, without entering into the final composition of minerals, render possible or facilitate their formation and crystalliza-

[1] *Lehrbuch der Allgemeinen Chemie*, vol. ii., part i. (1893), p. 515.

[2] A familiar example is the influence of manganese dioxide in the ordinary laboratory process of obtaining oxygen from potassium chlorate.

[3] 'Note sur les émanations volcaniques et métallifères,' *Bull. Soc. Géol. Fra.* (2), vol. iv. (1847), pp. 1249-1333.

tion.[1] There is, however, no reason for excluding the case in which the mineraliser, or part of it, enters into the mineral as finally constituted. Our mineralisers, then, are the same class of substances which have already been enumerated as employed by chemists in the synthesis of minerals which do not crystallize from simple igneous fusion. They are, in the first place, water; then other bodies, more restricted, perhaps, in their province, but at the same time more potent—viz., fluorides, chlorides, and borates, to which we may probably add others of less importance, such as tungstates, phosphates, vanadates, and carbonates.

Daubrée, Fouqué, Michel-Lévy, Lacroix, and others, have laid great stress upon the importance of the rôle played by mineralisers, not only in the crystallization of many rock-forming minerals, but also in magmatic differentiation, contact-metamorphism, and other geological processes. As regards their primary function, the agency of mineralisers has been invoked especially to explain the mineral constitution of the more acid igneous rocks. It is in these that the formation of the minerals actually found seems to be inexplicable without this hypothesis; and it is especially in the acid rocks that the presence of small amounts of fluorine, boron, etc., in certain of the constituent minerals affords in some degree direct evidence of the intervention of substances of the class postulated. Basic rocks are in general composed of minerals which can crystallize from 'dry' fusion; but, since it is part of the definition of mineralisers that they may be completely eliminated from the final products of crystallization, it is impossible to prove their absence from the parent-magma of any igneous rock. Water is probably present, though in varying amount, in all rock-magmas; and apatite, which is seldom wanting in igneous rocks, probably always contains either fluorine or chlorine.

[1] Elie de Beaumont, however, applied the term in the first place to such bodies as sulphur, arsenic, etc., which are found combined with the heavy metals in ordinary metalliferous lodes. He believed the contents of the lodes to have been introduced by sublimation, and his 'minéralisateur,' in this connection, was a body which, itself volatile, also forms volatile compounds with other bodies.

Since the mineralising agents are volatile substances, it is to be expected that they will in great measure escape from a magma which is extruded subaërially. Accordingly, we find that those minerals which are presumed to depend for their formation on the presence of mineralisers are, as a class, characteristic of plutonic rather than volcanic rocks. This is eminently true as regards micas, topaz, tourmaline, and others, which contain fluorine or boron as part of their constitution, though the occurrence of the sodalite group of minerals in lavas proves that the volatile constituents may not be wholly eliminated even from volcanic rocks. In general, having regard to both the physical and the chemical properties of these substances, it is clear that their escape, or partial escape, from an extravasated lava must have important consequences. Firstly, since they are powerful fluxes, their loss may be the immediate cause of consolidation, apart from any fall of temperature; secondly, since they are mineralisers, presumed to be necessary for the crystallization of certain minerals, their loss may be the immediate cause determining the consolidation of the magma as a glass.

A certain element of mystery has often attached to mineralisers and their functions, which has raised protests in some quarters.[1] This mystery arises partly from the practice of some French petrologists of ranking mineralisers with temperature and pressure as among the conditions which control the crystallization of rock-magmas; partly also from the conception of special mineralising agents, which seems to be entertained in some quarters, as gaseous emanations having an origin in some sense independent of the magmas on which they operate. From our point of view, however, the 'mineralising' substances are not to be regarded as in any sense adventitious or connected with external conditions, but as *an integral part of the rock-magma itself* (p. 47). There is clear evidence that they have been operative throughout the whole progress of consolidation of the magma. Further, as we shall see, the paragenesis of

[1] Morozewicz, *Tscherm. Min. Petr. Mitth.* (2), vol. xviii. (1898), p. 8.

the characteristic minerals shows very clearly that particular mineralisers are especially associated with particular kinds of rocks—boric acid with granites, fluorine with granites and syenites, chlorine with gabbros, etc.

Again, to preclude misunderstanding, it is to be remarked that the properties which we attribute to what are conveniently called mineralisers are not in any wise peculiar, but are shared by the silicates and other constituents, the difference being one *not of kind, but of degree.* This was clearly recognised by Elie de Beaumont himself. In one passage he remarks (*loc. cit.*, p. 1308): "According to the hypothesis proposed by M. Durocher, granite would have owed its liquidity to the fact that the felspar, quartz, and mica were in solution in one another, and formed a sort of fusible alloy." But he goes on to observe that this is not sufficient, since felspar and sometimes quartz are found to have crystallized in mass by themselves, and he therefore supposes the intervention of some more powerful flux.

The question of the part played by water in the formation of such rocks as granite is a very old one, and both Plutonists and Neptunists are still among us, though the issue now seems to be rather one of terminology.[1] To take a simple illustration, a salt such as sodium chloride has a definite melting-point; the addition of even the smallest proportion of water lowers the melting-point proportionately, more water causes further depression, and so forth. The distinction between fusion and solution has, from this point of view, no practical significance. To say that a granite magma is in a condition of 'hydato-pyrogenic,' or 'hydro-thermal,' or 'igneo-aqueous' fusion is to assert of it what is almost certainly true in its degree of all natural rock-magmas. So long as it is recognised that the distinction implied is one of degree rather than of kind, such terms may perhaps be conveniently applied to cases in which water (with other volatile substances) plays a very essential rôle. The chief

[1] It is impossible, however, to maintain the extreme position taken by Vogt, who believes that quartz may crystallize from dry fusion: *Tscherm. Min. Petr. Mitth.* (2), vol. xxv. (1906), p. 408.

objection to the terms themselves is that they ignore other mineralisers than water.

The physical and the chemical functions of mineralising agents may to some extent be considered separately. As regards the former, mineralisers are, before all, powerful *fluxes*. One of their most important offices is that of *reducing the viscosity* of a magma. For instance, the experiments of Day and Allen go to show that the reason why albite and orthoclase cannot be made to crystallize from a 'melt' of their own composition is merely the extreme viscosity which characterizes those compounds in the neighbourhood of their melting-points. It is easy to believe that the presence of a small amount of water may so reduce the viscosity as to enable the alkali-felspars to crystallize freely, perhaps even at temperatures not very much below their proper melting-points. Fluorides, as their very name imports, may be expected to produce an even greater effect. This consideration is relevant not only to the crystallization of rock-magmas, but also to magmatic differentiation; for any cause which promotes fluidity will thereby facilitate diffusion.

Mineralisers, however, act as fluxes in another sense, producing a general *lowering of freezing-points* in the constituent minerals of a magma. We have seen that the mixture of different silicates, etc., in a molten magma causes a mutual lowering of freezing-points, which may be measured in some cases by hundreds of degrees; but such substances as water and hydrofluoric acid must produce, in proportion to their amount, a much more marked effect of the same kind. This follows from Van 't Hoff's law, in view of the very low molecular weights of these substances (p. 191). Mineralisers, therefore, have the effect of deferring crystallization to a later stage, or to a lower temperature, as was clearly recognised by Elie de Beaumont.[1] This may have very important consequences, quite apart from any chemical

[1] "The presence of these substances seems to have had the effect of suspending the crystallization of the granite, and of suspending it to a more advanced degree of cooling in proportion as they were more concentrated" (*loc. cit.*, p. 1315).

reactions. For instance, Day and Shepherd put the inversion-point between tridymite and quartz at about 800°, and we may assume that quartz cannot crystallize above that temperature. Now, it may be doubted whether a slowly-cooled magma, consisting simply of such compounds as felspars, mica, etc., and free silica (without water, fluorides, or the like), could ever remain fluid down to 800°; but the presence of water or other mineralisers, by merely depressing the freezing-points of the minerals in the magma, may make the crystallization of quartz possible. As already pointed out, a flux in this sense is only another name for a solvent.

As regards the *chemical* functions of mineralisers, we have already made some remarks on the part which they appear to play in the formation and crystallization of many minerals; but it is to be noted further that their action may be destructive as well as constructive. This is *pneumatolytic* action, in the proper sense of the term, and will be considered below.

Minerals formed by the Agency of Mineralisers.—It may be taken as a general rule that minerals which have been artificially reproduced only by the aid of special fluxes and reagents (or have hitherto defied synthesis) owe their formation in nature also to the agency of mineralising agents. Some qualification of this general conclusion is, however, necessary. For example, zircon and sphene do not crystallize from 'melts' of their own composition; but there is no evidence from the laboratory to prove that they cannot crystallize from dilute solution in a silicate-magma, and we need not assume the agency of mineralisers to account for the minute crystals of early formation which occur sparingly in so many igneous rocks. Where, on the other hand, zircon or sphene occurs very abundantly in relatively large crystals of somewhat late formation, we may safely deduce the intervention of mineralising agents. Again, the alkali-felspars do not crystallize from fusion of the pure material in a crucible; but this seems to be due merely to extreme viscosity, and it is quite conceivable that in a mixed magma viscosity may be sufficiently reduced to permit the crystallization of these minerals without any special flux. The abundance of sanidine in many lavas seems, indeed, to necessitate this supposition. No such argument can avail, however, for quartz, micas, amphiboles, beryl, and many others, which must be regarded as essentially *low-temperature minerals*. They doubtless represent forms which are not stable at high magmatic temperatures; and this is especially true of those compounds in which water (or hydroxyl), boric acid, fluorine, or chlorine makes an essential part. Minerals rich in these constituents belong to the latest stages of crystallization, when, besides a low temperature, there was an increased concentration of the mineralising substances in the residual magma.

The minerals which may be presumed to owe their formation in igneous rocks to the agency of mineralisers are a very numerous class, though most of them are of rare occurrence. They possess, as a whole, certain chemical characteristics which may be briefly mentioned:

(i.) Many of these minerals have a *highly complex constitution*. Not only do the chemical analyses show a large number of constituents in a given mineral, but these vary in a manner which indicates a constrained isomorphism[1] of very unlike compounds. In such minerals as the amphiboles and micas, for instance, "such unlike constituents as H_2O, Na_2O, CaO, MgO, and Al_2O_3 are in some way brought into a molecule as isomorphous constituents or radicals, a result which we do not meet with nor expect to meet with in the case of simple chemical compounds."[2] More remarkable examples might be cited from the rarer minerals. Euxenite, for instance, appears to be a 'homœomorphous' mixture of an orthoniobate—(Y, Er) $(NbO_3)_3$—with metatitanates—$Y_2(TiO_3)_3$ and (U, Th) $(TiO_3)_2$—but contains also Ta, Sn, Ce, La, Di, Fe, Mn, Ca, Pb, Na, K, etc. To many minerals of this class no formula can be confidently assigned.

It is to be remarked that many minerals of complex con-

[1] Compare Brögger on 'homœomorphism,' *Die Mineralien der südnorwegischen Granitpegmatitgänge*, I. (1906), pp. 117-122.

[2] Penfield and Stanley, *Amer. Journ Sci.* (4), vol. xxiii. (1907), p. 25.

stitution, such as hornblendes and micas, have no melting-point in any intelligible sense, since they break up into other substances before or during fusion. This is true also of some simple compounds, such as quartz, which are dimorphous.

(ii.) To the agency of special mineralisers we owe a large proportion of the material which enriches the cabinets of mineralogical collections. A long list might be made of species which are practically confined in nature to pegmatite veins, undoubtedly crystallized under the influence of mineralising agents. In particular, minerals of this class are the repository of most of the *rarer elements*—lithium; beryllium; yttrium, cerium, lanthanum, didymium, etc.; thorium; niobium and tantalum; molybdenum, tungsten, and uranium; with others of still rarer occurrence. Some of the 'rare earths' are practically unknown except in this connection; and in Southern Norway, Ceylon, and elsewhere some of these occurrences have acquired in late years considerable economic importance.

(iii.) In many of the minerals of this class, as already pointed out, we find *direct evidence* of the co-operation of mineralisers in a certain content of hydrogen (usually as hydroxyl), fluorine (which, as Penfield has proved, often replaces hydroxyl in isomorphous fashion), chlorine, boric acid, or some other special constituent.

(iv.) Groups of minerals belonging to this class are found associated in pegmatites in a manner indicating a significant *paragenesis*. Further, the rarer elements which locally become important as constituents of minerals in pegmatite veins may be found in extremely small amount in the normal plutonic rocks of the neighbourhood; and this is true likewise of boron, fluorine, chlorine, etc. It is clear that both the rare elements and the mineralising substances were contained in the original rock-magma, though the direct evidence of their presence is mostly confined to the latest products of crystallization.

Pegmatites.—Since the volatile constituents of a rock-magma enter at most in very exiguous amount into any of the common minerals—at least, those which separate in the earlier stages—the progress of crystallization brings about an *increasing concentration* of these substances in the residual fluid magma. To this, as well as to the falling temperature, it is due that the influence of mineralising agents becomes more manifest in the later stages of the consolidation. It becomes most conspicuous in the pegmatite fringes and veins, a concomitant of most plutonic intrusions, which represent the very last products of magmatic crystallization.

The original *pegmatite* of Haüy (1822) was a coarse graphic intergrowth of felspar and quartz ('graphic granite'); but the name was extended by Delesse (1849) to all very coarse granitic rocks, whether graphic or not, these occurring commonly as veins or dykes traversing an ordinary granite, or as apophyses or a more or less continuous fringe on its border. Subsequent usage has further extended the term to cover corresponding modifications of other plutonic rocks (syenite, diorite, gabbro, etc.), and it is in this sense that the word pegmatite is here employed.

The origin and significance of pegmatite veins have been much discussed,[1] but it is not necessary here to enter into the subject at length. Setting aside the 'lateral secretion' theory of Credner and others, for which it is difficult to find any support, the question agitated seems to resolve itself into an antagonism between an igneous and an aqueous origin. From our point of view, as already sufficiently indicated, the antithesis marked by these terms finds no place. The magma or solution from which the pegmatites crystallized was igneous, in that it was the residual part of a granitic or syenitic or other igneous magma, of which the greater part had already crystallized under plutonic conditions. It was aqueous, inasmuch as it contained, perhaps very richly, magmatic water, concentrated (with other volatile constituents) in the residual magma by continued crystallization of anhydrous minerals. The pegmatites

[1] For a summary of the controversy, and an admirable discussion of the question, see Brögger, *Die Mineralien der Syenitpegmatitgänge der südnorwegischen Augit- und Nephelinsyenite*, *Zeits. Kryst.*, vol. xvi. (1890), pp. 215-234.

themselves represent this watery residual magma, except that the greater part of the water and other volatile substances was expelled in the final crystallization. This final crystallization from a solution rich in water and perhaps in other fluxes may have taken place at quite low temperatures, but genetically it is impossible to separate the resulting product from igneous rocks. Logically, indeed, we might include under the same head simple quartz-veins crystallized from solution in water (at perhaps 200° C.), if both quartz and water were of direct intratelluric origin, the final residuum of an igneous rock-magma.[1]

The *mode of origin* of pegmatites, regarded as the latest products of crystallization of igneous rock-magmas, has been discussed by Arrhenius[2] in a paper already cited. He considers a magma as a complex solution containing various silicates, etc., and also gases, the latter, of course, including in the first place water, which above its critical temperature of about 365° must be ranked as a gaseous body. He proceeds, however, on the supposition that, between the silicates and other minerals on the one hand, and the gaseous constituents on the other, there is only a limited mutual solubility in the fluid state, becoming more limited with declining temperature. On this supposition, the magma, as it cools, will separate progressively into two partial magmas, one containing most of the silicate and other minerals, and the other with a preponderance of water and other volatile substances. The latter, which we may call the aqueous solution, will collect especially in the marginal parts of an intrusive body, where the cooling and consequent separation begin; but smaller portions may subsequently detach themselves in the interior, giving rise to druses and veins. When the other part of the magma (the silicate solution) has become effectively solid, fissures in it may be injected with the still fluid aqueous solution, the normal plutonic rock thus being traversed by veins, sheets, and strings of pegmatite. Owing

[1] See Lomas, *Geol. Mag.*, 1903, pp. 34-36; Harker, *ibid.*, p. 95; Bonney, *ibid.*, pp. 138, 139.

[2] *Geol. Fören. Förh.*, vol. xxii. (1900), pp. 395-419.

to its great fluidity, the aqueous solution may be injected, even in the finest apophyses, into the contiguous sedimentary or other rocks, and may travel to considerable distances from the main body of the intrusion.

The various constituents of the original magma will be divided between the two partial magmas according to their relative solubility in one or other of these. In the aqueous solution will be collected, not only the bulk of the water, but the chief part of all those constituents which are more soluble (at the temperature) in water than in the silicate solution. These include boric, carbonic, and hydrosulphuric acids, the fluorides, chlorides, and borates of the alkali-metals and the metals of the rare earths, and other bodies. The silicates, with free silica and other free oxides etc., will be taken up in proportion to their solubilities. It may be presumed that the ionization of the various constituents in the aqueous solution will be more complete than in an ordinary igneous magma, and chemical activity will be augmented.

Most of Arrhenius's conclusions would have equal force without the assumption of a limited mutual solubility between water and molten silicates and the consequent separation into two immiscible solutions. The pegmatite may in any case be regarded as representing the residual 'mother-liquor' at the end of the process of crystallization.

The actual *manner of occurrence* of pegmatites in the field is found to accord perfectly with this conception of their nature and origin. They occur commonly as veins, dykes, or sheets, not far from the margin of the plutonic intrusion of which they are satellities, and traversing either the main plutonic rock itself or the country-rock adjacent. If in greater bulk, the pegmatite may form a more or less continuous border on one or more sides of the main body. Under certain conditions, as we shall remark later, the final residual magma may become separated from the main body of earlier consolidation, and may give rise to detached intrusions of pegmatite of considerable dimensions. Where dykes of pegmatite intersect sedimentary or other extraneous rocks, they have the same characteristic features as other dykes, being sharply bounded and sometimes enclosing fragments of the country-rock. They have often produced very considerable metamorphism. Where, on the other hand, the pegmatite occurs in the normal plutonic rock, it often forms veins, streaks, knots, or lenticles, more or less irregular and discontinuous, and the boundary may not be a very sharp one. This is easily intelligible if the pegmatite-magma has been exuded from a crystallizing mass in the last stage of consolidation.

Perhaps the most obvious peculiarity of pegmatites is their *very coarse texture.* This is a necessary consequence of the fluidity of the magma, rich in water and other fluxes, from which they are formed. Again, from the effect of these substances in deferring the crystallization of the silicate and other minerals, it results that these latter often show no very determinate order of crystallization.[1] Very characteristic is the simultaneous crystallization of two or more constituents, which, indeed, is what we must expect in pegmatites when we regard them as the final products of crystallization of rock-magmas. Graphic intergrowths are very common, the well-known 'graphic granite' being a typical example. The presence of water, modifying the eutectic proportions, may be one reason for the variation of composition among graphic pegmatites. In the syenite-pegmatites perthitic intergrowths of different felspars are a characteristic feature.

The *mineralogical composition* of pegmatites stamps them, not merely as igneous rocks, but as of the granitic, syenitic, or other family, according to the nature of the normal plutonic rocks with which they are associated. They have the same chief constituent minerals as these, but are in general somewhat richer in the later products of crystallization, and thus often more 'acid' in a generalised sense. Further, the pegmatites often, though by no means always, contain special minerals which are wanting in the ordinary plutonic rocks, or occur there only occasionally or sparingly. Such special minerals, as well as other evidences of the influence of mineralisers, are much more frequently found

[1] Brögger, *Syenitpegmatitgänge* (1890), p. 158.

in connection with acid than with basic intrusions. It may be remarked that we have here still another count in the parallelism already pointed out between the progress of crystallization in a magma and the course of magmatic differentiation (p. 131). Just as the crystallization of any magma causes a concentration of the volatile constituents in the final residuum, so the ordinary sequence of a series of plutonic intrusions from more basic to more acid, attended by declining temperature (p. 139), is accompanied by enrichment of the later magmas in the volatile constituents.

From what has been said above, it follows that each family of pegmatites has its own train of special minerals, which are more or less distinctive and sometimes confined to that association. Different occurrences, however, may be characterized by different groups of special minerals, so that it may be possible to distinguish *different types of pegmatites* belonging to the same family of rocks. Among the granite-pegmatites, for instance, some in Southern Norway[1] are rich in niobates and tantalates, with phosphates of the yttrium-metals and many other minerals. Another type is characterized especially by tourmaline, and another by beryl, topaz, and fluor. In various districts there are granite-pegmatites carrying numerous lithia-bearing minerals, such as lepidolite, spodumene, petalite, etc. The Ivigtut type, in Southern Greenland,[2] has fluorides as the characteristic minerals, including cryolite, fluor, and others. The pegmatites of the alkali-granites carry other groups of minerals. The best-known pegmatites of the syenite and nepheline-syenite families are those of Southern Norway.[3] The larvikite-pegmatites of the Fredriksvärn district often contain zircon, and sometimes the zircono-titanate polymignite. The nepheline-syenite-pegmatites of the Langesundsfjord present a much richer assemblage of special minerals, some of very

[1] Brögger, *Die Mineralien der südnorwegischen Granitpegmatitgänge*, I., *Vid.-selsk. Skr.*, *Math.-Nat. Kl.* (1906), No. 6.

[2] Flink, *Meddelelser om Grönland*, xiv. (1898), pp. 221-262; and Böggild, 'Mineralogia Groenlandica,' *ibid.*, xxxii. (1905).

[3] Brögger, *Syenitpegmatitgänge* (1890), pp. 121-200.

restricted distribution. Many are rich in zirconium and titanium, and others contain beryllium. Some contain fluorine and others boron. The pegmatites of the diorites, gabbros, essexites, and peridotites do not in general carry special minerals.

Pneumatolytic Action.—With the completion of crystallization in a rock-magma the contained water and other volatile substances, excepting such portion as has been incorporated in some of the crystallized minerals, must be disengaged. Since the critical temperature for water is about 365° C., and for the other substances lower, they must be in the gaseous state, however great the pressure to which they may be subjected. Some part of these residual gases may be imprisoned in cavities in crystals, and possibly 'occluded' or dissolved in the substance of the crystals; but the bulk is collected in druses and geodes, or occupies fissures opened in the now solid rock-body. Such cavities and fissures may be sufficient to permit of a general circulation through the rock-mass. The compressed gases probably carry a certain amount of the silicates and other constituents in solution, and deposit them on the walls of druses as cooling proceeds. We must suppose a certain leakage by diffusion into and through the contiguous country-rocks; but it is probable that this escape is not important until the temperature has fallen considerably.

The active rôle of the volatile constituents does not terminate with the completion of crystallization. Having fulfilled the constructive office of mineralisers, they now enter upon a new activity which is partly of a *destructive* kind. As cooling proceeds, some of the compounds crystallized at higher temperatures cease to be stable in presence of the concentrated gaseous residuum, and are decomposed with the production of new minerals. The volatile substances themselves may or may not enter into the composition of these new minerals, but in general they do so to a greater extent at this stage than during the magmatic crystallization. It is to destructive action of this kind that Bunsen[1] applied

[1] *Pogg. Ann.*, vol. lxxxiii. (1851), p. 238.

the term *pneumatolytic*, although he had in view rather the analogous processes in volcanic rocks. Brögger[1] has extended the term to embrace also the constructive action of mineralisers during crystallization of the magma; but, although he has been followed by some other petrologists, it seems better to retain the original usage. As thus limited, pneumatolysis is essentially an '*after-action*.' Nevertheless, it is closely bound up with igneous intrusion, and must be regarded as belonging to igneous action. The pneumatolytic agents themselves are merely the final residuum of an igneous rock-magma, and the processes are conducted at temperatures more or less elevated, though progressively declining.

Pneumatolytic action in plutonic rocks is not confined to their pegmatoid modifications. In many cases it clearly follows joint-fissures in the body of the rock, a sufficient proof that it belongs to a stage distinctly posterior to the completion of consolidation. Since in different kinds of rocks the destructive action is exerted on different minerals, and the volatile agents themselves are not always the same, we may recognise *different types of pneumatolysis*. In the granite family several types may be distinguished. In one the characteristic change is the conversion of alkali-felspar to alkali-mica and quartz, or even its replacement as a final stage by quartz alone. There may be no direct evidence of any agent other than water, though the fact that the rock has sometimes been depleted of alumina as well as alkalies argues some more potent agency. The kaolinisation of alkali-felspars is also attributed to pneumatolytic action.[2] Here the frequent occurrence of some tourmaline suggests that boric acid as well as steam has participated. In the greisen type of change felspar is replaced by quartz, topaz, and white mica, often of a lithia-bearing variety, and the action must be ascribed to fluorides. Closely bound up with the greisens are the tinstone veins,[3] the cassiterite probably

[1] *Syenitpegmatitgänge* (1890), p. 213.

[2] Butler, *Min. Mag.*, vol. vii. (1886), pp. 79, 80; Flett, *Geol. Land's End*, *Mem. Geol. Sur.* (1907), pp. 58-60.

[3] Collins, *Min. Mag.*, vol. iv. (1880), pp. 1-20, 103-116, vol. v. (1882), pp. 121-130; reprinted, with additions, under title *Cornish Tin-stones*

resulting from reaction between the volatile tin fluoride (SnF_4) and water. The destructive action of fluorides is exceedingly energetic. At Geyer, in Saxony,[1] granite is locally converted to a rock containing more than 90 *per cent.* of topaz. A well-known type of pneumatolysis in granites is tourmalinisation, in which boric acid or some borate is the agent. Tourmaline may be formed by mineralising agency

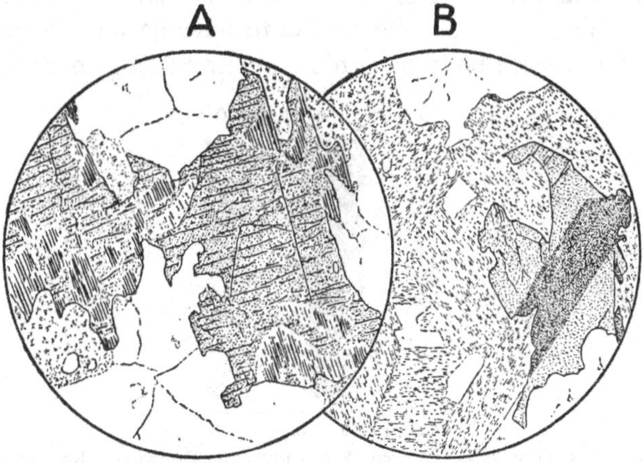

FIG. 95.—TOURMALINISATION IN GRANITE. × 20.

A, Replacement of biotite by brown tourmaline, Busava, Cornwall. Some patches of biotite, only partly destroyed, show the original cleavage.
B, Replacement of felspar by brown and blue tourmaline, Ivybridge, Dartmoor. The single crystal of tourmaline is mostly brown, but with a sharply-defined blue part, which replaces a plagioclase crystal.

in the stage of magmatic crystallization, but more usually it is produced at the expense of mica and felspar (Fig. 95). In the nepheline-syenite-pegmatites of Norway Brögger notes various transformations attributable to pneumatolytic action in the strict sense, and shows how they fall into distinct stages, corresponding with declining temperature. A late stage is marked by the production of zeolites, the chief agent being water, now probably in the liquid state, though there is still evidence of the co-operation of fluorine

and Tin-capels (1888, Truro); Flett, *Summary of Progress Geol. Sur.* for 1903, pp. 156-158, and *Geol. Land's End.*

[1] Salomon and His, *Zeits. deuts. geol. Ges.*, vol. xl. (1888), pp. 570-574.

(in apophyllite). Still later carbonic acid became an efficient agent, producing certain fluo-carbonates and carbonates. In the gabbro family, the most important type of pneumatolytic action is that which has given rise to the valuable apatite veins of Norway, Canada, and elsewhere.[1] The characteristic minerals after apatite are rutile, ilmenite, hæmatite, and metallic sulphides. Titanium plays somewhat the same part here as tin in the greisens and zirconium in the syenitic rocks. Vogt supposes the apatite to have been concentrated from the general body of gabbro by circulating solutions, in

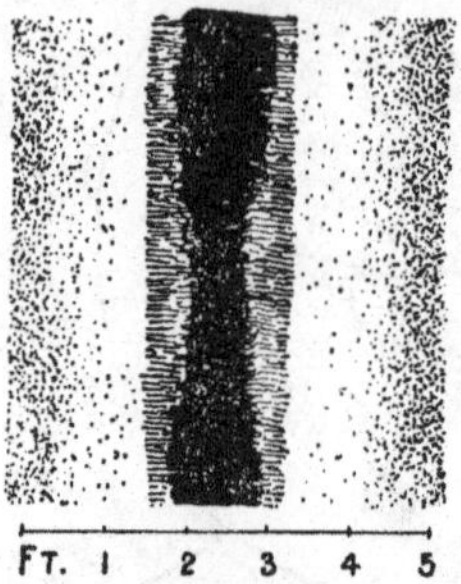

FIG. 96.—APATITE VEIN, WITH SCAPOLITIZED BORDERS, REGAARDSHEI, NORWAY. (AFTER VOGT.)

A, Apatite; H, hornblende, mica, etc.; S, scapolite-hornblende-rock; O, olivine-hyperite.

which chlorides played the chief part as solvents. As a pneumatolytic effect, in the stricter sense, the gabbro immediately bordering the apatite veins has its felspar transformed to scapolite by the addition of chloride.[2] The pyroxene is at the same time converted to hornblende.

Pneumatolysis in Relation to Metamorphism.—It is safe to assume that the water and other volatile constituents actually found in igneous rocks, whether chemically combined or mechanically enclosed, represent only a fraction of what was originally contained in the parent rock-magmas. The rest has been lost, and, in the case of intrusive rocks, which we are more particularly considering, must have

[1] Vogt, *Zeits. prakt. Geol.*, 1895, pp. 367-370, 444-459, 465-479.

[2] Judd, *Min. Mag.*, vol. viii. (1889), pp. 186-198.

passed into or through the surrounding country-rocks. Probably some leakage goes on throughout the process of consolidation, but not with equal freedom at different stages. High temperature, which in liquids diminishes viscosity and so promotes diffusion, has the opposite effect in gases; and, in view of all the conditions, it is likely that a large part of the volatile constituents is in general retained down to a late stage. Nevertheless, more or less of the water and other gases must pass into the neighbouring rocks while these are still heated by the intrusion, and this has an important bearing on the conditions of thermal metamorphism in sedimentary and other rocks.

Metamorphism at high temperatures has in general the effect of actually expelling volatile constituents even from combination. For instance, in the amygdaloidal basalts of Skye, when metamorphosed against the plutonic intrusions, the lime- and lime-soda zeolites are converted to plagioclase felspars. Carbon dioxide is expelled from calcium carbonate in the presence of silica which can replace it, and may be expelled from magnesium carbonate even in the absence of silica. These effects prove at least that the rocks are pervious to gases. Very many of the most characteristic transformations in thermal metamorphism, however, cannot have proceeded at very elevated temperatures. The minerals most commonly produced—micas, amphiboles, epidotes, alkali-felspars, quartz, etc.—are indicative of relatively low temperatures and of the co-operation of water. Whether the 'ground water' of sedimentary rocks would suffice for this office, or was reinforced by water of magmatic origin, is a debatable question. In many cases the phenomena, studied in detail, lend no support to the supposition that water has acted to any sensible extent as a carrier of material in solution.[1] Moreover, the phenomena of thermal metamorphism are sometimes found far from any igneous intrusion.

It is none the less true that locally the *volatile constituents*

[1] Harker and Marr, *Quart. Journ. Geol. Soc.*, vol. xlix. (1893), pp. 368-370.

given out from an igneous intrusion may contribute in an important degree to the metamorphism of the rocks within a certain distance of the contact. This is proved in numerous instances by the production in the metamorphosed rocks of special pneumatolytic minerals, containing boron, fluorine, chlorine, etc., in different cases, and related in this way to the nature of the intruded magma as well as to the material undergoing metamorphism. The special minerals are not always the same in the metamorphosed as in the igneous rock, since the pneumatolytic agents operate on different materials. The metamorphosed slates bordering the tourmaline-bearing granites of Cornwall often contain abundant tourmaline; but in the metamorphosed 'greenstones,' richer in lime, axinite appears instead, and this is produced also in connection with the impure Devonian limestones. Again, it is interesting to note that proof of the presence of mineralising agents may be preserved in the metamorphic aureole surrounding an igneous rock, which itself furnishes no such evidence. Lacroix points out that the lherzolites and ophites of the Pyrenees are of quite simple mineralogical composition; but gaseous emanations from them have given rise to abundant dipyre (scapolite) and tourmaline in the limestones which they have metamorphosed.

This pneumatolytic metamorphism, which is also metasomatism, does not extend far from the actual contact, and might with propriety be termed contact-metamorphism, had not this term been unfortunately applied by many geologists to thermal metamorphism in general, which is not confined to contacts nor necessarily related to igneous intrusions.

More closely limited to the immediate vicinity of igneous contacts are certain metasomatic changes of a more radical kind, which point not merely to gaseous emanations, but to a transfusion of material by the medium of *aqueous solutions.* The clearest example is the conversion of slate to adinole at the contact of diabase intrusions in the Harz[1] (Fig. 97). The adinole consists essentially of albite and quartz; and

[1] Kayser, *Zeits. deuts. geol. Ges.*, vol. xxii. (1870), pp. 103-172.

analyses show a large addition to the rock of silica and soda, with a loss of iron oxide, potash, and water. Similar effects are seen near Padstow in Cornwall, where slate is transformed to an almost pure albite-rock.[1] Near San Francisco[2] radiolarian cherts, as well as felspathic sandstones, are in like manner converted to glaucophane-schists, composed of albite, quartz, glaucophane, biotite, etc.

All these effects are strictly local, extending at most to a very few feet from the actual contact. By the French school of petrologists a much wider province is assigned in thermal metamorphism to metasomatic processes of this order. Michel-Lévy[3] has maintained that metamorphosed sedi-

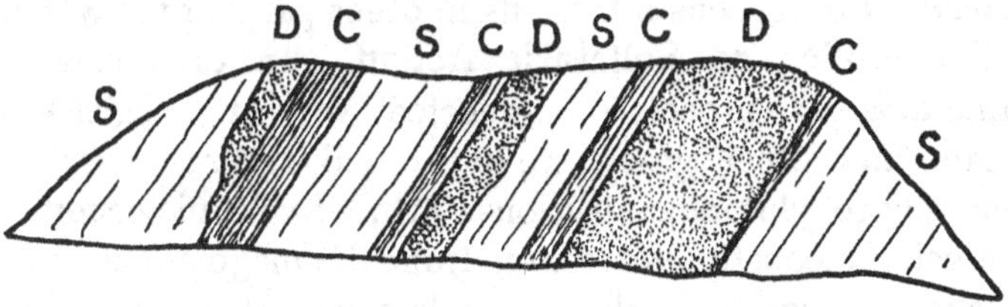

FIG. 97.—SECTION AT ALLRODE, HARZ. (AFTER KAYSER.)
S, Slate; D, diabase sills; C, contact-rocks.

ments near a granite contact may be materially altered in total chemical composition by the introduction of felspar and other minerals, and he would extend this idea to explain the production of crystalline schists on a large scale. This process of 'feldspathisation' he apparently attributes, not merely to injection along a network of veins and strings, but to some kind of transfusion. Lacroix[4] goes farther, for at

[1] Fox, *Trans. Roy. Cornw. Geol. Soc.*, vol. xi. (1894), pp. 687-724, and *Geol. Mag.*, 1895, pp. 13-20; McMahon and Hutchings, *ibid.*, pp. 257-259.

[2] Ransome, *Bull. Dep. Geol. Univ. Cal.*, vol. i. (1894), pp. 193-240.

[3] See, *e.g.*, *Contribution à l'étude du granite de Flamanville*, *Bull. No.* 36 *Carte géol. Fra.* (1893), pp. 18-24.

[4] *Les phénomènes de contact de la lherzolite et de quelques ophites des Pyrénées*, *Bull. No.* 42 (1895); and *Le granite des Pyrénées et ses phénomènes de contact*, *Bull. No.* 64 (1898) and *No.* 71 (1900) *Carte géol. Fra.* For summary see *Notice sur les travaux scientifiques de M. A. Lacroix* (1903), pp. 24-40.

the contact of lherzolites in the Pyrenees he describes the Mesozoic limestones as transformed partly, or even wholly, into silicate and other minerals, including dipyre, felspars, micas, amphiboles, pyroxenes, tourmaline, etc. These large additions, amounting sometimes to total replacement of the limestone, he ascribes, not to the introduction of the substance of the intrusive rock itself (which, indeed, consists of quite other minerals), but to emanations from the magma. In the absence of full local details it is not possible to form any independent judgment upon questions which turn entirely on field-evidence;[1] but the French geologists are here at issue with most of those who have studied the behaviour of igneous intrusions in other parts of the world.

Fumerolic and Solfataric Action.—In discussing the mineralogical processes connected with the agency of steam and other gases, we have had regard almost exclusively to plutonic intrusions. Under volcanic conditions the volatile constituents escape from the magma much more freely, and consequently play a less important part in the constructive rôle of mineralisers during the primary or magmatic consolidation. The partially destructive processes, which are the analogue here of pneumatolysis in the plutonic intrusions, may, however, have far-reaching effects; and they possess an added interest in the fact that they can be, to some extent, witnessed in actual progress in modern volcanic vents and in the fumeroles and solfataras of active or recently active volcanic districts.

It results from the difference of conditions that these processes are analogous, rather than identical, with those which affect plutonic rocks. The temperatures in the volcanic focus during the maximum of activity must be considerably higher than in a plutonic rock in an advanced stage of consolidation or wholly solid. Further, there are *differences of temperature* between different parts of the volcanic apparatus. It is in consequence of this that many of the special minerals in an active volcano appear as *sublimates*, or are *formed in place by reactions* involving substances

[1] Compare Adams, *Journ. Geol.*, vol. ix. (1901), pp. 28-46.

in the gaseous state. The common minerals of Vesuvius,[1] for instance, include rock-salt, sal-ammoniac, cotunnite ($PbCl_2$), and other chlorides, besides sodalite; fluor; various sulphates, vanadates, etc.; native sulphur, probably formed by reaction between sulphuretted hydrogen and sulphur dioxide; and various metallic oxides and ferrates, such as melaconite (CuO) and cuprite (Cu_2O), hæmatite, magnetite, and magnesioferrite ($MgFe_2O_4$). Most of these are unknown as constituents of the Vesuvian lavas. The list leads to another remark, which is of general application. These minerals, *rapidly formed at atmospheric pressure*, have not the highly complex chemical constitution which characterizes so many of the pneumatolytic products in plutonic and metamorphosed rocks, but are comparatively simple compounds.

Different types of solfataric action might be distinguished, and these are in some degree characteristic of different kinds of lavas. But it is also to be remarked that different volatile constituents may figure prominently at a given volcanic centre at *different stages in the decline* of activity. This is in part a matter of direct observation; for it has frequently been remarked that only the hottest fumeroles emit hydrochloric and hydrofluoric acids, while sulphurous and hydrosulphuric acids are connected with lower temperatures of emission, and water and carbonic acid with the lowest temperatures. Some indications of a like sequence have already been noted in pneumatolysis under plutonic conditions. The later stages of solfataric activity may be continued long after the cessation of vulcanicity in the narrower sense. Thus, the rhyolites of the Yellowstone Park, of Pliocene age, have been in some localities rather extensively decomposed by the long-continued passage of steam and heated waters, and this process has not yet ceased. Doubtless more potent agents than water have played a part here, and what these were may be inferred from analyses of the geyser-waters of the vicinity, which contain alkaline chlorides

[1] Scacchi, *Riv. min. e crist. ital.*, vol. v. (1889), pp. 34-38; Lacroix, *Bull. soc. fra. min.*, vol. xxx. (1907), p. 219, vol. xxxi. (1908), p. 259.

and carbonates, besides smaller amounts of borates and arsenates.

Certain common alterations in volcanic rocks, which have sometimes been ascribed to 'weathering,' belong more properly to this place, the water, carbonic acid, etc., which have been the principal agents, having a magmatic, not an atmospheric, source, and the processes being effected at a late stage in the cooling down of the rocks. This is certainly true in many cases of the infilling of the steam-cavities in lavas with zeolites and other secondary minerals. In the amygdaloidal Tertiary basalts of the British area the commonest secondary minerals are various zeolites, with chloritic minerals, calcite, chalcedony, etc., clearly derived from partial decomposition of the minerals of the basalt; and there is every reason to believe that the water and carbonic acid which have been added were of direct volcanic origin.[1] Here we are able to contrast these solfataric effects with those which result from true weathering—*i.e.*, from the action of the atmospheric gases and water at ordinary temperatures. Each lava-flow, where it was exposed to the atmosphere before being covered by a later flow, presents a layer of red ferruginous clay, the residue of basalt decomposed in place; and it is clear that the processes concerned in this change were totally different from those by which the amygdaloidal cavities were filled.

[1] *Tertiary Igneous Rocks of Skye*, *Mem. Geol. Sur.* (1904), pp. 42-46. See also an interesting paper by Strachan, *Ann. Rep. Belfast Nat. Field-Club*, (2) vol. vi. (1908), pp. 92-98.

CHAPTER XIII

MAGMATIC DIFFERENTIATION

The problem of magmatic differentiation.—Possible causes of differentiation in a still fluid magma.—Fractional crystallization.—Subsidence of crystals in a partly fluid magma.—Expulsion of the residual fluid magma.—Some geological aspects of differentiation.

The Problem of Magmatic Differentiation.—The most fundamental problem of modern petrology, that of the origin of the great diversity of rock-types actually found, has engaged the attention of numerous writers, and their essays towards its elucidation have taken various lines.[1] In general they aim at providing some physical explanation of the derivation of diverse rocks from one parent-magma. Some attempts in this direction have taken the form of very large speculations relative to the differentiation of a general cosmic magma. In other cases appeal is made to some known physical law, the application of which is suggested as one factor in a confessedly complex evolution. These endeavours to reduce the problem to a more definite shape have been facilitated during recent years by the rapid expansion of physical chemistry; but the application of general principles to the particular case of rock-magmas is still greatly hampered by the deficiency of quantitative data.

[1] For a review of opinion concerning this subject see Iddings, 'The Origin of Igneous Rocks,' *Bull. Phil. Soc. Wash.*, vol. xii. (1892), pp. 91-127; Zirkel, *Lehrbuch der Petrographie*, 2nd ed. (1893), vol. i., pp. 658-671; Löwinson-Lessing, 'Studien über die Eruptivgesteine,' *Compte Rendu VII. Congr. Géol. Internat.* (1899), pp. 308-401; Vogt, *Zeits. prakt. Geol.*, 1901, pp. 327-340.

That the actual diversity met with among igneous rocks and the varying composition of many single rock-bodies are in the main attributable to processes of differentiation is a thesis which needs no formal discussion. It has been the common ground of almost all speculations on this subject during the last sixty years—that is, since the date of Darwin's *Geological Observations on the Volcanic Islands*—and, as already remarked, it must be established by the inductive, not the deductive, method. The evidence for magmatic differentiation, which seemed sufficient to the earlier petrologists, has been enormously strengthened by increasing knowledge of the distribution and mutual relations of different kinds of igneous rocks. We shall briefly notice below the possible modifications which rock-magmas may undergo by absorption of foreign material, upon which Lyell and some more recent writers have laid stress; but it is clear that no process of admixture can afford a substitute for differentiation, or enter into consideration as more than a subsidiary factor. The only practical alternative to magmatic differentiation, as accounting for the observed facts, is the doctrine of countless special creations.

It is to be remarked that there are two orders of facts to be explained: firstly, the differences between rocks constituting distinct intrusions (or extrusions), but giving evidence of derivation, more or less direct or remote, from a common source; and, secondly, variation between different parts of a single rock-body, presumed to have been intruded as a homogeneous magma. The distinction is that already drawn between differentiation prior to intrusion (or extrusion) and differentiation in place (p. 133). It does not necessarily imply two different kinds of differentiation as regards the mechanism of the processes concerned. The distinction is that in the former case we see only the finished result, the causes which have brought it about being a matter of conjecture, or at most of inference; while in the latter case the stages of the variation, and in some measure the nature of the processes involved, may be directly studied.

How far conclusions based on a study of differentiation in

place can be applied to the more obscure question of the differentiation of magmas in an unknown deep-seated reservoir is perhaps open to debate. The only intelligible basis of discussion is to regard rock-magmas as complex solutions, with properties differing only in degree, not in kind, from those of ordinary solutions which are the subject of laboratory treatment. If, now, a chemist desires to separate the constituents of an aqueous solution of mixed salts, or to prepare from it other solutions containing different proportions of the several constituents, he has at command various '*fractionating*' processes ; but in general, unless the mutual solubility be of the limited kind, he cannot arrive at any very effective separation without proceeding to *actual crystallization*. We shall see good reason for believing that the variation observed in a single igneous rock-body is due to processes of diffusion closely bound up with progressive crystallization. Here, as in the laboratory, the causes tending to bring about differentiation in a solution wholly fluid appear to be negligible in their effects. It is possible that they may be more efficient in a very large subterranean body of rock-magma, but this is a question concerning which we possess no real knowledge. If, then, we are to apply to the earlier unknown differentiation the same principles which appear to be valid in relation to differentiation in place, it seems necessary to assume that crystallization and re-fusion occur, sometimes repeatedly, in the intercrustal magma-reservoirs. This assumption is a legitimate one. We have already seen that these primary reservoirs must be situated in a zone of the earth's crust where solid and liquid rock are in approximate equilibrium, and where, consequently, crystallization or fusion will readily be brought about in response to changes of pressure or temperature.

Possible Causes of Differentiation in a Still Fluid Magma.—If we consider what causes may conceivably set up or maintain differentiation in a continuous body of fluid magma *without proceeding to crystallization*, two factors suggest themselves as possibly efficient—viz., gravity and differences of temperature. Since appeal has been made to both, we will discuss them briefly in turn.

Theoretically a partial separation should be brought about in a mixed solution in consequence of the *differential action of gravity*, the denser constituents being more concentrated in the lower layers and the lighter constituents in the upper. For an aqueous solution of a salt, the question was first investigated on thermodynamic principles by Gouy and Chaperon,[1] who found that in such a case a column 100 metres high would be necessary to produce any sensible difference of concentration between top and bottom. The physical constants are wanting for any such calculation as applied to silicate-magmas; but there is positive evidence from the experimental side to suggest that separation under gravity may be much more effectual in this case than in that of an aqueous solution. The technical processes in the manufacture of glass are instructive in this respect.[2]

The conception of a large body of magma stratified in accordance with varying density underlies the speculations of such geologists as von Waltershausen, Durocher, von Richthofen, and others, who have sought a general cosmic origin for all the igneous rocks of the world. On the scale contemplated by them, granted that the interior of the globe is or has been fluid throughout, the stratification under gravity may be freely conceded; and the high mean density of the earth, about double that of the superficial rocks, affords strong support to such a hypothesis. We are, however, more immediately concerned with the question whether this stratification can attain importance within the limits of such intercrustal magma-basins as can reasonably be postulated. To this question we can give no answer from certain knowledge, but there are various classes of facts which might be cited as throwing light on the subject. We have clear proof that two very different magmas may

[1] *Ann. chim. et phys.* (6), vol. xii. (1887), pp. 384-394.

[2] Morozewicz, *Tscherm. Min. Petr. Mitth.* (2), vol. xviii. (1898), p. 233. It is necessary to discriminate between stratification of the fluid mass itself and the sinking of crystals in it.

be intruded almost or quite simultaneously in circumstances which seem to prove conclusively that they must have coëxisted in the same reservoir. It is difficult to see how this could be unless the denser magma (in general the more basic) underlay the lighter (more acid); and that this was, in fact, their relative situation is sometimes indicated by the occurrence in the basic rock of xenocrysts representing the beginning of intratelluric crystallization in the overlying acid stratum.

Within the smaller vertical limits of a single intruded body, such as a sill or laccolite, any stratification under gravity will necessarily be less marked; and, so far as our knowledge goes, it is in general negligible, though closer observation bearing on this point is desirable. We are, of course, referring to the possibility of differentiation from this cause after intrusion: it would be easy to cite stratified laccolites and sills in which this character results from successive intrusions of different magmas.

We have tacitly assumed that there is perfect miscibility among the constituents of the fluid magma. If there were only a *limited miscibility* between two portions of the magma, a stratification in accordance with their relative densities would necessarily follow, and some petrologists have sought in this principle an explanation of the phenomena of magmatic differentiation. Bäckström,[1] for instance, would "give to liquation and not to diffusion its place as the working hypothesis, upon which the theory of differentiation is to be constructed." On his hypothesis a rock-magma is comparable with a mixture of aniline and water, which is homogeneous only above 166° C., and, when cooled below that temperature, separates into a lower layer of aniline with a diminishing proportion of water, and an upper layer of water with a diminishing proportion of aniline. It is clear that if the assumption of limited miscibility, decreasing with fall of temperature, were admissible, it would be applicable to differentiation in place as well as in an intercrustal magma-reservoir, and the results should be apparent in *discontinuous*

[1] *Journ. Geol.*, vol. i. (1893), pp. 773-779.

variation between different parts of a single rock-body. The variation observed is, in fact, continuous, and many phenomena familiar to petrologists go to confirm Vogt's conclusion that natural rock-magmas are in general freely miscible with one another in any proportions at all temperatures at which they remain fluid. The most important exception is the case of the sulphides, which appear to have only a very limited degree of miscibility with silicates in the molten state (p. 199). Since there is also a great difference of density, we should expect a magma of silicates and sulphides to divide into two layers with only a very partial commingling, the separation becoming more complete with falling temperature. This seems to be, in fact, illustrated in the most important occurrence of this kind which we know—the famous nickel-bearing intrusion of Sudbury in Ontario. This is a vast sheet-formed mass, estimated to have a thickness of about $1\frac{1}{4}$ miles and an original volume of about 1,000 cubic miles. The greater part consists of a micropegmatitic rock varying between granite and quartz-diorite, and this passes downward into a norite. The norite in its lower part contains some of the sulphide minerals, pyrrhotite with pentlandite, (Fe,Ni)S, and the amount of these rapidly increases, until in places the rock is composed essentially of sulphides, with only a very little of the norite minerals. These masses of ore occupy pockets or irregular offshoots from the base of the sheet, as if accumulated at the lowest points of the large body of magma.[1] The relations as described seem to accord well with the hypothesis that a sulphide-magma has separated under gravity from a silicate-magma, with which it was only very partially miscible. Whether the separation of the norite from the overlying acid rock falls under the same head is a much more doubtful question.

Rosenbusch's *Kern hypothesis*,[2] in so far as it is apprehended by the present writer, seems to postulate limited miscibility not as an exception, but as the rule. It is an

[1] See especially Coleman, *Journ. Geol.*, vol. xv. (1907), pp. 759-782.
[2] *Tscherm. Min. Petr. Mitth.* (2), vol. xi. (1889), pp. 144-178.

attempt to construct a natural classification of igneous rocks or rock-magmas on the basis of magmatic differentiation; but the differentiation appealed to is apparently a *spontaneous division* of a homogeneous magma into partial magmas, characterized by a certain chemical individuality and having only a limited mutual solubility. It seems to imply also that differentiation must always proceed on the same lines, for it assumes a tendency towards the separation of definite types of magmas in an approximately pure state, though various qualifying factors are introduced to account for the non-stochiometric composition of actual rocks.

Rosenbusch discusses a collection of bulk analyses of igneous rocks, which are reduced to atomic proportions and grouped about the particular types which he recognises. Thus, one set of analyses, belonging to nepheline-syenites, phonolites, etc., is rich in alkalies and alumina but very poor in the dioxides. These are referred to the 'foyaite-magma,' in which there is a great predominance of a certain 'Kern' or nucleus $(Na,K)AlSi_2$. This, by the crystallization of its oxides, gives rise to such minerals as leucite, nepheline, and orthoclase, the last two resulting presumably from a splitting up of the Kern:

$$2(Na,K)AlSi_2=(Na,K)AlSi+(Na,K)AlSi_3.$$

In like manner other types of magmas are characterized—the granitic, the granito-dioritic, the gabbro, and the peridotite magmas. The last, like the foyaite magma, is an approximately pure one, with the two Kerne R_2Si and RSi—*i.e.*, orthosilicate and metasilicate—but the others are more or less mixed; and a considerable degree of mutual solubility must be assumed in order to find place in the scheme for the analyses considered. Thus, as Brögger points out, the lardalite of Norway, instead of falling under the foyaite type, in which the characteristic Kern has only a small admixture of others, gives very nearly the formula:

$$3(Na,K)_2Al_2Si_4O_{12}+FeSiO_3+MgSiO_3+CaSiO_3.$$

The Kern hypothesis has been criticized from more than

one side by Roth,[1] Brögger,[2] and others. Apart from the difficulties which arise when we try to reduce the analyses of actual igneous rocks into correspondence with the scheme propounded, the continuous gradation observed in rock-masses where variation has been set up by differentiation in place seems to be an insuperable objection to any theory which supposes a limited degree of miscibility between different rock-magmas.

Some petrologists have attributed a leading part in the differentiation of a still fluid rock-magma to *differences of temperature* between different parts of the magma-basin, or, again, between the central and marginal parts of an intruded body of magma. Teall[3] first cited in this connection the experiments of Soret, who had proved that in an aqueous solution of a salt, different parts of which are maintained at different temperatures, there is a stronger concentration of the salt in the cooler region. On this Teall formulated the rule that "the compound or compounds with which the solution is nearly saturated tend to accumulate in the colder parts"; and Brögger and others have followed him in laying great stress on 'Soret's principle' as a factor in the differentiation of rock-magmas. Its application in this connection has, however, been criticized,[4] and a consideration of it from the quantitative point of view shows that it must be of very trifling importance. The true rule—a simple corollary of Van 't Hoff's theory of osmotic pressure—is that, when a continuous body of a dilute solution is at different temperatures in different parts, equilibrium will be attained only when the concentration at every point is inversely proportional to the absolute temperature. If, for instance, the hottest part be at 1200° C. (1473° absolute), and the coolest at a temperature 100° lower—an extravagant sup-

[1] *Zeits. deuts. geol. Ges.*, vol. xliii. (1891), pp. 1-42.

[2] *Eruptivgesteine des Kristianiagebietes*, III. (1898), pp. 302-333.

[3] *British Petrography* (1888), pp. 394-403.

[4] Bäckström, *Journ. Geol.*, vol. i. (1893), pp. 774, 775; Harker, *Geol. Mag.*, 1893, pp. 546, 547; Becker, *Amer. Journ. Sci.* (4), vol. iii. (1897), pp. 23, 24.

position—then the greatest difference of concentration that can be set up from this cause will be represented by the ratio 1473 : 1373, or 1·07 : 1. It appears, therefore, that, even apart from any consideration of convection, no sensible degree of differentiation can be brought about by differences of temperature in a still fluid magma. In particular, the conception of a great magma-reservoir stratified from this cause, the upper layers basic[1] and the lower layers acid, in defiance of gravity, is seen to be wholly chimerical.

Fractional Crystallization.—We come next to discuss the manner in which differentiation is brought about *concurrently with, and as a consequence of, the progress of crystallization* in a rock-magma. Here we stand on firmer ground, since the appeal is to familiar principles concerning which there can be no doubt. It is to be noted, in the first place, that the crystallization of any constituent from a mixed solution necessarily implies differentiation. When, for instance, a crystal of olivine is formed in a basaltic magma, it fills a space which was formerly occupied by a mixture of olivine, augite, felspar, etc. Its growth is fed by diffusion-currents, which maintain a certain degree of supersaturation in the immediate vicinity of the crystal, carrying olivine substance to it and augite and felspar away from it. Here, then, we have differentiation in miniature. If many crystals of the same mineral are forming instead of one, and if the conditions are such that they form wholly or principally in one part of a body of magma, initially homogeneous, it is clear that there is no theoretical limit to the differentiation set up by crystallization alone. It is even conceivable that, the requisite conditions being maintained, there may be formed a rock consisting wholly of one mineral, the remaining fluid magma being entirely deprived of that constituent. The requisite conditions are, firstly, that the earlier minerals shall separate successively in some definite order, and, secondly, that one part of the body of magma shall be constantly at a lower temperature than the rest. The difference

[1] Brögger, *Syenitpegmatitgänge* (1890), p. 85.

of temperature need not be great, for it is not here (as in the Soret action) a measure of the differentiation.

To fix ideas, we will consider the case of a laccolite or a large sheet or dyke intruded among solid rocks at a lower temperature than the invading magma. Cooling then takes place by conduction, and, so long as it is in progress, the marginal parts of the intruded body will be constantly at a lower temperature than the interior.[1] Suppose now that, for the earlier minerals, there is a definite order of crystallization—*A*, *B*, *C*, . . . When cooling has proceeded so far as to initiate the crystallization of *A*, this will begin at the margin of the mass, where the temperature is lowest, and where the Soret action has already caused a slight enrichment in that constituent. This crystallization at the margin would, in the absence of any compensating factor, cause an impoverishment of that part of the magma in the constituent *A*, but the concentration is maintained by diffusion throughout the whole body of magma. In so far as this diffusion can keep pace with the crystallization, there will be an impoverishment of the whole body of magma in the constituent *A*, while crystallization is confined mainly to the marginal region. When the temperature has fallen so far as to occasion the crystallization of B, this, in like manner, will, in the act of crystallization, become concentrated towards the margin. The concentration, however, will not be so marked as in the case of *A*, for diffusion will be gradually checked; partly because the temperature-gradient is lessened by conduction and by the heat set free in crystallization; partly, perhaps, because viscosity is heightened by falling temperature and by the increasing acidity of the residual magma, though the concentration of water, etc., in the latter will modify this effect. The same qualifying considerations will apply with more force to a third constituent, *C*, so that the concentration of successive minerals in the

[1] Harker, *Geol. Mag.*, 1893, pp. 546, 547, and *Quart. Journ. Geol. Soc.*, vol. l. (1894), pp. 327, 328. The argument is now restated without allusion to Berthelot's principle, which has proved a stumbling-block to some critics.

margin becomes less pronounced. Since the diffusion of some constituents towards the margin implies diffusion of the others in the opposite direction, the later-crystallized minerals will be concentrated in the interior.

This process is what Becker[1] appropriately styles *fractional crystallization;* but he denies the importance of diffusion,[2] and would substitute convection-currents. To the present writer it appears that convection can play at most a minor part, since there is abundant evidence that diffusion proceeds freely in a cooling rock-magma long after the increasing viscosity has rendered bodily movement impossible. It is to be remembered, also, that the rate of diffusion is accelerated by pressure.[3]

Fractional crystallization seems to afford a complete explanation of differentiation in place as exemplified in some of the commonest phenomena of this order, and especially in the relatively *basic marginal modifications* of intrusive rock-bodies. (i.) Under favourable conditions—*i.e.*, if there be a sufficient initial difference of temperature between the intruded magma and the contiguous rocks, and if the volume of magma be so large that there is no rapid chilling of the whole mass—the resulting body of igneous rock may show a considerable difference in composition between different parts. (ii.) The variation will be continuous, showing a gradual transition from one extreme type to the other. (iii.) Since the differentiation is conducted with reference to the surface of contact, the variation will be disposed in an orderly manner—with bilateral symmetry in a dyke or sheet, and with an arrangement of concentric zones in a boss or laccolite. (iv.) The process consists in a concentration of the earlier-crystallized minerals in the marginal region, and, further, a differential concentration of these according to their order of crystallization. The composition, arrangement, and order of consolidation of the different varieties of rock

[1] *Amer. Journ. Sci.* (4), vol. iv. (1897), pp. 257-261.

[2] *Ibid.*, vol. iii. (1897), pp. 21-40, 280-286. Compare Bäckström, *Journ. Geol.*, vol. i. (1893), p. 773.

[3] Röntgen, *Wiedem. Ann.*, vol. xlv. (1892), pp. 98-107.

will therefore stand in direct relation to the order of crystallization of the constituent minerals. If the minerals follow more or less closely Rosenbusch's 'order of decreasing basicity,' there results the familiar case of a relatively basic margin. (v.) Since the process depends upon the existence of this or some other definite order of crystallization, differentiation from this cause will be more or less effective according to the nature of the original magma. (vi.) The nature of the magma will be of consequence also as regards the degree of fluidity or viscosity, which necessarily controls in some measure a process depending on diffusion. For example, we should expect differentiation of this kind to be more marked in a gabbro than in a nepheline-syenite. Specific viscosity is, however, greatly modified by the presence of water or other fluxes, and granite intrusions frequently show a more or less marked basic modification at the margin.

It is reasonable to suppose that the process of fractional crystallization, which appears to be of prime importance in the differentiation in place of intruded rock-magmas, plays a part likewise in that intratelluric differentiation by which these magmas had their origin. Its most obvious application is to a series of rocks which are composed of the same minerals in different relative proportions. We have already seen that in such cases the distinction between differentiation prior to intrusion and differentiation in place may have but little significance (p. 128).

Subsidence of Crystals in a Partly Fluid Magma.—We have seen in the preceding section how differentiation may be brought about concurrently with crystallization by the latter being more or less localised at a given stage in a particular region of the magma. We have now to remark that, even when crystallization proceeds uniformly throughout a body of rock-magma, important consequences may result from a partial mechanical separation between the crystals already formed at a given stage and the residual fluid magma. Such separation may conceivably be effected by more than one cause; but the result in any case will be variation of such a kind that one extreme type is relatively enriched in the earlier and the other extreme in the later products of crystallization. This implies a degree of correspondence with the variation discussed in the foregoing section, but the intervention of the mechanical element imports certain differences, and may give rise to *discontinuity* in the variation set up.

One process which readily suggests itself is the *sinking of crystals* in a fluid magma. This was long ago maintained by Darwin[1] as a principal cause of differentiation. After noticing instances of lavas in which porphyritic crystals have accumulated at the bottom, he remarked that such facts "throw light on the separation of the trachytic [*i.e.*, acid] and basaltic series of lavas." He clearly connected it with progressive crystallization in the magma. On the one hand, he considered that a separation due to gravity could not be effective, as had been supposed by Scrope, in a magma still wholly liquid; on the other hand, there would be no sinking of crystals if all the minerals crystallized simultaneously, which he believed to be the case in plutonic rocks. He further applied his conception of differentiation to explain the sequence of different types of lava at one centre. We find, he says, that "where both trachytic and basaltic streams have proceeded from the same orifice, the trachytic streams have generally been first erupted, owing, as we must suppose, to the molten lava of this series having accumulated in the upper parts of the volcanic focus."

The sinking of crystals in a liquid lava-stream has been witnessed by von Buch, Darwin, Green, Dutton, and others; but it is safe to assume that these lavas were of very exceptional fluidity. Analogous records,[2] in which variation in a vertical direction can plausibly be ascribed to such a cause, are certainly rare, either in lava-flows or in intrusive sills. Porphyritic crystals, in a rock-body representing a single homogeneous intrusion, appear to be in general

[1] *Geol. Obs. Volc. Is.* (1844), p. 118.

[2] See, *e.g.*, Stock, *Tscherm. Min. Petr. Mitth.* (2), vol. ix. (1888), pp. 429-469; Lane, *Rep. Geol. Sur. Mich.*, vol. vi. (1898), pp. 143-148; Clarke, *Bull.* 228 *U.S. Geol. Sur.* (1904), p. 105.

equably distributed; and it has not been shown that, for instance, the heavy iron-ore minerals tend to accumulate at the base even of a thick laccolite. Since it can scarcely be doubted that crystals are almost always denser than the magma from which they form, the general absence of any evidence of their sinking can be explained only by the viscosity of rock-magmas in the temperature-range of crystallization; and we have here another proof that a degree of viscosity sufficient to check molar movement does not prevent diffusion.

It is necessary to observe, however, that viscosity, if it follows the laws laid down by physicists, cannot inhibit movement, but can only retard it. The time-element therefore becomes an all-important factor. We are left free to conjecture that the settling down of crystals, which seems to be generally ineffective in a sill or laccolite, may give rise to very important differentiation in a large intercrustal magma-basin, cooling at an extremely slow rate. Various special features observable in igneous rocks are susceptible of interpretation on this hypothesis, and serve in a measure to support it. The dark basic secretions or 'clots,' which occur sporadically in many granites and other rocks, may be taken as an example. These consist in general of the same minerals as the normal rock, but are much enriched in the darker and denser minerals, or in those of earlier crystallization. It seems reasonable to regard them as portions picked up from a lower stratum of the magma-reservoir, where crystals of these minerals accumulated by settling down in the magma. In the Shap granite of Westmorland[1] the dark patches enclose porphyritic crystals of orthoclase like those in the matrix, but always corroded and frequently converted at the margin of each crystal to plagioclase and quartz. The dykes and sheets of mica-lamprophyre in the district surrounding the Shap granite contain scattered crystals of quartz and alkali-felspar, always deeply corroded. It is suggested

[1] Harker and Marr, *Quart. Journ. Geol. Soc.*, vol. xlvii. (1891), pp. 280-282.

that these lamprophyres were derived from the basic lower layers of the magma-reservoir, into which some quartz and felspar crystals sank from the overlying acid magma, where they were beginning to form. Scattered quartz-grains are a highly characteristic feature of many groups of lamprophyres, and are found also in certain basaltic lavas. It is scarcely to be believed that the quartz crystallized, at an early stage, from a highly basic magma, and the explanation offered in the case of the Shap rocks has possibly a wide application.[1] We shall recur to the subject in the following chapter.

Expulsion of the Residual Fluid Magma.—Any differentiation which depends on the sinking of crystals under gravity belongs necessarily to a somewhat early stage of crystallization, when the bulk of the magma was still in a liquid condition. At a later stage, when the crystals formed are so numerous or so large as to touch and support one another, the condition may be likened to that of a sponge full of water; and it is easy to picture a partial separation being effected by the *straining off or squeezing out of the residual fluid magma* from the portion already crystallized. That such a process does in fact take place is amply proved by the phenomena of pegmatites, which represent the final residual magma of plutonic intrusions. This not only collects in fissures to form veins in the plutonic rocks themselves, but is often intruded as dykes and sheets in the neighbouring country-rock. Moreover, dykes and more considerable masses of pegmatite may occur at various distances from the main plutonic intrusion to which they belong, and not in visible connection with it; a fact suggesting that in the lower depths separation of the kind in question may be carried out on an important scale.

The application of the principle is by no means confined to plutonic intrusions and their extrusions of pegmatite. We will take as an example the dykes and sheets of younger augite-andesite and pitchstone, which represent one of the latest episodes in the British Tertiary cycle. The mutual

[1] Harker, *Geol. Mag.*, 1892, pp. 199-206, 485-488.

relations of these two groups of rocks are, as Judd[1] has observed, of special interest, and can be explained only on the supposition that the two are complementary products of differentiation of one magma. Moreover, this differentiation took the form of a partial separation between the crystallized elements at a given stage and the residual magma; and the gradations of the process are almost as clearly marked as if it could be witnessed in progress. (i.) The less basic augite-andesites contain a considerable amount of glassy residue, and this, in a thin slice, shows a very pronounced tendency

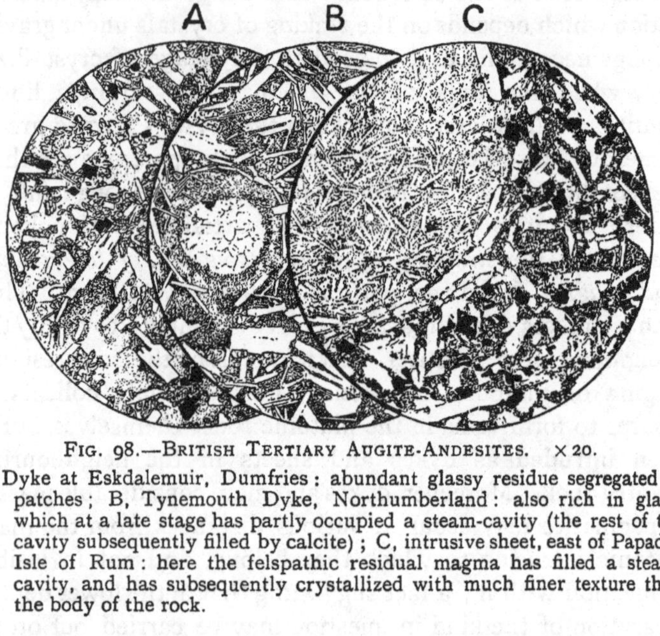

FIG. 98.—BRITISH TERTIARY AUGITE-ANDESITES. × 20.

A, Dyke at Eskdalemuir, Dumfries: abundant glassy residue segregated in patches; B, Tynemouth Dyke, Northumberland: also rich in glass, which at a late stage has partly occupied a steam-cavity (the rest of the cavity subsequently filled by calcite); C, intrusive sheet, east of Papadil, Isle of Rum: here the felspathic residual magma has filled a steam-cavity, and has subsequently crystallized with much finer texture than the body of the rock.

to segregate into little patches. A very characteristic feature of this group of rocks throughout the whole region is the oozing of the residual magma into vesicular cavities, which thus become round spots of glass or of devitrified glass (Fig. 98). (ii.) In some of the dykes there occur kernels,

[1] *Quart. Journ. Geol. Soc.*, vol. xlvi. (1890), pp. 371-381, and vol. xlix. (1893), pp. 536-564. Compare Harker, *Tertiary Igneous Rocks of Skye*, *Mem. Geol. Sur.* (1904), p. 402, with references.

streaks, or more considerable masses of highly vitreous rock in the less glassy andesite, these being constantly of much more acid composition than the rock in which they are found. (iii.) In Arran and elsewhere occur composite dykes, in which a central band of pitchstone (sometimes devitrified) is flanked on each side by augite-andesite. The two rocks have been intruded successively, but their intimate relationship is beyond question. (iv.) A complete separation gives rise to distinct intrusions of augite-andesite on the one hand and pitchstone on the other, sometimes, as in Ardnamurchan, in close association.

In magmas of andesitic composition, as Teall and Lagorio have pointed out, the general order of crystallization is such that the residual magma becomes progressively more acid. Thus, by the process of separation outlined, a magma with the composition of an intermediate andesite gives rise to relatively basic rocks (the more crystalline andesites) and relatively acid rocks (pitchstones and felsites). This is a discontinuous variation. But for the same reason there may also be variation between rather wide limits in each of the two complementary groups. On the one hand, the composition of a pitchstone derived in this way will depend upon the stage to which crystallization had progressed before the residual magma was drafted off. The British Tertiary pitchstones are of andesitic, trachytic, dacitic, or rhyolitic composition in different examples. On the other hand, the bulk-composition of the complementary rock will depend on the same factor, and also on the extent to which the rock has been deprived of its residual fluid magma. That the latter consideration is of importance appears from the fact that these augite andesites are more basic in proportion as they are more nearly holocrystalline.

The same idea is developed in another paper by Judd,[1] in which he discusses the hypersthene-andesite lavas, of various geological ages, found in many parts of the world. The examples cited have in all cases the same crystallized

[1] 'The Natural History of Lavas as illustrated by the Materials ejected from Krakatoa,' *Geol. Mag.*, 1888, pp. 1-11.

minerals in similar relative proportions, together with a glassy residue, which is perhaps of not very different composition in different cases; but bulk-analyses show a silica-percentage ranging from 52 to 70. This variation depends on the proportion of glassy residue present, which varies from 10 to 90 *per cent.* Judd concludes that "after the partial separation of a magma into crystals and a colloid residue, the two may be separated by a process of liquation, and subsequently become mingled again in varying proportions."

It is easy to conceive various modes of liquation and decantation, straining and filtration, by which a partial separation may be brought about between the already crystallized elements and the more mobile liquid residuum at different stages during the consolidation of a rock-magma. As in the simple case of fractional crystallization first considered, the differentiation thus effected belongs to the period of consolidation, or to some epoch during that period; but the purely molecular processes of diffusion and crystallization are here complicated by the intervention of the mechanical element. The special interest lies in the fact that magmatic differentiation is brought into *direct relation with crustal stresses.* The importance of this will be rendered clearer by a concrete illustration.

The 'older granites,' which play so prominent a part in the geology of the Eastern Highlands of Scotland, have been studied especially by Barrow,[1] and such of his results as have been made public possess a high interest in connection with our present subject. The rocks in question were intruded and consolidated during a period of great crust-movements; and this has been a prime factor in determining, not only their manner of intrusion and their frequent gneissic structure,

[1] 'On Certain Gneisses with Round-grained Oligoclase and their Relation to Pegmatites,' *Geol. Mag.*, 1892, pp. 64, 65; 'On an Intrusion of Muscovite-biotite Gneiss in the South-Eastern Highlands of Scotland, and its accompanying Metamorphism,' *Quart. Journ. Geol. Soc.*, vol. xlix. (1893), pp. 330-356. See also *Summary of Progress Geol. Sur.* for 1902, p. 94.

but also their varying mineralogical and chemical composition. As exposed at the surface, they do not usually occur as large continuous bodies, but tend to an infinitude of small intrusions permeating extensive areas of highly metamorphosed sediments. It may be inferred that the visible intrusions are offshoots from large continuous masses at a deeper level.

This region affords most decisive evidence of differentiation effected by *mechanical stress in conjunction with progressive crystallization.* This took the form of a squeezing-out of the residual magma, which was pressed forward, and made a fringe of pegmatite in advance of the direction of crust-movement. The packing together of the already crystallized elements, with a minimum of interstitial residuum, gave rise to a rock with more or less pronounced gneissic foliation; but the pegmatite, which might remain fluid until movement had ceased, is often non-foliated. As a consequence of the usual order of crystallization in granitic magmas, the pegmatite was enriched in silica and potash, and the complementary gneissic granite impoverished in these constituents. In a typical case, more particularly described by Barrow, the gneiss is composed of oligoclase, muscovite, biotite, quartz, and microcline, the last in subordinate amount. While the micas have sharp angles, the oligoclase shows a rounded outline, due to the draining off of the magma before the growth of the crystals was completed. Traced southward—*i.e.*, in the direction of movement—the rock changes character, the proportion of oligoclase diminishing, and muscovite becoming predominant over biotite. Pegmatite veins become abundant, and on the southern edge of the gneiss is a massive fringe of pegmatite, up to 700 yards in breadth, composed essentially of microcline, quartz, and muscovite.

Some Geological Aspects of Differentiation.—Magmatic differentiation is of fundamental importance in relation to the partition of petrographical provinces, the variety of rock-types developed within a single province, and the sequence in time of these different associated rocks. The considerations advanced in the foregoing paragraphs may therefore serve to link together the geological matters touched on in the earlier part of this volume, and the more purely petrological subjects which have occupied the later chapters. Accordingly a few remarks may properly be offered under this head, notwithstanding the speculative character which necessarily attaches to them.

The existence of extensive intercrustal magma-reservoirs, from which successive extrusions and intrusions are fed, seems to be a plain inference from general considerations, and is the only rational explanation of that family likeness among associated igneous rocks which is implied in the expression 'consanguinity.' Considering a single reservoir, supposed, for simplicity, of homogeneous magma, enclosed by solid rock, it is clear that, if this condition is to have any element of permanence, the temperature of the magma cannot be far from that at which crystallization would begin, and the rocks which enclose it must be at a similar temperature. This is a necessary condition for equilibrium between a liquid and a solid of like composition, and it must remain true with proper qualification for any difference in composition which can be postulated. Now, we have seen that the temperature-range of crystallization is higher for basic than for acid magmas. We may infer, therefore, that, in so far as concerns magma-reservoirs of the kind contemplated, ***basic rocks are of more deep-seated origin than acid rocks.*** This conclusion is in accord with the relative densities of the rocks, as well as with thermal considerations, and it seems to be borne out by such geological evidence as we possess. In applying like reasoning to the different layers of a *stratified magma-reservoir,* supposed of graduated composition from basic at the bottom to acid at the top, we raise questions which cannot be resolved so decisively. It is clear that the supposed arrangement will always be stable so far as concerns the relative densities of the different magmas[1]—*i.e.*, a

[1] This is easily seen from the very different densities of basic and acid glasses at ordinary temperatures (about 2·7 and 2·4). The coefficient of dilatation at high temperatures may be taken at about ·00005.

basic magma will be denser than an acid one, notwithstanding any difference of temperature that can reasonably be postulated. Whether the arrangement is stable with reference to diffusion is a question which cannot yet be answered from the physical side, but the geological evidence tells strongly in the affirmative. The stratified magma-basin, with a temperature-gradient comparable with that usually assumed for the solid crust, may therefore rank at least as a working hypothesis.

Assuming this, we must see in it the natural starting-point for any discussion of magmatic differentiation as applied to a single petrographical province. The data do not yet suffice for following out all the consequences of this hypothesis, and interpreting on the lines indicated the actual succession of events in a given case; but certain clues are afforded even by the imperfect survey of the facts which we have been enabled to make. We have seen that a normal cycle of igneous action opens with volcanic outbursts, and that the sequence of different rock-types in this phase is in general one of increasing divergence in opposite directions from the initial type. These facts suggest that the volcanic phase corresponds with the period during which the parent magma-reservoir, supposed originally of uniform composition, is undergoing progressive differentiation and stratification. If the correspondence is a strict one, the earliest erupted volcanic rocks will represent more or less closely the undivided magma of the province; but in some cases there is evidence that differentiation was already well advanced before the beginning of any overt action.[1] Types of lava belonging to the two diverging lines may alternate, or may even be erupted contemporaneously at different centres, the succession in this particular being determined by conditions of the second order of importance.

The pause which divides the volcanic from the plutonic phase probably corresponds with a freezing of the main magma-reservoir; and the localisation of plutonic intrusions

[1] Harker, *Tertiary Igneous Rocks of Skye*, *Mem. Geol. Sur.* (1904), p. 417.

will then be determined by re-fusion beginning at particular centres. However this may be, the order of decreasing basicity, which is so general a rule in a series of plutonic intrusions, may be taken to prove that the primary differentiation is completed before the beginning of this phase, and that the earlier intruded magmas are drawn from the lower levels of the reservoir, and later magmas from successively higher levels. This order may perhaps be connected with a steady rise of the isothermal surfaces within the local crust during the plutonic phase. Such, apparently, is the view adopted by Barrow,[1] with special reference to the Caledonian complexes of Scotland and to the succeeding minor intrusions, in which the law of sequence is reversed. He suggests that the whole of the phenomena may be due to "the remelting of an old magma," and supposes that "the irruptions commenced with the re-fusing of the base, and ended, in the dyke-stage, with its reconsolidation."

There is another order of facts which we may hope to see elucidated by a fuller comprehension of the nature of magmatic differentiation—viz., the geographical distribution of rock-magmas, as illustrated by the separation of the Atlantic and Pacific regions and the delimitation of different petrographical provinces within one region. Here we have evidence of *extensive differentiation and separation in the horizontal sense.* The fact that regions and provinces are often divided by important orographic lines points to *mechanical forces of the nature of lateral thrust* as the essential factor in differentiation of this order. Such agency can bring about actual separation only when it operates on a body of rock which is partly crystalline and partly liquid; and, according to the conception which we have adopted, this condition is realised in a nascent magma-reservoir at a period anterior to the beginning of overt activity. The differentiation, then, is effected by a deep-seated displacement of a greater or less proportion of the mobile magma, which is carried forward in the deeper levels of the crust, leaving the mainly crystalline residuum correspondingly altered in total composition.

[1] *Quart. Journ. Geol. Soc.*, vol. xlviii. (1892), p. 121.

This action is illustrated, on a relatively small scale, by such phenomena as those described by Barrow in the Eastern Highlands; and at the same time the geological relations of that region warrant us in supposing that the observed effects are only an index of more extensive operations of the same kind conducted in the deeper levels of the Earth's crust. It is to be noted, however, that the requisite conditions for this differentiation by filtration—viz., a partially solidified body of magma (or partially fused body of rock) and appropriate mechanical forces—may be realised at more than one stage in a cycle of igneous action. This factor, therefore, to which we assign an important part in the delimitation of different petrographical provinces, may be supposed to intervene also at later epochs in the evolution of a given province, its operation being perhaps restricted then to isolated reservoirs of smaller dimensions.

The consequences of such differentiation, from the petrographical point of view, will evidently depend on the order of crystallization of the minerals in the magma. Where this approximates to Rosenbusch's 'order of decreasing basicity,' as in the old granitic magmas of the Eastern Highlands, the separation will be one between more 'basic' and more 'acid' derivatives. Where the order is that of the ideal case discussed in Chapter VIII., the fluid magma expelled under pressure will approximate to eutectic composition ('anchi-eutectic' of Vogt; see below, Chapter XV.).

In the direct connection here contemplated between lateral thrust and magmatic differentiation we see at least one reason for the fact, already observed, that there is in general greater diversity among the associated igneous rocks in a mountain than in a plateau region (p. 104). Another question arises when we regard orographic lines as the boundaries between petrographical provinces. In a given folded mountain-chain there is a definite direction in which the thrust has operated, in general from the concave towards the convex side of the curving line. Where such a chain divides two petrographical provinces, we should expect the broad petrographical differences between the two provinces to stand in some definite relation to the direction of thrust. The question cannot, however, be adequately stated in such simple terms. Indeed, it often happens in a region of complex folding that a given province lies on the inner side of one mountain-arc and on the outer side of another (compare map, Fig. 1). The manner in which the rocks composing a mountain-chain are folded and overthrust indicates only the *relative* direction of movement in the outer portion of the crust, and there is no real criterion for discriminating between an overthrust from south to north and an underthrust from north to south. The magmatic differentiation, to which we attribute the establishment of different petrographical provinces, must be related rather to the *absolute* direction of displacement at deeper levels, and to this we have no direct clue. Again, the apparent connection which we have remarked between the Pacific and Atlantic petrographical regions and the two contrasted types of coast-line suggests that vertical, as well as horizontal, crust-movements may play a part in magmatic differentiation; but no useful purpose would be served by venturing farther into the realm of conjecture.

CHAPTER XIV

HYBRIDISM IN IGNEOUS ROCKS

Commingling of two magmas.—Absorption and assimilation of solid rocks by a magma.—Hybridism in plutonic complexes.—Hybridism in composite sills and dykes.—Cognate xenoliths and xenocrysts.—Hybridism as a factor in petrogenesis.

Commingling of Two Magmas.—In the preceding chapter we have considered the origin of different igneous rocks by various processes of differentiation. We have also to recognise, however, the possibility of new varieties of rocks arising from what is in some sense the reverse of differentiation—viz., admixture. The phenomena of *mixed or hybrid rocks* possess an intrinsic interest; and, since certain geologists have laid much stress on this element in petrogenesis, it will be proper to examine briefly the kind and degree of effects which may be expected from such a cause.

Two possibilities may be distinguished—viz., commingling of two fluid rock-magmas and reactions between a fluid magma and solid rocks. The former idea is identified especially with comprehensive theories designed to account for all the igneous rocks of a region, or of the whole globe, by a process of simple admixture on a large scale. Thus Bunsen,[1] principally from a chemical study of the volcanic rocks of Iceland, enunciated the hypothesis that all intermediate varieties have arisen from the admixture in different proportions of two extreme magmas, to which he assigned definite compositions. The variation which could arise in

[1] *Pogg. Ann.*, vol. lxxxiii. (1851), pp. 197-272. For a wider application of Bunsen's hypothesis, see Roth's *Gesteinsanalysen* (1861).

this way is represented graphically in Fig. 99. Durocher[1] subsequently put forward his well-known theory that all igneous rocks have been derived from two permanent magmas, which coëxist beneath the crust of the Earth. The variation which would result is exhibited in Fig. 100.

In the light of later knowledge concerning the great diversity of igneous rocks which may occur even in one

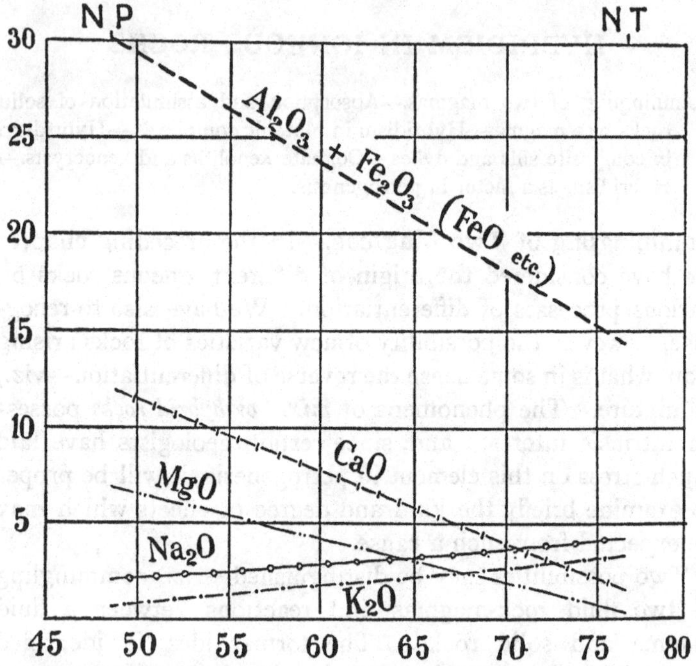

FIG. 99.—VARIATION-DIAGRAM FOR MIXTURES OF BUNSEN'S TWO MAGMAS, THE 'NORMAL PYROXENIC' (NP) AND THE 'NORMAL TRACHYTIC' (NT).

small district, such theories have only a historic interest. Moreover, it is evident that no hypothesis of the commingling of two magmas can afford a complete explanation of the variation even in a simple series of igneous rocks. The original two magmas remain to be accounted for, and it appears that admixture of the kind supposed is only a partial

[1] 'Essai de pétrologie comparée,' *Ann. des Mines*, vol. xi. (1857), pp. 217-259. A translation is given in Haughton's *Manual of Geology*.

undoing of the differentiation which has still to be postulated. On this point a remark may be made which has a wider application. It is easy to see[1] that the processes of magmatic differentiation are of too complex and subtle a nature to be reversible by so crude a method as that of mixture between two extreme terms. Any variation arising from the latter cause is necessarily of the *linear* kind (Figs. 99, 100); but whether we regard particular series of rocks (p. 120) or igneous rocks in general (Figs. 43, 44), it is clear that the

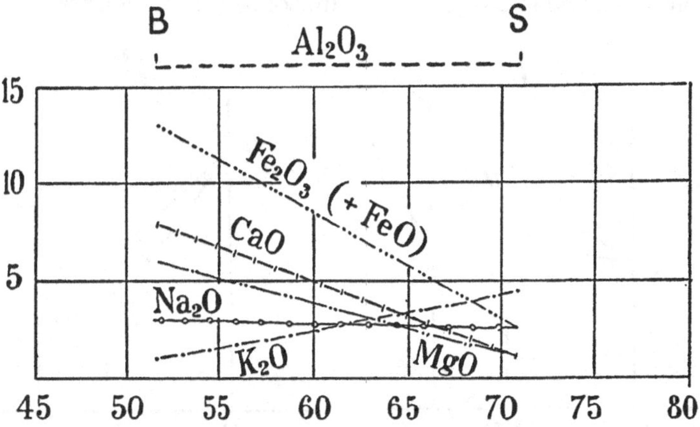

FIG. 100.—VARIATION-DIAGRAM FOR MIXTURES OF DUROCHER'S TWO MAGMAS, THE BASIC (B) AND THE SILICEOUS (S).

variation met with in nature is not of the linear kind. It follows from this that, theoretically at least, *hybrid rocks are abnormal in composition.* Consider a mixture of two rocks belonging to any natural series (Fig. 101). It is evident that the hybrid rock will differ from a member of the series with like silica-percentage, there being a deficiency in those constituents which have convex curves on the variation-diagram and an excess of those which have concave curves. By averaging, *e.g.*, the analyses of a peridotite and a granite, we obtain a result much richer in magnesia and poorer in lime and alumina than any natural intermediate rock. If such a mixture of magmas were to take place, there would

[1] Harker, *Journ. Geol.*, vol. viii. (1900), pp. 389-399.

be considerable chemical rearrangement (*e.g.*, olivine combining with silica to form abundant enstatite), and the resulting product would have a very peculiar mineralogical composition. The same consideration applies to reactions between a magma and a solid rock, and it may often be used as a test of hybrid origin.

Absorption and Assimilation of Solid Rocks by a Magma.—Not infrequently we find evidence that an intruded rock-magma has to a certain extent permeated and impregnated contiguous solid rocks; or, again, solid rock-fragments have

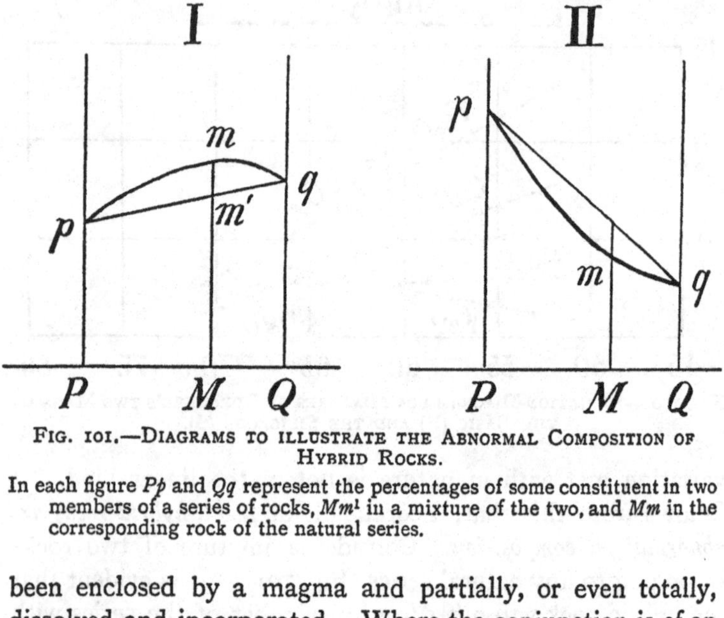

FIG. 101.—DIAGRAMS TO ILLUSTRATE THE ABNORMAL COMPOSITION OF HYBRID ROCKS.

In each figure *Pp* and *Qq* represent the percentages of some constituent in two members of a series of rocks, Mm^1 in a mixture of the two, and *Mm* in the corresponding rock of the natural series.

been enclosed by a magma and partially, or even totally, dissolved and incorporated. Where the conjunction is of an *accidental* kind—*i.e.*, with no genetic relationship between the magma and the rocks which it invades or envelopes—such action is—so far as our positive knowledge from observation extends—always local and limited. Reactions between *cognate* igneous rocks, crystalline and liquid, may give rise to hybridism on a somewhat larger scale.

The distinction just drawn is of some importance. In the former case the solid rock assimilated by the magma may be, and often is, of sedimentary origin. Now, the great variation in composition among sedimentary and other stratified rocks follows laws very different from those which control the variation in igneous rocks. Chemical degradation and mechanical sorting may produce on the one hand a deposit almost purely siliceous, and on the other a sediment consisting largely of aluminous silicates with little alkali. Organic agency builds up non-silicated rocks containing

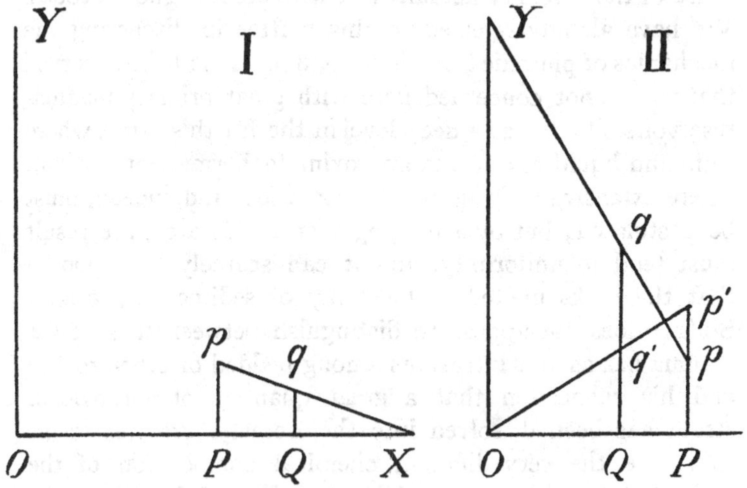

FIG. 102.—DIAGRAMS REPRESENTING THE CHANGE OF COMPOSITION IN A MAGMA DUE TO AN ADMIXTURE (I) OF SILICA OR (II) OF LIME.

Here $OX = OY = 100$. OP represents the original and OQ the modified silica-percentage of the magma. In I the silica-percentage is raised and the percentage of any other constituent reduced (from Pp to Qq). In II the silica-percentage is reduced, the percentage of lime is raised (from Pp to Qq), and that of any other base reduced (from Pp^1 to Qq^1).

no base but lime. In general the chemical analysis of a sedimentary rock is enough to distinguish it at once from any igneous product. Its dissolution in an igneous magma gives rise to a hybrid rock much more markedly abnormal in composition than one of which both components were igneous. For illustration we represent graphically in Fig. 102 the result of dissolving in an igneous magma a quantity (I) of quartzite, or (II) of limestone, the carbonic acid in the latter case being eliminated. It is easy to see that the modified magmas will produce rocks of peculiar mineralogical composition. We find, in fact, that such reactions between igneous magmas and sedimentary rocks give rise to various minerals (wollastonite, sillimanite, cordierite, zoisite, and many others) which, well known in metamorphosed sediments, are foreign to normal igneous rocks.

Certain geologists have made very large demands on 'assimilation,' and have even regarded it as the principal cause of the diversity actually met with among igneous rocks. We have already referred to this matter in discussing the mechanics of plutonic intrusions (p. 83). It is to be observed that we are not concerned here with great primary magma-reservoirs situated at a deep level in the Earth's crust, where solid and liquid rock are in approximate thermal equilibrium. There extensive melting may be conceded, and, indeed, must be postulated; but by averaging over a wide area, the result must tend to uniformity, and it can scarcely be supposed that the rocks melted include any of sedimentary origin. Sollas[1] does not appear to distinguish between these great magma-basins and intrusions among bedded or other rocks; and his suggestion that a great quantity of Ordovician strata has been dissolved into the Leinster granites seems to ignore the very different chemical composition of the sedimentary and igneous rocks in question. Johnston-Lavis[2] has put forward, to replace differentiation, a theory of transmutation which he calls 'osmotic reciprocal reaction,' in which all consideration of chemical composition is frankly thrown over. He ascribes the essexites of Gran to the modification of an acid magma by absorption of basic sedimentary rocks, without inquiring what sort of sedimentary rocks, and in what amount, would be required to produce such a change, and without observing that the essexites are in fact more basic than the stratified rocks of their district. Apart from merely fanciful suggestions, the extreme position is represented, among British geologists, by Cole.[3] As a

[1] *Trans. Roy. Ir. Acad.*, vol. xxx. (1894), p. 505.
[2] *Geol. Mag.*, 1894, pp. 252-254.
[3] *Sci. Trans. Roy. Dubl. Soc.* (2), vol. vi., 1897, p. 246.

result of his study of Slieve Gallion, in Londonderry, he concludes "that the underlying magmas of the earth's crust may be of far simpler character than has been commonly supposed," and suggests "that plutonic rocks, as we ordinarily know them, are phenomena of contact, produced in what are, comparatively speaking, the upper layers of the earth's crust." He further concludes that, "by a combination of absorption and concomitant or subsequent differentiation, an invading igneous rock may come to occupy the place of a pre-existing rock, and may, in fact, represent it as a pseudomorph, the absorbed matter being drawn off through the molten mass to lower levels."

The insuperable objection to any such theory is that it demands an enormous amount of heat to raise the solid rocks to the point of melting and to melt them, and no source for this heat is indicated. We have already pointed out that no noteworthy superheating can be assumed in the intruded magma, and this consideration sets a very moderate limit to its capability for absorbing solid rock-material. The limit will be wider if the solid rock is already at a temperature not far below that of melting; for in that case only its latent heat of fusion is to be reckoned with, not its thermal capacity. Accordingly the most striking phenomena of mutual reactions between an invading magma and a solid rock are found where the two belong to successive cognate intrusions, and one has followed the other after no long interval of time.

Hybridism in Plutonic Complexes.—We have already had occasion to remark that, where different rocks have been intruded successively at the same place with no long intervals, a certain degree of intermixture is a common incident (p. 127). Other conditions being the same, it must clearly be more efficient in proportion as the interval is shorter, so that the earlier rock is still hot when the later magma invades it. Another rule is that, the temperature conditions being the same, the mutual reactions will be more pronounced the greater the difference in composition between the two rocks concerned. Especially instructive in this connection is the

British Tertiary suite, in which types of mean acidity are scarcely represented, and basic and acid rocks are brought into close conjunction in a variety of circumstances. Not only do many of the rocks carry what Sollas terms xenoliths (quasi-foreign inclusions) or xenocrysts (isolated crystals of extraneous origin), but, in more than one series, basic and acid rocks in bulk enter into hybrid products, usually only locally, but in favourable circumstances on a somewhat larger scale.

Among the plutonic rocks, granites are found in intimate

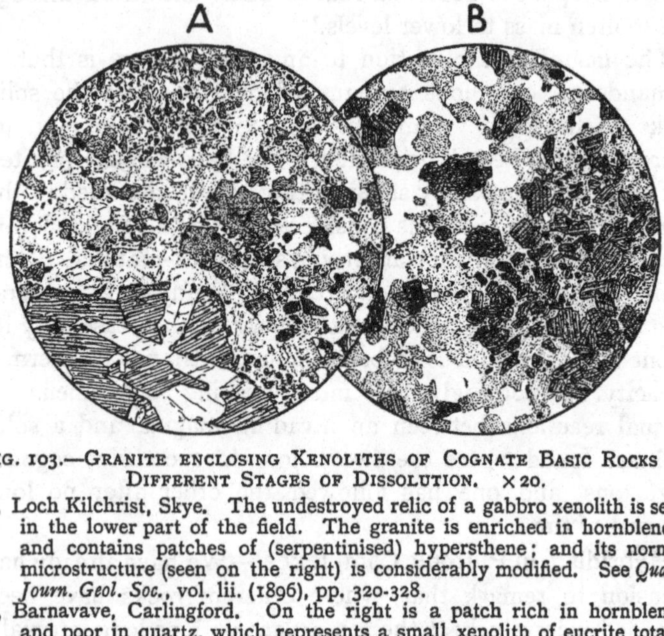

FIG. 103.—GRANITE ENCLOSING XENOLITHS OF COGNATE BASIC ROCKS IN DIFFERENT STAGES OF DISSOLUTION. ×20.

A, Loch Kilchrist, Skye. The undestroyed relic of a gabbro xenolith is seen in the lower part of the field. The granite is enriched in hornblende, and contains patches of (serpentinised) hypersthene; and its normal microstructure (seen on the right) is considerably modified. See *Quart. Journ. Geol. Soc.*, vol. lii. (1896), pp. 320-328.

B, Barnavave, Carlingford. On the right is a patch rich in hornblende and poor in quartz, which represents a small xenolith of eucrite totally dissolved and recrystallized. (Compare Sollas, *loc. cit.*)

association with eucrites in Rum[1] and in the Carlingford[2] and Dundalk district, with gabbros in Skye[3] and Mull, and

[1] Harker, *Geology of Small Isles, Mem. Geol. Sur.* (1908), pp. 105-114.

[2] Sollas, *Trans. Roy. Ir. Acad.*, vol. xxx. (1894), pp. 477-510.

[3] Harker, *Tertiary Igneous Rocks of Skye, Mem. Geol. Sur.* (1904), pp. 169-196. Compare also Carrock Fell in Cumberland, *Quart. Journ. Geol. Soc.*, vol. li. (1895), pp. 130-139.

with both gabbros and norites in Arran.[1] The minute veining of the basic rock by the acid, with reactions producing an acidification of the former and a correlative basification of the latter, the inclusion of xenoliths of the basic rock in the acid, and their progressive dissolution, can be studied in every stage (Fig. 103). The resulting hybrid rocks often betray their origin by the presence of corroded xenocrysts, or by an ill-defined patchy appearance due to scarcely obliterated veins or xenoliths; but, even when these direct evidences are lost, there are peculiarities in the composition of the rocks, which are elucidated by the field-evidence proving their essentially hybrid nature. They are in fact abnormal products, in the sense indicated, and they cannot be correctly designated by names, such as quartz-diorite, which belong to products of magmatic differentiation.

The most interesting effects are found where a basic rock has been broken up, but only partially dissolved or fused by an invading magma, and the heterogeneous mass has been drawn out by contemporaneous differential movement, so as to produce *gneissic banding*. For the production of a primary gneissic banding in plutonic rocks, two conditions are requisite—viz., a heterogeneous constitution and a sufficient differential deformation while the mass is still effectively fluid. The heterogeneity may arise either from differentiation or from admixture. In the banded peridotites, eucrites, and gabbros of the Scottish isles, the heterogeneous character was the result of incomplete differentiation—*i.e.*, various partial magmas had already segregated, but had not been separated into different places. The other case, that of imperfect admixture, is illustrated by the remarkable Tertiary gneisses of the Isle of Rum, which have resulted from the partial destruction of a eucrite by granitic intrusions. The gneisses occur in several detached areas, the most instructive one measuring about a mile by half a mile (Fig. 104). Along its northern and western border is a strip of eucrite, which is clearly the attenuated relic of a more considerable mass. It has been invaded by the acid magma in an intricate net-

[1] *Geology of North Arran*, *Mem. Geol. Sur.* (1903), pp. 105-108.

work of veins, and in places its substance has been more intimately permeated, producing a certain degree of acidification. Detached portions of the eucrite, up to 50 yards in

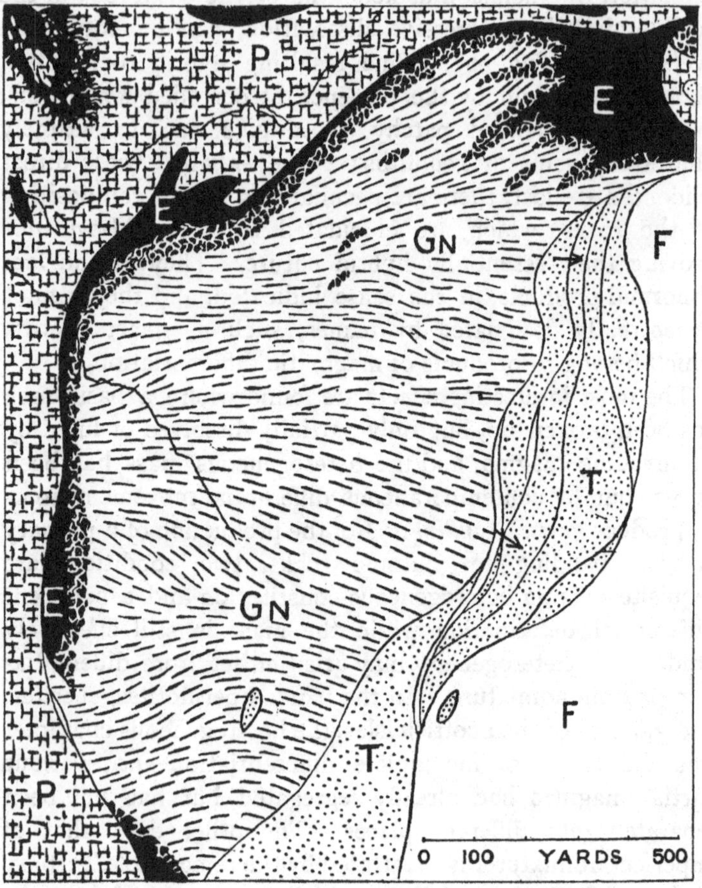

FIG. 104.—SKETCH-MAP OF A PART OF THE ISLE OF RUM. SCALE, 6 INCHES TO A MILE.

T, Torridonian strata, overthrust and partly brecciated, in the slopes of Ashval.
P, peridotite. In the north-west corner it is invaded by the eucrite in a very intricate fashion, but without any noteworthy mutual reactions.
E, eucrite, veined and largely destroyed by the succeeding granitic intrusion.
GN, gneiss, of hybrid origin from eucrite and granite, making the floor of the corrie Fiadh-innis.
F, porphyritic quartz-felsite, a later intrusive sheet, making the ridge of Ashval and Sgùrr nan Gillean.

length, with similar veining and permeation, are enclosed in the contiguous gneiss; and smaller blocks are seen in every stage of dissolution, their relics being traceable finally into the dark streaks of the gneiss itself, the significance of which is thus conclusively demonstrated. These processes may be studied step by step to distances of some 200 yards from the main eucrite border; but beyond this the earlier stages, which furnish the clue, cease to be recognisable. Without the connecting links the true origin of the gneiss could scarcely have been surmised, for the relics of the eucrite are transformed beyond recognition by metamorphism, by fusion

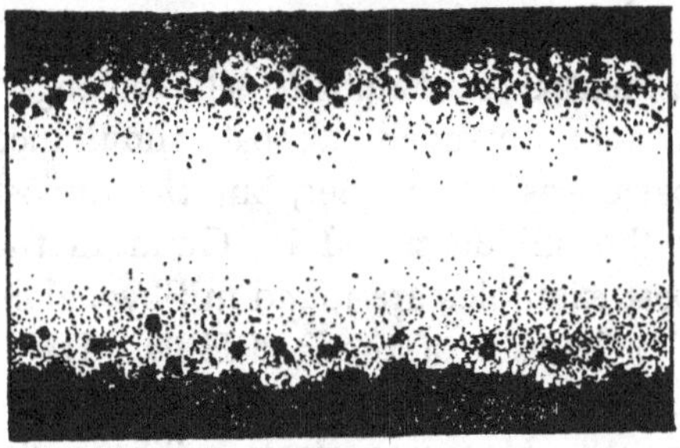

FIG. 105.—SECTION OF A REGULAR TRIPLE COMPOSITE SILL (OR DYKE).

and recrystallization, and by partial impregnation. The recrystallization following upon partial fusion gives sometimes a thoroughly coarse texture, probably because inoculation by relics of original crystals precluded supersaturation. The close resemblance of these rocks to the Lewisian gneisses suggests that the principles here illustrated may have a wide application.

Hybridism in Composite Sills and Dykes.—Minor intrusions tend to have a much more regular habit than plutonic masses. Hence, when different rocks have been successively intruded along the same channels, there may arise, instead of a 'complex' with little sign of orderly disposition, a composite sill or dyke in which the different members show a regular stratiform arrangement. The typical arrangement is that in which the later rock is flanked on each side by the earlier (Fig. 105), the arrangement being ABA, or, if three different rocks are involved, ABCBA. The interpretation of this is that the first magma consolidated from the margin inwards, and the interior was still hot, and, in some cases, not yet wholly crystallized, when the second magma followed, so that the latter naturally found its way along the middle of the sill or dyke.

Very interesting examples are found among the Tertiary intrusions of the western isles of Scotland,[1] where they followed immediately after the plutonic phase of action. We have already pointed out that the sequence of decreasing basicity in that phase was reversed in the succeeding phase of minor intrusions. At a certain intermediate epoch, to which these remarkable composite sills belong, basic and acid magmas were intruded almost simultaneously. In general the basic was the earlier, but the opposite case is illustrated by the sill described by Gunn in the south of Bute.[2] The best examples are found in Skye, along a belt of country lying to the east of the Red Hills. Here numerous composite sills, from a few feet to more than 200 feet thick, are intruded in the Mesozoic strata. The earlier rock of each may be called a basalt, while the later and central member is a more acid rock, usually a 'granophyre.' There are also composite dykes, the feeders of the sills. In the dykes, owing to more prolonged flow and probably somewhat higher temperature, the acid magma has made greater inroads upon the earlier basic rock than is usual in the sills, and in places has totally destroyed it.

In sills and dykes alike the junction between the two rocks is more or less sharp, or nearly or quite obliterated, in different cases. It has a very irregular form, and the later rock contains abundant partly digested xenoliths of the earlier. The extent of the mutual reactions is doubtless related to the longer or shorter interval of time which

[1] Harker, *Tertiary Igneous Rocks of Skye*, *Mem. Geol. Sur.* (1904), chaps. xii., xiii., with references.

[2] A recent examination of this sill by the present writer does not confirm this statement.

separated the two intrusions; and comparison of different occurrences enables us to trace out the stages of the process:

(i.) Boundary of basalt and granophyre irregular, and xenoliths of former enclosed in latter. These are partly acidified by impregnation, and have some interstitial quartz. Granophyre reciprocally modified near the junction, being enriched in the dark minerals.

(ii.) Basalt in place begins to be acidified by impregnation. Xenoliths more acidified, and enclose xenocrysts introduced from the granophyre[1]—viz., corroded alkali-felspars and rounded quartz-grains with reaction-border of granular augite; outlines of xenoliths still quite distinct. Granophyre more basified, and its felspar phenocrysts beginning in consequence to be corroded at the angles.

(iii.) Basalt in place more acidified, and encloses introduced quartz-grains with reaction-border and corroded alkali-felspars full of secondary glass-cavities. Xenoliths still outlined, though extremely altered; the quartz-xenocrysts, now more nearly in equilibrium, have lost the reaction-border.

(iv.) Basalt xenoliths not merely impregnated but totally recrystallized, with andesine in place of labradorite. The introduced xenocrysts of alkali-felspars have grown a new border of clear substance.

(v.) 'Basalt' xenoliths and inner border of 'basalt' in place assimilated to the 'granophyre' by acidification of the one and basification of the other; the junctions fading out only when practical identity is attained between the two.

(vi.) In one sill only is seen an insensible gradation from the basalt of the margins to the more acid rock of the middle, here not a granophyre but a bostonite. In this case the interior of the basic sill was still fluid when the newer felspathic magma was intruded.

(vii.) Near the preceding are two thin sills perhaps representing a final stage, in which all distinction of basalt and dolerite has disappeared. Presumably the two magmas

[1] Compare Cole, *Sci. Trans. Roy. Dubl. Soc.* (2), vol. v. (1894), pp. 239-248.

mingled freely, but not perfectly, for considerable differences may be found between specimens from the same sill.

It is interesting to note that, independently of these reactions in place, the basalt of these sills contains occasional xenocrysts of quartz and alkali-felspars, sharply contrasted with the fresh phenocrysts of labradorite, and undoubtedly brought up by the basalt magma. We have thus direct proof that the basic and acid magmas, prior to their intrusion, coëxisted in subterranean reservoirs; that the lighter acid magma overlay the denser basic one; and that crystals which formed in the upper acid stratum sometimes sank into the lower basic one. We have not direct proof, but at least a strong suggestion, that the two magmas were complementary differentiation-products, and that the differentiation which produced them was effected in these same local reservoirs. That they were distinct local reservoirs, and not one general one, is indicated by the special characteristics which link together in some instances the two very different rocks associated in one composite sill—*e.g.*, where the acid rock is a bostonite, the scattered xenocrysts of early intratelluric derivation in the associated basalt are of alkali-felspars only, without quartz. It would be easy to show in other ways how a study of xenocrysts and xenoliths may be made to throw light on the origin and mutual relationships of associated igneous rocks, and some considerations under this head will find a place in the following section.

Cognate Xenoliths and Xenocrysts.—In his exhaustive memoir on the xenoliths ('enclaves') of volcanic rocks, Lacroix[1] distinguishes 'enclaves énallogènes' and 'enclaves homœogenes.' In the first there is no relation as regards mineralogical composition or origin between the enclosed and the enclosing rock—*e.g.*, limestone enclosed in trachyte. In the second there is a greater or less analogy of composition and origin between the two—*e.g.*, olivine-nodules in basalt. This division does not appear quite satisfactory; for similarity of mineralogical composition cannot be accepted as a sufficient criterion, either positively or negatively, of community

[1] *Les enclaves des roches volcaniques* (1893).

of origin. Lacroix gives granite enclosed in basalt as an example of his first category, but there can be no doubt that in some occurrences of that association the two rocks have a real genetic relationship, notwithstanding their great difference of composition. We shall therefore prefer the terms already used above (p. 336), and distinguish *accidental* and *cognate* xenoliths. The former, although they afford much of interest in the metamorphic and metasomatic changes connected with them, are of very little importance from the

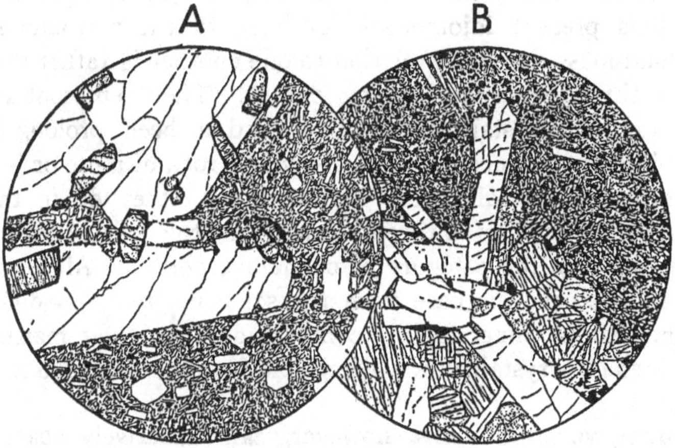

FIG. 106.—GLOMEROPORPHYRITIC STRUCTURES. × 10.

A, Hypersthene-andesite, Aspronisi, Santorin; with coarsely crystalline aggregate of felspar and hypersthene.
B, Basalt, Ovifak, Disco, Greenland; with coarse aggregate of felspar, augite, and (serpentinised) olivine. This rock contains also metallic iron of intratelluric origin.

petrogenetic point of view; but the latter have a significance which demands some further consideration.

On general grounds it seems evident that crystallization (often followed by re-fusion) must take place in intercrustal magma-reservoirs; and we have seen reasons for believing that magmatic differentiation is mainly dependent on this intratelluric crystallization. From the consequences of differentiation we can draw only general inferences relative to the crystallization which caused it, but in the cognate xenoliths of igneous rocks we seem to have a more direct clue.

Most igneous rocks afford evidence more or less clear of some crystallization in their magmas prior to intrusion or extrusion. In the volcanic rocks it is very evident, for we cannot doubt that the porphyritic crystals were formed at a relatively early stage in an intercrustal reservoir. There is often a marked tendency of these crystals to aggregate in groups, giving a structure for which we may conveniently use Judd's term *glomeroporphyritic* (Fig. 106). The grouped crystals present idiomorphic outlines to the surrounding ground-mass, but their relation to one another is rather that of a 'hypidiomorphic' plutonic rock. They represent an intratelluric crystallization which, had it been prolonged, would have produced a relatively coarse rock of more or less deep-seated consolidation. Occurring as they do in the heart of a rock of different characters (the ground-mass), they fall into the category of cognate xenoliths. Although crystallized in this case from the same magma, they differ somewhat in mineralogical composition from their matrix, progressive crystallization being itself a process of differentiation.

Many volcanic rocks, however, carry relatively coarse-textured inclusions which differ more notably in composition from the general mass, being considerably more basic or, speaking precisely, richer in the earlier-crystallized elements. These inclusions, often called *basic secretions*, are sharply bounded, and must be classed as xenoliths, doubtless of the cognate kind. Of the same general nature are the ovoid dark patches common in many plutonic rocks, especially in the more acid families (Fig. 107). Here the matrix itself is more or less coarse-grained, and the inclusion is usually of finer texture, owing to the fact that the earlier minerals of which it is largely composed build relatively small crystals. The same rocks sometimes contain xenoliths of the accidental kind, representing metamorphosed or partly fused fragments of slate, etc. These are, as a rule, easily distinguished,[1] by

[1] Phillips, *Quart. Journ. Geol. Soc.*, vol. xxxvi. (1880), pp. 1-21.

their angular or subangular shape and schistose or foliated structure, from cognate xenoliths. The true nature of the latter comes out clearly in comparative analyses of them and their matrix, which show that the two are consanguineous, but present a greater or less difference, always of the same kind. Regarding these basic secretions, then, as the products of an early crystallization in an intercrustal reservoir, we have still to ask how the earlier-formed minerals come to be so closely aggregated or concentrated as is here seen. The most probable explanation appears to be that

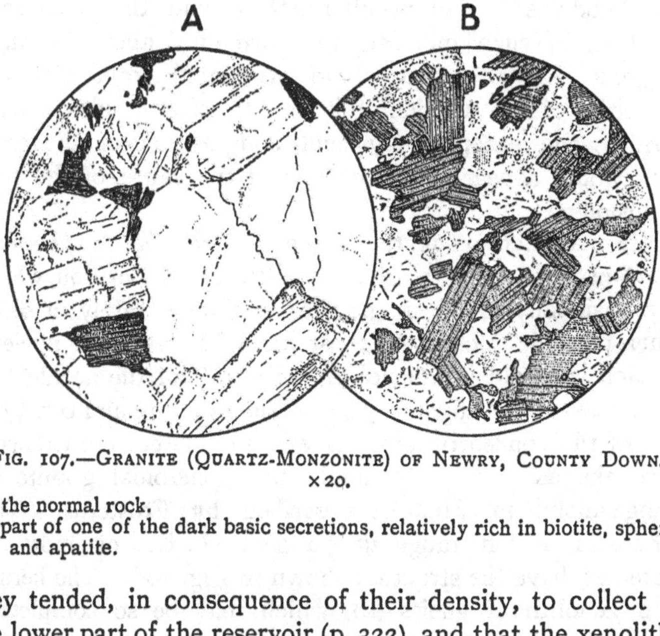

FIG. 107.—GRANITE (QUARTZ-MONZONITE) OF NEWRY, COUNTY DOWN. ×20.

A, the normal rock.
B, part of one of the dark basic secretions, relatively rich in biotite, sphene, and apatite.

they tended, in consequence of their density, to collect in the lower part of the reservoir (p. 322), and that the xenoliths are clots or patches carried up from the lower levels as an incident of intrusion.

Vogt[1] treats these inclusions in granite merely as the products of early crystallization in the granite magma, without considering the significance of their aggregation; and he illustrates from comparative analyses how the separation of relatively basic secretions reduces the magma more

[1] *Tscherm. Min. Petr. Mitth.* (2), vol. xxv. (1906), pp. 363-366, 374-376.

nearly to eutectic composition. This is, of course, in accordance with the general laws of crystallization; although it does not seem quite justifiable to assume that the bulk-analysis of the rock (secretions and matrix together) represents the composition of the magma from which the secretions formed. Vogt, however, goes farther, and supposes that a magma notably more acid than the eutectic type may give rise conversely to relatively acid secretions, thus again reducing the residuum more nearly to eutectic composition. The only two examples cited[1] are not free from ambiguity. The 'nodules' are of peculiar nature, and the occurrence in them of such minerals as tourmaline and sillimanite suggests that they may have originated from dissolved xenoliths of the accidental kind. If such suspicion were verified, the exceptionally acid composition of the total rock would appear as the effect, not the cause, of the acid inclusions.

We may mention in this place the *orbicular or spheroidal structure*, with concentric shells of different composition, which is occasionally found in granites and more rarely in some other plutonic rocks. In most cases at least[2] the nucleus of each spheroid is a xenolith, sometimes of the accidental, sometimes of the cognate kind. The structure and constitution of the concentric growths vary in different occurrences. We take as a single example the spheroidal granite of Kangasniemi in Finland, described by Frosterus.[3] The spheroids, which range up to eight inches or more in diameter, have the structure shown in Fig. 108. The kernel is a xenolith of gneiss (*a*), which may be so completely recrystallized as to have a granitoid structure. This is surrounded by a coarse-textured zone (*b*) composed principally of andesine, with quartz, biotite, etc. It contains

[1] Adams, 'Nodular Granite from Pine Lake, Ontario,' *Bull. Geol. Soc. Amer.*, vol. ix. (1898), pp. 163-172. Brögger has described a similar occurrence on Kragerö.

[2] See especially Von Chrustschoff, 'Ueber holokrystalline makrovariolithische Gesteine,' *Mem. Acad. Sci. S. Pet.* (7) vol. xlii. (1897), No. 3, 244 pp. and three plates.

[3] *Bull. No.* 4 *Comm. Géol. Finl.*, (1896).

zircons derived from the gneiss, and doubtless represents a hybrid zone resulting from the partial dissolution of the xenolith in the granitic magma. This 'andesine-zone' passes into a finer-grained one with partial concentric and radiate structure, and so into the 'innermost fine-grained plagioclase-zone' (*c*), in which this structure is more pronounced. The constituents are plagioclase, quartz, and biotite. To this succeeds a thick, pale 'microcline-zone' (*d*), in which predominant microcline, with a radiate arrangement, is micrographically intergrown with quartz. Lastly comes the 'outermost fine-grained plagioclase-zone' (*e*), with only imperfect radiate and concentric structure, composed of plagioclase and quartz with some potash-felspar and biotite. The mineralogical composition of the several zones and of the matrix between the spheroids can be roughly calculated from chemical analyses, and is quoted in the table. The composition of the matrix is shown in column *f*.

	a.	*b.*	*c.*	*d.*	*e.*	*f.*
Plagioclase	Ab_2An_1 63·64	Ab_4An_3 73·22	Ab_1An_1 56·07	Ab_3An_2 16·52	Ab_2An_1 44·28	Ab_2An_1 35·49
Microcline and orthoclase ...	—	—	—	57·44	6·79	16·88
Biotite... ...	18·92	8·00	5·60	—	4·00	12·27
Quartz... ...	19·60	16·50	39·72	26·74	42·82	32·53
	102·16	97·72	101·39	100·70	97·89	97·17

The crystallization of the plagioclase in zones *b* and *c* was probably promoted by inoculation from the xenolith of gneiss. Vogt[1] considers that, after the crystallization of the andesine-zone (*b*) felspar and quartz were in eutectic proportions, and the succeeding zones represent supersaturation with orthoclase and plagioclase alternately. In other words, making abstraction of the biotite, the crystallization of the andesine-zone brings the indicating-point to the line EE_1 (Fig. 80), and it then oscillates about E in the line E_1E_2.

A given igneous rock may contain cognate xenoliths of

[1] *Loc. cit. sup.*, pp. 396-403.

more than one type. In the simplest case they differ from the enclosing rock in the same manner (as being more basic in a generalised sense), but in different degrees. This suggests that they represent different stages in an early intercrustal crystallization, which was also a differentiation. Lacroix,[1] discussing in particular the lavas of Mont Dore and their xenoliths, supposes that the 'enclaves homœogènes' of a given volcanic rock *constitute a series*, of which one extreme type is an ultrabasic rock and the other is the deep-seated equivalent of the volcanic rock itself. In this way the

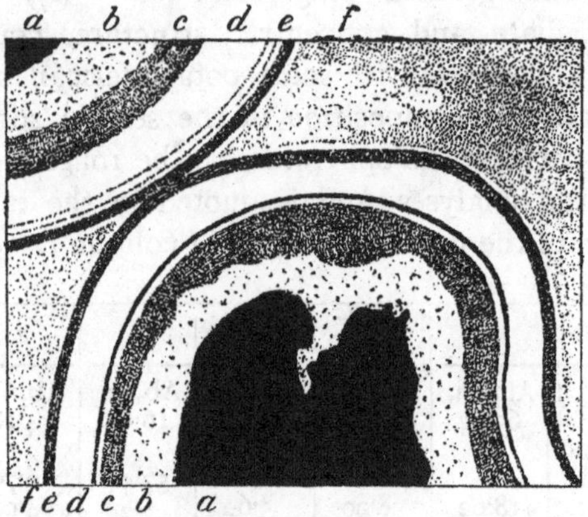

FIG. 108.—SECTION OF SPHEROIDAL GRANITE OF KANGASNIEMI, FINLAND. ONE-FOURTH NATURAL SIZE. (AFTER FROSTERUS.)

xenoliths may give a picture, more or less complete, of the genetic relations of the rock. This, indeed, seems to follow, if we can assume that all the types in question have been derived from one magma by differentiation along the same lines. The actual conditions may, however, be of a more complex kind. Fig. 109 is the variation-diagram of the nepheline-syenite complex of the Serra de Monchique, in Portugal,[2] with its cognate xenoliths. The different shape of each curve in the left and right parts of its course shows

[1] *Comptes Rendus*, vol. cxxxiii. (1901), pp. 1033-1036.

[2] Kraatz-Koschlau and Hackman, *Tscherm. Min. Petr. Mitth.* (2), vol xvi. (1896), pp. 197-306.

that in this case the laws of differentiation are not the same in the xenoliths as in the intruded rocks.

If in the intercrustal reservoir, where we suppose cognate xenoliths to have crystallized, different magmas coëxisted,

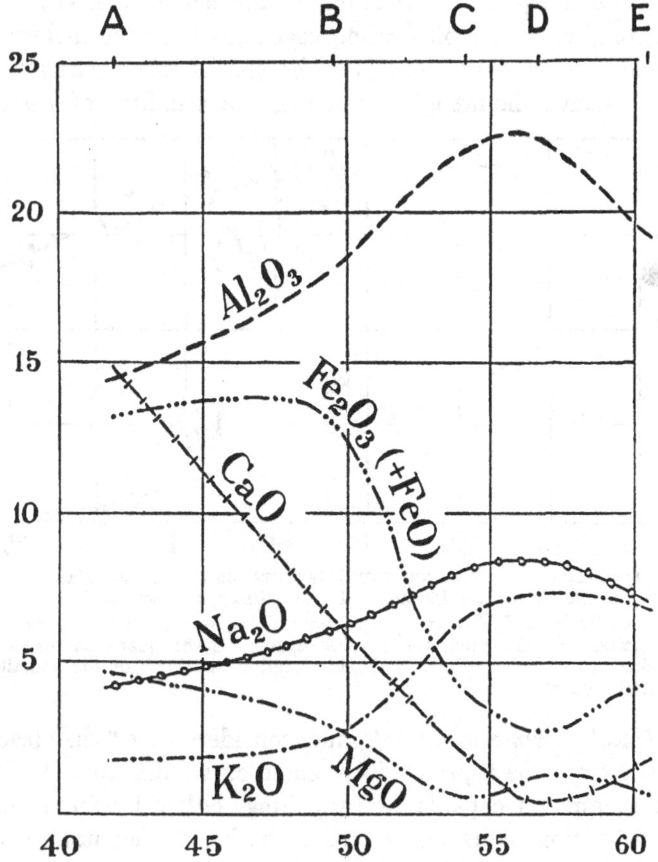

FIG. 109.—VARIATION-DIAGRAM OF THE NEPHELINE-SYENITE COMPLEX OF SERRA DE MONCHIQUE, PORTUGAL, WITH ITS COGNATE XENOLITHS.

A, Xenolith of teschenite.
B, Xenolith of essexite.
C, Nepheline-syenite, Picota.
D, Fine-textured 'schliere' in nepheline-syenite.
E, Pulaskite, Foya.

presumably with a stratified arrangement in accordance with their densities, we cannot assume that the magma from which a given xenolith crystallized is represented by its present matrix. The xenoliths must be considered with reference, not merely to the particular rock which encloses them, but to the whole suite. The true relations seem, therefore, to be more adequately expressed in another generalisation of Lacroix,[1] in a memoir dealing especially with the inclusions in the lavas of Santorin. Here he speaks more generally of the 'enclaves homœogènes' or cognate xenoliths of a *petro-*

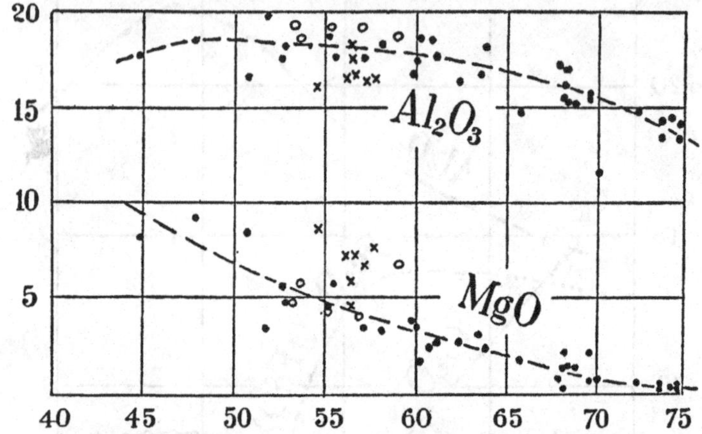

FIG. 110.—DIAGRAM SHOWING THE PERCENTAGES OF ALUMINA AND MAGNESIA IN THE LAVAS OF LASSEN PEAK, CALIFORNIA.[2]

The dots correspond with thirty-four normal lavas—viz., basalts, andesites, dacites, and rhyolites; the crosses represent seven quartz-basalts; and the small circles belong to basic secretions in some of the andesites and dacites.

graphical province as a whole, and considers that "they teach us what are the types of deep-seated crystalline rocks which the magma is capable of furnishing, either by the bodily consolidation of its mean type or by basic differentiation or original heterogeneity."

It is to be observed that crystalline aggregates more basic than the rock which encloses them are not the only kind of cognate xenoliths. In a district where basic and acid rocks occur in intimate association, we not infrequently find *acid*

[1] *Comptes Rendus*, vol. clx. (1905), pp. 971-975.

[2] From analyses made for the U.S. Geological Survey. See *Bull.*, *No.* 148 (1897), pp. 192-200.

xenoliths, or xenocrysts of the acid minerals, in a basic rock. Their constant occurrence in a particular group of rocks, often far from any possible source of accidental fragments, compels us to regard them as cognate xenoliths. The basic dykes of the British Tertiary region sometimes enclose xenoliths or isolated xenocrysts in considerable quantity in the neighbourhood of a plutonic centre, and more rarely at greater distances.[1] They are sometimes of gabbro or the minerals of gabbro, but altered xenoliths of granite and corroded xenocrysts of quartz and alkali-felspars seem to be quite as common. The explanation offered is that the basic magma which fed these dykes was drawn from a reservoir, in which it underlay an acid magma. Some crystallization was in progress in the latter, and crystals, singly or in aggregates, sometimes sank from the upper into the lower strata. We have already made the same suggestion in connection with the cognate xenocrysts found in many mica-lamprophyres, satellites of granitic intrusions (p. 323). Another interesting case is that of the basalts carrying quartz-grains, which occur as lava-flows in some districts. They are found at several localities in California, Nevada, Arizona, New Mexico, and Colorado, and examples have been described by Iddings[2] and Diller.[3] These authors believe that the quartz crystallized from the basalt magma itself in consequence of special physical conditions. Iddings, indeed, considers that the quartz-bearing basalts do not differ in chemical composition from those without quartz; but an examination of the analyses shows certain differences which must certainly be regarded as significant. The lavas of the Lassen Peak district afford a good illustration. Here the normal basalts, andesites, dacites, and rhyolites constitute a regular series (p. 125); but the quartz-basalts do not fall into place in this series, as is very apparent on attempting to plot the curves of variation. For the sake of simplicity, we

[1] Harker, *Tertiary Igneous Rocks of Skye*, *Mem. Geol. Sur.* (1904), chap. xx.

[2] *Amer. Journ. Sci.* (3), vol. xxxvi. (1888), pp. 208-221.

[3] *Bull. No.* 79 *U.S. Geol. Sur.* (1891).

show in Fig. 110 the relations for alumina and magnesia only. It will be seen that the points for the quartz-basalts (indicated by small crosses) fall below the convex curve of alumina and above the concave curve of magnesia; which strongly confirms the supposition that they are hybrid rocks (p. 335). It may be remarked that this is not the case with the basic secretions enclosed in some of the lavas. The positions of the small circles, which correspond with these, seem to indicate that they have resulted from a differentiation not very different from that which produced the series of lavas.

Hybridism as a Factor in Petrogenesis.—In a complete theory of the origin of igneous rocks by magmatic differentiation, hybridism would appear as in some sense a disturbing factor; and it is therefore necessary to inquire whether the considerations briefly discussed in the present chapter import any serious modification of our earlier conclusions. We may conveniently distinguish (*a*) admixture effected in an intercrustal magma-reservoir, and (*b*) effects due to the invasion of an older rock by a newer magma.

There seems to be clear evidence that, in a differentiated magma-basin with stratiform arrangement of the different partial magmas, crystals formed in one magma may become enclosed in a different one, either by settling under the effect of gravity, or by disturbance incident to the act of intrusion. This we may take to be the general source of those cognate xenoliths which are of intratelluric origin. Applying the same idea to isolated crystals, it appears that the distinction between phenocrysts and xenocrysts may be in some cases not very significant, the former having crystallized from the magma of the enclosing rock itself and the latter from an allied magma. The partial resorption or transformation, which is usually the sign of a xenocryst, will in general be proportional to the difference of composition between the magma from which the crystal formed and that in which it became involved. Consequently xenocrysts which imply any admixture important from the petrogenetic standpoint will easily be recognized as such, and the same reasoning applies to xenoliths. This criterion may fail if the dissolution has been complete; but in that case, either the foreign material was of trifling amount, or the difference of composition was so great that the resulting rock will be of a markedly abnormal variety. In no case can the hybrid origin of a rock be concealed.

The evidence goes to show that any important modification of a rock-magma by the inclusion of foreign elements of intratelluric origin is only *an exceptional incident*. The British Tertiary province offers, as has been remarked, specially favourable conditions, basic and acid rocks being found there in very intimate relation. Nevertheless there is, so far as is known, only one rock forming distinct intrusions which is so much modified by partly dissolved xenolithic material as to be of medium acidity, and this has only a restricted occurrence. It is found at certain localities in Skye, and has been described for convenience under the name 'marscoite.[1] It occurs as dyke-like intrusions between gabbro and granite, being also intermediate between these in age, and it likewise makes independent sill-intrusions. The rock clearly represents a basic magma much modified by enclosed acid material. In addition to fresh phenocrysts of labradorite, it contains corroded xenocrysts of quartz and alkali-felspars, and the ground-mass consists chiefly of hornblende and a felspar of the oligoclase-andesine series with more or less interstitial quartz. It is interesting to observe that the marscoite sills have been partly enveloped and destroyed by a later granitic intrusion, and there results in places a xenolithic rock of sub-acid composition, which is thus a *hybrid of the second order*.

In the invasion of an older rock by a newer magma the amount of mutual reaction depends, as before, upon a more or less marked difference of composition between the two, any important fusion of a rock by a magma of its own composition being precluded by the fact that the latter carries little, if any, excess of heat. There is then the further condition that, for any considerable reaction to be possible, the

[1] Harker, *Tertiary Igneous Rocks of Skye*, *Mem. Geol. Sur.* (1904), chap. xi.

solid rock invaded must itself be at a high temperature. This may be realised in some circumstances even in sedimentary rocks—*e.g.*, near a channel which has served for the prolonged passage of molten magma; but the most important case is that of an intruded igneous rock which has not had time to cool before it is invaded by a new intrusion of a different magma. Remembering the very low conductivity of rocks, we should expect this latter case to be more common than the former, and this is confirmed by observation. In any case effects of this kind are naturally *of local occurrence*, and are often confined to the neighbourhood of junctions.

Here again the British Tertiary plutonic rocks are instructive. Where granite invading gabbro shows a chilled margin, there is no sign of mutual reactions. Any important destruction of the basic rock is found only in places where it was followed after no great interval of time by the acid magma; and the more remarkable effects are usually connected with special situations, as where strips of gabbro have been wholly enveloped by the granite magma. Even where rock of hybrid origin is in considerable bulk, the admixture has not often been so complete as to produce a perfectly homogeneous appearance; and here the hybrid origin is still betrayed by an abnormal composition, chemical and mineralogical.

As regards chemical composition, these rocks illustrate another point, not yet mentioned. We have supposed that a hybrid rock has a composition which can be calculated from the analyses of the two component rocks, taken in proper relative proportions. While this is necessarily true of the rocks in bulk, it does not always hold good of every specimen, for admixture may be complicated by unequal diffusion of the different constituents. This is exemplified by the following figures for a granite much modified by dissolving gabbro material:

	SiO_2.	CaO.
Gabbro	46·39	15·29
Granite	70·34	1·24
Calculated mixture ...	64·71	4·54
Hybrid rock (found) ...	64·72	2·98

A mixture of gabbro and granite in the proportions 23·5 to 76·5 would give the silica-percentage of the hybrid, and would give 4·54 *per cent.* of lime, an amount not unusual in such rocks as quartz-diorites. The actual amount of lime, however, is only two-thirds as much; and this appears to indicate that the lime-compounds (viz., the anorthite molecule and the metasilicate which enters into hornblende) diffuse less freely than free silica and the alkali-felspars. It seems, therefore, that the composition of a hybrid rock may sometimes be more abnormal than that calculated, sometimes perhaps less so.

Reviewing what has been written in this chapter, we see no good ground of apprehension that any generalisations based on the doctrine of magmatic differentiation are invalidated by the consequences of admixture of different igneous rocks. The conditions which limit this process, and minimise its practical results, serve thereby to keep pure the stock of magmatic differentiation. Finally, it should be remarked that, when admixture takes place, it belongs to a late stage in the genetic history of a rock-magma, either to the epoch of intrusion (or extrusion) or to the time immediately following that. Any peculiar variety of rock which may result does not make a starting-point for the elaboration of new varieties. In short, like other hybrids, these *hybrid rocks are barren.*

CHAPTER XV

CLASSIFICATION OF IGNEOUS ROCKS

Principles of classification.—The 'Quantitative Classification.'—Is a natural classification possible for igneous rocks?—Some essays towards a natural classification.—Desiderata in an ideal system.

Principles of Classification.—Our attitude towards any classificatory scheme, or our opinion concerning the principles which may properly enter into such schemes, must depend upon the view which we hold of the object of classification in general. On the one hand is the position of Pinkerton (1803), who pronounced that "The only advantage of every methodic system in Natural History is to assist the memory." On the other hand is the aspiration of many petrologists at the present day towards a natural system, which shall represent the true relationships of the various rock-types. Between these extremes lie the numerous schemes which are or have been current, which we may characterize from this point of view as artificial systems. They are based on the likeness or unlikeness of rocks in respect of particular characters, and they therefore differ greatly in value according to the kind of characters selected.[1]

In the early years of the last century, when petrography had not yet detached itself from a mineralogy founded on 'external characters,' the classification of rocks was necessarily of a very crude kind. In the hands of Werner and

[1] For a historical account of the classification of igneous rocks see Whitman Cross, *Journ. Geol.*, vol. x. (1902), pp. 331-376, 451-499; reprinted in *Quantitative Classification of Igneous Rocks* (1903).

his school it was further hampered by having grafted on it an unreal stratigraphy; and, as regards any true advance in the systematic study of igneous rocks, the union was a singularly unfruitful one. Petrography on a modern basis may be considered as beginning with von Leonhard (1823-24) and Brongniart (1827). Most of the systems propounded since their time have followed them in making mineralogical composition and structure the principal criteria, one or other factor taking the first place in different schemes. Notwithstanding the pioneer work of Cordier (1815), the more fine-textured igneous rocks were not completely brought within the compass of a mineralogical classification until the preparation of thin slices by Sorby (1858) made the minute study of rocks comparatively easy. *Zirkel* (1873), in the first systematic work making use of microscopical methods, based his most important divisions and subdivisions upon the chief constituent minerals of the rocks. Most later classifications have proceeded on the same general lines, though with some modifications and with elaboration consequent upon increased knowledge of different rock-types. Structure has been employed in conjunction with mineral compositon as a factor in classification. The distinction of 'older' and 'younger' rocks, a belated survival of Wernerian theories, has not yet quite disappeared, but will scarcely be seen again in standard works.

Ideally it would be possible to make chemical instead of mineralogical composition a basis of classification; but the alternative is scarcely a practicable one, since we cannot always wait for a chemical analysis of a rock before assigning its classificatory position. Chemical composition accordingly has not figured prominently in any comprehensive scheme, except, in a disguised form, in the new American classification to be noticed below. It has, however, been used more or less explicitly, especially by Roth (1861), as a controlling factor, importing a certain quantitative element into classification. In the various schemes based mainly on mineralogical composition the criteria are mainly of the qualitative kind, the relative proportions of the constituent minerals being recognised only in a general way, as, for instance, in the distinction between essential and accessory minerals. This is not in itself to be regarded as a defect in a classification which aims at representing the facts of nature, for one of the most fundamental characters of igneous rocks is their variability.

Speaking generally, the various classificatory systems to which we have referred are artificial, in that they divide rocks into mutually exclusive categories on the ground of particular selected characters, without any direct reference to natural—*i.e.*, genetic—relationships. Nevertheless the development of systematic petrography, as embodied, for instance, in the successive editions of Rosenbusch's great work (1877-1908), shows an advance on definite lines. The tendency is to define groups of rocks not by particular characters, but by the aggregate of their presumably essential characters, and to regard as essential those which may be supposed of genetic significance. Indeed, the classification in the later editions of Rosenbusch is influenced in some measure by his theory of magmatic differentiation, propounded in 1889. Petrologists pay much more attention now than formerly to the mutual relationships of igneous rocks, and there is a general feeling that we have almost within grasp a fundamental principle analogous with that of descent, which lies at the root of natural classification in the organic world.

The 'Quantitative Classification.'—It is necessary to offer a few remarks on the system put forward in 1902 by four American petrologists.[1] Having regard to the trend of opinion in recent years, as indicated above, we cannot but

[1] *Journ. Geol.*, vol. x. (1902), pp. 555-690, separately published under title *Quantitative Classification of Igneous Rocks* (1903), by Whitman Cross, J. P. Iddings, L. V. Pirsson, and H. S. Washington; with supplementary article, 'The Texture of Igneous Rocks,' *Journ. Geol.*, vol. xiv. (1906), pp. 692-707. For collections of analyses of rocks arranged according to this system see Washington, *U.S. Geol. Sur.*, *Prof. Papers* No. 14 (1903) and No. 28 (1904). For illustrations of the application of the classification see Washington, *The Roman Comagmatic Region* (1906), and recent papers by other American petrologists.

view the 'Quantitative Classification' as a retrograde movement, for here the artificial element is applied to the complete exclusion of the natural. The system is too elaborate to be described in a few words. It is based primarily upon mineralogical composition, but upon a hypothetical mineralogical composition, which is not in general that of the actual rock, but is computed from a chemical analysis. A division into five classes depends upon the relative proportions of light and dark minerals—*i.e.*, of quartz, zircon, corundum, felspars, and felspathoids on the one hand, and non-aluminous ferromagnesian and calcic silicates, with iron-ores and various minor accessory minerals, on the other hand. Subclasses, orders, and suborders are formed by further consideration of the relative proportions of the minerals; rangs and subrangs are made to depend directly on chemical characters—viz., the molecular ratios of certain oxides; and grads and subgrads are constituted with reference to the relative proportions of subordinate mineral constituents. All divisions are made on *a priori* arithmetical principles. Thus, for the five classes, where the dividing criterion is the proportion of light ('salic') to dark ('femic') minerals, the points of division are the ratios $\frac{7}{1}$, $\frac{5}{3}$, $\frac{3}{5}$, and $\frac{1}{7}$. For the working-out of the system, and the new terminology in which it is expressed, the reader must be referred to the original source.

Regarded on its merits as a confessedly artificial system, the Quantitative Classification is open to serious criticism.[1] The design of any artificial classification should be to bring together things which are like and separate those which are unlike. But, to take an example, the subrang Toscanose in Washington's tables includes rhyolites, toscanites, trachytes, latites, dacites, and hypersthene-andesites, with an equal diversity among abyssal and hypabyssal rocks. It would be difficult to say wherein these resemble one another and differ collectively from the assemblage placed in any of the neighbouring compartments of the scheme. The authors begin by asserting the chemical composition of a rock to be its most

[1] Evans, *Science Progress*, vol. i. (1906), pp. 259-280.

fundamental character, and lay it down that "all rocks of like chemical composition should be classed together"; but in the analyses under Toscanose we find that the percentages of the eight principal oxides vary as follows:

SiO_2	...	76·48 to 62·20	MgO	...	2·39 to 0·05
Al_2O_3	...	20·82 to 12·04	CaO	...	5·07 to 0·85
Fe_2O_3	...	4·69 to 0·02	Na_2O	...	4·92 to 2·14
FeO	...	4·64 to none.	K_2O	...	7·09 to 3·28

Such is the range of composition found within a single subrang, apparently the smallest subdivision used in ordinary practice. The larger divisions of course embrace rocks with still wider diversity of chemical composition. To take an extreme case, the first of the five classes, called Persalane, would include a pure quartz-rock or a pure corundum-rock. Consequently, as Evans points out, the silica-percentage of rocks in this class ranges from 100 to 0, and the alumina-percentage between the same limits.

This anomaly results from the dichotomous principle and the fact that each division or subdivision is made with reference to a single factor, expressible as a ratio. Thus, the rock-forming minerals are divided into two groups, salic and femic, and the division of rocks into five classes is based on the ratio of total salic to total femic constituents. If corundum had not been thrown into the former group, with quartz and felspars, it would have been found in no less strange company with olivine and pyroxenes in the second group. Clearly the arrangement of so diverse an assemblage as the rock-forming minerals in two groups only must violate natural relationships; and a like objection applies in greater or less degree to the factors used in making the subdivisions of the system.

The mineralogical composition employed in the Quantitative Classification is, as we have remarked, not the real composition, but an ideal one, called the 'norm.' This is calculated from a chemical analysis of the rock on the supposition that the only minerals which enter are those of a certain artificially selected 'standard' list. The micas, the garnets, the common aluminous hornblendes and augites, melilite, spinel, and others are denied admittance, and in their place we find kaliophilite, sodium and potassium metasilicates, wollastonite, and åkermanite—compounds which are foreign to igneous rocks, and some of which are not known in nature. As an extreme instance, an almost pure hornblende-rock, from Gran in Norway, is assumed to consist of diopside, anorthite, magnetite, nepheline, ilmenite, olivine, orthoclase, leucite, and hæmatite. The reason given for this arbitrary treatment is that the rejected minerals, such as hornblende, have not such simple compositions as would lend themselves to arithmetical manipulation; but it would seem more reasonable to adapt the classification to the rocks, not the rocks to the classification.

It will be seen that this system, though expressed for the most part in mineralogical terms, is in reality a chemical one, the 'norm' being only a circuitous manner of representing a chemical analysis. The manner in which the various oxides are actually combined in the minerals—in the author's terminology, the 'mode' of the rock—finds no place in the classification. The underlying idea seems to be that the constituents exist in a rock-magma in the form of free oxides, a conception which necessarily obscures natural relationships. Thus, in another place, Pirsson and Washington[1] endeavour to minimise the distinction, formerly urged by their colleague Iddings, between the alkali and sub-alkali branches. It is true that this distinction is less strongly marked in acid rocks than in intermediate and basic; but this fact is not correctly expressed by saying that "the alkalic nature . . . is to a great extent masked by dilution with silica," a statement which wholly ignores that the rock is composed of minerals. The authors take the chemical analysis of a granitic rock, subtract the greater part of the silica, and recalculate from the analysis thus mutilated a 'norm' which shows a large amount of alkali-minerals. But in so doing they have not only disregarded the quartz of the original rock, but have also abstracted most of the silica from the soda-felspar, reducing it to

[1] *Amer. Journ. Sci.* (4), vol. xxii. (1906), pp. 442, 443.

nepheline. The process here performed arithmetically has no counterpart in nature, and can throw no light on the actual relationships of rocks.

The strongest objection to the Quantitative Classification is, however, that it is planned entirely on *a priori* lines. In the development of the system (as distinct from the preliminary discussion) there is scarcely anything to show that the authors are aware that igneous rocks actually exist and have been made the subject of study. Starting from a list of standard minerals which corresponds only very imperfectly with the actual rock-forming constituents, the scheme is built up without any consideration of how known rocks will fit into the rows of compartments so constructed. This is the antithesis of the ideal to which petrological research points.[1] As von Richthofen aptly remarked, "in order to establish a more natural system, we have not to *make* groups, but to *find* them." In other words, the proper place of petrology is with the inductive, not the deductive, sciences. The rounded completeness, the measured precision, the finality, which would be admirable in a mathematical treatise, serve only to condemn a classification of igneous rocks, since they make more evident its aloofness from the scheme of nature, based not on arithmetical but on physical and chemical principles.

Is a Natural Classification possible for Igneous Rocks?—It has often been held an obstacle to any natural classification of rocks that there are in petrography no species. Pinkerton[2] long ago, criticizing Dolomieu, urged that "in the other branches of natural history a species produces a similar progeny," and protested against the application of such terms to "dead and inert matter." But in the last hundred years the conception of a species in botany and zoology has become very materially modified; and it is misleading to speak of rocks as dead and inert, when they exhibit that continual response to environment which gives rise to the evolution of new types. The recurrence of the

[1] Harker, *Geol. Mag.*, 1903, pp. 173-178.

[2] *Petralogy: a Treatise on Rocks* (1811), vol. i., pp. i, ii.

same types in different regions and at different geological periods gives them an individuality which cannot be ignored. It does not imply an absolute constancy of composition, variability being a characteristic of the rock-type as it is of the organic species. The question, however, does not turn upon the definition of a species or upon any supposed parallel between the organic and inorganic worlds. If the actual characters of an igneous rock depend upon the stock from which it is derived and the conditions under which its evolution has been brought about, there is here a theoretical basis for a natural classification comparable in a broad sense with that employed for animals and plants. The question remains whether the system thus indicated can be realised in practice, and to this question the future must give an answer.

In a memoir published in 1903, Iddings[1] endeavours to prove that no natural subdivisions of igneous rocks can be recognised on the basis of chemical composition. Taking silica-percentage as abscissa and the molecular ratio of alkalies to silica as ordinate, he plots 2,000 bulk-analyses of igneous rocks, and remarks that there is no clustering of the points in particular parts of the diagram. This test is clearly fallacious. It might serve as applied to rocks containing no other minerals than quartz and alumino-silicates of the alkali-metals, but is not applicable where lime-bearing felspars, pyroxenes, olivine, and other minerals are important constituents. There are at least seven or eight variables to be considered, of which the method employed takes account of two only. Moreover, as has already been insisted, it is the mineralogical, not the chemical, composition that should enter into such a comparison. Nevertheless, as Lane[2] has pointed out, there is on Iddings' diagram a very marked clustering, not about points, but, as might be expected in the circumstances, about a line. This clustering is seen in the acid half of the diagram, and follows a horizontal line

[1] 'Chemical Composition of Igneous Rocks expressed by Means of Diagrams,' *U.S. Geol. Sur.*, *Prof. Paper* No. 18 (1903).

[2] *Journ. Geol.*, vol. xii. (1904), pp. 83-93.

corresponding with the alkali-silica ratio ·083, or one-twelfth. This doubtless represents a tendency to eutectic proportions of felspar and quartz. Vogt's estimate of the eutectic proportions for orthoclase and quartz gives a ratio ·105, and the lower figure is due presumably to the presence of other minerals in the rocks. The line of clustering declines a little as it passes from acid to intermediate rocks and alkali-felspars give place to others containing lime. The clustering may be shown in another way (Fig. 111). Of the rocks

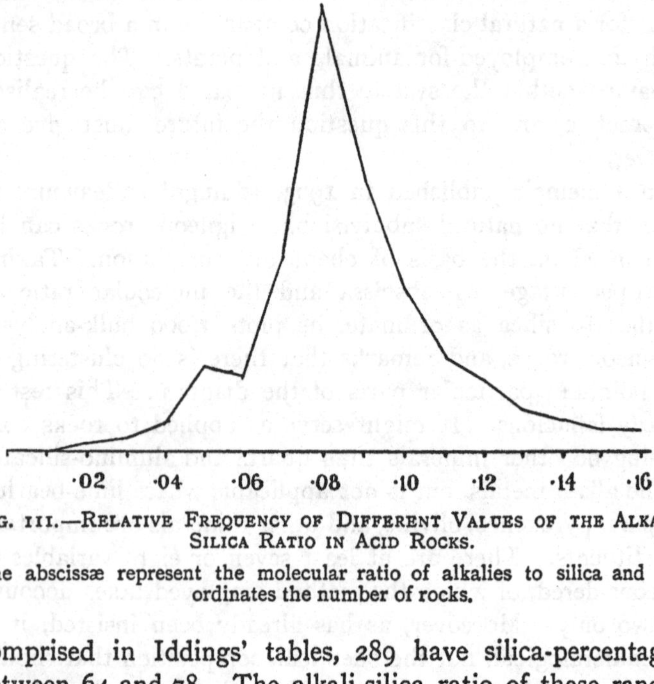

FIG. 111.—RELATIVE FREQUENCY OF DIFFERENT VALUES OF THE ALKALI-SILICA RATIO IN ACID ROCKS.

The abscissæ represent the molecular ratio of alkalies to silica and the ordinates the number of rocks.

comprised in Iddings' tables, 289 have silica-percentages between 64 and 78. The alkali-silica ratio of these ranges from ·02 to ·16, but for nearly three-fourths of the number it lies between ·06 and ·10, with a strongly marked maximum of frequency a little beyond ·08.

Iddings again seeks to minimise the importance of petrographical provinces as possessing distinctive chemical characteristics, quoting Brögger's statement[1] that "the

[1] *Quart. Journ. Geol. Soc.*, vol. l. (1894), p. 36.

same group of differentiated rocks can be produced by separation from mother magmas of quite different chemical composition." To illustrate this he takes seriatim a number of rocks from the Christiania district, and sets beside them for comparison rocks from other petrographical provinces. For example, a quartz-porphyry from Kroftkollen is paralleled with an aplite from Butte, Montana; but, turning to the actual analyses, we find that soda is (in molecules) the preponderant alkali in the former rock, while potash is largely preponderant in the latter; and these are, in fact, the most marked characteristics of the two petrographical provinces in question. Brögger's generalisation, quoted above, was prompted by the observation that, while camptonites and bostonites are in several American districts associated with nepheline-syenites, they are found in the neighbourhood of Gran in close association with the plutonic rock which he first styled olivine-gabbro-diabase. This rock, however, is no gabbro but an essexite, a type associated in Norway and elsewhere with nepheline-syenites. The bostonites of Gran are quite different from those found, in Norway as in America, in close association with nepheline-syenites, and Brögger has since constituted for them the new type mænaite. Clearly, any profitable inquiry into the true relationship of igneous rocks must proceed by a closer characterization of types, not by a laxity of grouping.

It is true that, where differentiation has reached an advanced stage, a wide range of variety in composition may be found among the rocks of a single petrographical province, and exceptionally aberrant types may occur which closely resemble rocks elsewhere derived from a different stock. The reasonable deduction from these facts is that a natural classification must have regard to the mutual associations of rocks as well as to their descriptive characters. A rigidly artificial classification may lead to much confusion even where differentiation has not proceeded so far as to cause any real complexity. The volcanic rocks of the Andes represent, as Iddings himself has remarked (see above, p. 103), a comparatively early stage of evolution, and it cannot be doubted that the andesites (including dacites) of this region constitute a natural group with close relationship. But in the Quantitative Classification we find them scattered through two classes, four orders, and nine rangs. They are, in different instances, bracketed in the same subrang with granites, granodiorites, syenites, nepheline-syenites, grorudites, teschenites, trachytes, tephrites, and many other rocks, with which they have no sort of relation founded in nature.

Some Essays towards a Natural Classification.—Although no comprehensive scheme of classification based on the genetic relationships of different igneous rocks has yet been put forward, and the problem is, as we have said, one for future solution, it will be of interest to glance at some of the suggestions which have been made touching the broad principles of such an ideal system. The 'Kern' theory of Rosenbusch has already been mentioned (p. 315). His conception of the spontaneous partition of a mixed magma into partial magmas, tending to stochiometric composition, seems to demand a limited mutual solubility of the latter, which is not easily reconciled with the phenomena of igneous rocks. Brögger's[1] ideal genetic classification follows other lines. There is a primary division of igneous rocks into comprehensive families, based on broad chemical and mineralogical characteristics, such as the granite family, the diorite family, etc. The granite family, for example, includes not only granites, but also all igneous rocks presumed to be derived from the same stock. Besides the normal abyssal types there are marginal facies, pegmatoid and other modifications, and various hypabyssal types. The latter include non-differentiated rocks, such as granite-porphyrites, quartz-porphyries, pitchstones, etc., but also differentiated derivatives, such as minette and aplite, which may differ widely in composition from granites. In addition there are the volcanic equivalents of the granites, with their tuffs. Other families are constituted in like manner, and a certain parallelism of evolution is recognised in the several families, analogous members of different families falling into place in

[1] *Eruptivgesteine des Kristianiagebietes*, I. (1894), pp. 87-95, 177, etc.

'rock-series' (p. 144). Brögger, however, recognises transitional types (Uebergangstypen) connecting the different families. The scheme appears to be conceived with special reference to the evolution of types in one petrographical province, and for that of Christiania an attempt is made to represent the mutual relations of the rocks by a kind of genealogical tree. It is not necessary to emphasize the element of hypothesis in such tentative schemes, for it is of the essence of a natural classification that there can be no finality in it. Indeed, its changes from time to time will be in some sense a measure of the progress of petrology.

It is to Becker[1] that we owe the first suggestion of a classification of igneous rocks based on *eutectics*, as the only mixtures which possess definite compositions and properties, and can therefore be used as standards of reference. He considers eutectic mixtures to play the part of solvents in rock-magmas, the excess of particular constituents in any case being dissolved in the eutectic. The ground-mass of any porphyritic rock may be supposed to tend to eutectic composition. In so far as this represents the facts, the logical plan will be "to group together in one genus all the rocks which have the same ground-mass and to regard the phenocrysts as minor or specific characteristics." Becker remarks, as an incidental advantage of such a scheme, that it will stand in relation to the geological behaviour of intruded and extruded magmas, which is largely controlled by viscosity and temperature. The eutectic mixture of given constituents has a definite melting-point and a definite viscosity at any given temperature. So long as it remains liquid, "the phenocrysts are, mechanically speaking, mere flotsam and jetsam in the stream; and the behaviour of the molten magma, in so far as it is dependent on specific properties, is determined by the eutectic or quasi-eutectic mixture." From Becker's point of view, 'fractional crystal-

[1] 'Report on the Geology of the Philippine Islands,' *Twenty-first Ann. Rep. U.S. Geol. Sur.*, III. (1901), pp. 519, 520; also Introduction to Day and Allen's *Isomorphism and Thermal Properties of the Felspars*, (1905), pp. 7-9.

lization' appears as a disturbing factor, and those rocks which result from differentiation in place would strictly find no place in the system except as varietal forms.

One of Vogt's more recent contributions[1] is an attempt to base a natural classification of igneous rocks on the actual *course of magmatic differentiation*, in so far as this may be supposed known. He takes as starting-point the parallelism which has been pointed out by Brögger, Teall, and others between the order of differentiation of cognate rock-magmas and the order of crystallization of the minerals in one magma. Vogt had formerly expressed the opinion[2] that the final result of magmatic differentiation carried to the extreme would be the separation of the several constituent minerals by themselves. He now remarks that, if the analogy with the course of crystallization holds good, the tendency must be to the separation on the one hand of magmas representing single minerals and on the other hand of eutectic magmas. He considers that very many rocks do actually approximate to these two ideal categories: firstly, 'anchi-monomineralic' rocks, composed essentially of a single mineral; and secondly, 'anchi-eutectic' rocks, with more or less closely eutectic proportions of two or more minerals.

The minerals which are found constituting simple rocks, with very little admixture of other constituents, are olivine, rhombic and monoclinic pyroxenes, hornblende, and the felspars; also, in less considerable masses, ilmenite, magnetite, titanomagnetite, chromite, pyrrhotite, etc. The dunites and anorthosites are the most important rocks of this kind. The olivine, which is the essential mineral of the dunites, is an isomorphous mixture of magnesian and ferrous orthosilicates, of which the former has the higher melting-point, and the mixed crystals belong to Type I. of Roozeboom (p. 234). Crystals forming from a magma are accordingly enriched in magnesium relatively to iron. If

[1] 'Ueber anchi-eutektische und anchi-monomineralische Eruptivgesteine,' *Norsk. Geol. Tidskr.*, vol. i. (1905).

[2] *Zeits. prakt. Geol.*, 1893, p. 277, and 1900, p. 235.

a like rule holds in the differentiation of peridotite magmas, the ratio Mg : Fe should be higher in proportion as differentiation has progressed towards a pure olivine-magma. Vogt shows from a comparison of chemical analysis of peridotites that, as they approach the composition of pure olivine-rock, the atomic ratio Mg : Fe steadily rises. We may illustrate this by the following table, including thirty-eight analyses, taking the diminishing percentage of alumina as indicating approach to the dunite type :

Number of Analyses.	*Percentage of Alumina.*	*Atomic Ratio. MgO : FeO.*
5	10 to 11	2·6
5	7 to 10	3·4
5	6 to 7	4·6
5	5 to 6	4·1
5	4 to 5	4·2
8	1 to 4	7·0
5	0 to 1	9·0

It is probably a general rule that, with concentration of a ferro-magnesian mineral, there is also an enrichment in magnesium relatively to iron. The general distribution of the members of isomorphous series seems to accord with this. Hedenbergite, like fayalite, belongs to acid rather than basic rocks. Magnesia, which is negligible in most analyses of ilmenite, reaches 4 or 5 *per cent.* in segregations of ilmenite-ore.

The plagioclase felspars, in crystallizing from a magma, are enriched in anorthite relatively to albite. If differentiation follows similar lines, it should be found that, in proportion as plagioclase makes up the greater part of an igneous rock, it tends to varieties rich in lime. For 25 analyses of plagioclase-rocks collected by Vogt, the felspar is oligoclase in 1 case, andesine in 3, labradorite in 13, bytownite in 7, and anorthite in 1. Löwinson-Lessing has since described a pure albite rock ; but other examples might be added of rocks composed essentially of the more calcic felspars, and these, unlike the rocks of more sodic felspars, occur in some regions in great force. Again, from considerations already set forth, we should expect the content of potash-felspar in a magma chiefly of plagioclase to diminish in proportion to the completeness of the differentiation. In accordance with this, it is found that there is less potash in anorthosites rich in lime than in those rich in soda. We may remark, indeed, that this is true of the plagioclase felspars in general, albite excepted. The analyses collected in Hintze's "Handbuch" give the following result:

	Number of Analyses.	*Mean Percentage of Potash.*
Albite	101	0·75
Oligoclase	153	1·59
Andesine	127	1·00
Labradorite	139	0·88
Bytownite	34	0·74
Anorthite	68	0·53

The exception in the case of albite seems to accord with the suggestion that the plagioclase felspars belong to Type III. of Roozeboom; but it is to be remembered that very many of the albite crystals analysed were not formed from igneous rock-magmas.

Among rocks which approximate to eutectic composition, Vogt reckons in the first place the ordinary types of granite, quartz-porphyry, and rhyolite, which correspond more or less nearly with a felspar-quartz eutectic or some ternary or quaternary eutectic mixture. It would appear that in a eutectic of felspars, quartz, and a ferromagnesian (or an iron-ore) mineral, the last enters only in small amount. The tendency to eutectic composition in the ground-mass of porphyritic rocks chiefly of felspars and quartz has already been pointed out. The phenocrysts are often in such small relative amount as to imply no very considerable excess of one or other constituent; and, when both felspar and quartz phenocrysts are present, the real excess may be much less than that represented by the total amount of porphyritic elements. Granitic rocks with silica-percentage about 73 to 75 have an exceedingly wide distribution. It is possible, as Vogt supposes, that a curve constructed like Fig. 42, but taking account of the actual masses of the rocks instead of the number of specimens analysed, would show a strong maximum in this part of the range (Fig. 112). Since it appears that in general the progress of magmatic differentiation, like that of crystallization in a single magma, terminates in the direction of the more acid products in an approximately eutectic mixture, anything approaching a pure quartz-rock is not to be expected. The range of normal igneous rocks, in fact, comes to an end quite abruptly on the acid side at a silica-percentage near 80, the excess of quartz over eutectic proportions never surpassing a moderate limit.

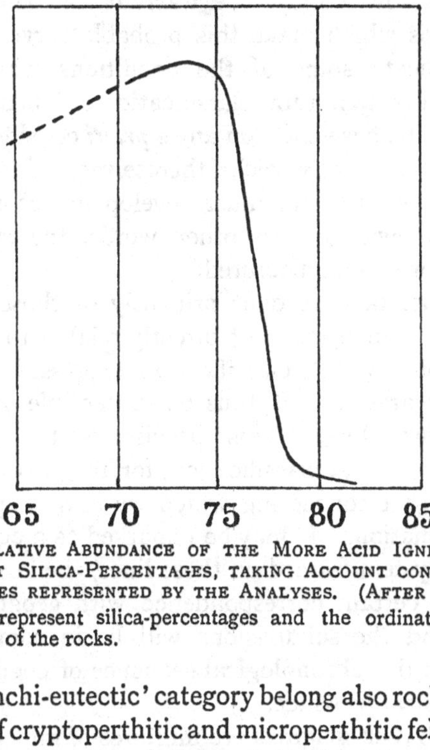

FIG. 112.—RELATIVE ABUNDANCE OF THE MORE ACID IGNEOUS ROCKS OF DIFFERENT SILICA-PERCENTAGES, TAKING ACCOUNT CONJECTURALLY OF THE MASSES REPRESENTED BY THE ANALYSES. (AFTER VOGT.)
The abscissæ represent silica-percentages and the ordinates the relative abundance of the rocks.

To the 'anchi-eutectic' category belong also rocks composed essentially of cryptoperthitic and microperthitic felspars, which are a binary eutectic of Or and Ab (+An), and probably other rocks of the nature of ternary and quaternary eutectics into which perthitic felspars enter. It is scarcely possible at present to discuss the basic rocks from the same point of view; but it may be conjectured that the best-marked types among these represent with some degree of approximation binary eutectics of felspar and augite, ternary eutectics of felspar, augite, and olivine or magnetite, or more complex mixtures of the same order.

Desiderata in an Ideal System.—Having regard to the imperfection of present knowledge in the domain of petrogenesis, it is clear that a systematic treatment of igneous rocks on these lines is not to be expected in the immediate future. That the establishment of a genetic classification is dependent only on a fuller knowledge of facts and principles, we have already found some warrant for believing. The considerations which make this probable serve at the same time to indicate some of the conditions which must be satisfied by a truly natural classification of igneous rocks.

(i.) It must be based, not on any *a priori* considerations, but upon actual study of the rocks themselves. Consequently it can have no finality, but must develop in accordance with increasing knowledge. In other words, the system must grow, and not be manufactured.

(ii.) It must take account primarily of those features of igneous rocks which are most directly related to the manner of their evolution. A classification expressed in terms of descriptive characters will thus be susceptible of interpretation on genetic lines. This proviso is the crux of the problem of a natural classification, for magmatic differentiation is a subject concerning which we have at present very scanty information. It may be remarked as a corollary that, in an ideal system framed on these lines, the larger divisions will show a certain correspondence with geographical distribution, and the subdivisions will be in some degree in relation with the chronological sequence of cognate rocks in a petrographical province.

(iii.) Since rock-magmas are mixtures of minerals, and the variation met with in igneous rocks is a variation in the associations and relative proportions of minerals, it follows that a natural classification must be in its expression a mineralogical one, not a chemical. It is true that the actual mineralogical constitution of a given magma may vary to some extent in accordance with temperature and other conditions, and this would seem to import some complication where successive differentiations have been effected at higher and lower temperatures. If it should be found necessary to represent complex low-temperature minerals by simpler compounds, these must at least be compounds which actually exist in the rock-magmas.

(iv.) As regards rock-structures, it is to be remarked that some, such as the pegmatoid, the graphic, and (in part) the porphyritic, have a real significance from the petrogenetic standpoint, and may therefore find a place, together with mineralogical composition, in a systematic scheme on a natural basis. Structures which depend on the special conditions under which a magma consolidates, and are not directly related to its composition and origin, are clearly of a lower order of importance.

(v.) Where the quantitative element enters, in fixing dividing-lines, regard must be had, not to arithmetical symmetry, but to chemico-physical principles. In the first place, as claimed by Becker, come eutectic ratios; and the determination, by whatever means, of the binary, ternary, and quaternary eutectic proportions for different rock-forming minerals will be an important step towards the realisation of a rational system of classification of rocks.

(vi.) Finally, an established system must be provided with an appropriate nomenclature; and, since a natural classification will traverse the subdivisions of all existing schemes, the nomenclature must be a new one. Classificatory terms must be made quite distinct in form from the names of rock-types: the want of such distinction is responsible for much of the confusion in petrography at the present time. The existing practice of naming a rock-type from its original locality, with a uniform termination, can scarcely be bettered, though the particular termination *-ite* might perhaps with advantage be surrendered to the mineralogists. Classificatory names might conveniently bear different terminations for divisions of different magnitude or status.[1]

[1] This plan is adopted in the Quantitative Classification—*e.g.*, a Vesuvian lava belongs to class Dosalane, order Italare, rang Vulturase, and subrang Braccianose.

INDEX

Printed in the United States
By Bookmasters